Population Genomics

Editor-in-Chief

Om P. Rajora, Faculty of Forestry and Environmental Management,
University of New Brunswick, Fredericton, NB, Canada

This pioneering *Population Genomics* Series deals with the concepts and approaches of population genomics and their applications in addressing fundamental and applied topics in a wide variety of organisms. Population genomics is a fast emerging discipline, which has created a paradigm shift in many fields of life and medical sciences, including population biology, ecology, evolution, conservation, agriculture, horticulture, forestry, fisheries, human health and medicine.

Population genomics has revolutionized various disciplines of biology including population, evolutionary, ecological and conservation genetics, plant and animal breeding, human health, genetic medicine, and pharmacology by allowing to address novel and long-standing intractable questions with unprecedented power and accuracy. It employs large-scale or genome-wide genetic information across individuals and populations and bioinformatics, and provides a comprehensive genome-wide perspective and new insights that were not possible before.

Population genomics has provided novel conceptual approaches, and is tremendously advancing our understanding the roles of evolutionary processes, such as mutation, genetic drift, gene flow, and natural selection, in shaping up genetic variation at individual loci and across the genome and populations, disentangling the locus-specific effects from the genome-wide effects, detecting and localizing the functional genomic elements, improving the assessment of population genetic parameters or processes such as adaptive evolution, effective population size, gene flow, admixture, inbreeding and outbreeding depression, demography, and biogeography, and resolving evolutionary histories and phylogenetic relationships of extant and extinct species. Population genomics research is also providing key insights into the genomic basis of fitness, local adaptation, ecological and climate acclimation and adaptation, speciation, complex ecologically and economically important traits, and disease and insect resistance in plants, animals and/or humans. In fact, population genomics research has enabled the identification of genes and genetic variants associated with many disease conditions in humans, and it is facilitating genetic medicine and pharmacology. Furthermore, application of population genomics concepts and approaches facilitates plant and animal breeding, forensics, delineation of conservation genetic units, understanding evolutionary and genetic impacts of resource management practices and climate and environmental change, and conservation and sustainable management of plant and animal genetic resources.

The volume editors in this Series have been carefully selected and topics written by leading scholars from around the world.

Julian R. Dupuis • Om P. Rajora
Editors

Population Genomics: Insects

Springer

Editors
Julian R. Dupuis
Department of Entomology
University of Kentucky
Lexington, KY, USA

Om P. Rajora
Faculty of Forestry and Environmental Management
University of New Brunswick
Fredericton, NB, Canada

ISSN 2364-6764 ISSN 2364-6772 (electronic)
Population Genomics
ISBN 978-3-032-10102-0 ISBN 978-3-032-10103-7 (eBook)
https://doi.org/10.1007/978-3-032-10103-7

This Springer imprint is published by the registered company Springer Nature Switzerland AG
The registered company address is: Gewerbestrasse 11, 6330 Cham, Switzerland

Population Genomics Book Series
This Population Genomics book series is dedicated to my (late) parents, and my wife Malti and children Apoorva, Anu, and Maneesha.

Om P. Rajora

Population Genomics: Insects
This book is dedicated to the world of beneficial insects providing enormous ecosystem and environmental services, vital for our sustenance, and insect population genomics researchers.

Om P. Rajora

Preface

High-throughput sequencing technologies and the ever-decreasing cost of generating large genomic datasets have revolutionized access to population genomic datasets for diverse groups of organisms. Insects, being the most diverse eukaryotes on the planet, have been a frequent target for these datasets and population genomics studies. Their diversity brings with it varied life history strategies and unique evolutionary and ecological phenomena; these characteristics have led insect species to be instrumental case studies for fundamental concepts of population genetics, from Peter Buri's experiments with mutant *Drosophila melanogaster* demonstrating genetic drift to Bernard Kettlewell's accumulated works on natural selection of melanism in the peppered moth. Insects also include some of the world's most dangerous animals and those with great economic impact; agricultural pests, be it those affecting fruits, vegetables, or animal production, and insects of medical importance impact human life in massive ways. In these systems, the most basic questions of evolutionary biology and population genetics often take on applied importance. Finally, in a world of increasing human impact, from climate change to habitat loss and phenomena like pesticide resistance, understanding the population genomics of these diverse organisms takes on a renewed importance.

As part of the pioneering Population Genomics book series, this volume explores various aspects of insect population genomics. Genomics research is inherently interdisciplinary, and population genomics is no exception, combining aspects of classical genetics, evolutionary and molecular biology, and bioinformatics. Topics covered in this volume include natural selection in insect systems, transposable elements, methodologies for population genomics, genomic exploration of pesticide resistance, pest management, landscape genomics, and treatments of heavily studied insect species. Chapters, written by leading and emerging experts in the field of insect population genomics, are divided into three sections. First, approaches, and evolutionary forces and dynamics in insect population genomics includes species-specific exploration into linked selection and transposable elements and a general overview of landscape genomics of insects. The second section focuses on insects as pests, and includes a methodological review of reduced representation sequencing methods for insect pest management and multiple chapters on population genomics of insecticide resistance. Finally, the third section contains a handful of "deep-dives" into insect species with long histories of population genomics study.

We envision this book to be suitable for a wide readership, including undergraduate and graduate students, research scholars and professionals, and any entomologists with an interest in evolutionary biology and population genetics. We hope it will serve as a valuable resource for future research on insect population genomics, and we thank all the authors who have contributed to this volume.

Lexington, KY, USA Julian R. Dupuis
Fredericton, NB, Canada Om P. Rajora

Contents

Contributors

Erin O. Campbell Canadian Food Inspection Agency, Ottawa, ON, Canada
Department of Biological Sciences, University of Alberta, Edmonton, AB, Canada

Lindsay P. Campbell Florida Medical Entomology Laboratory, Institute of Food and Agricultural Sciences, University of Florida, Vero Beach, FL, USA

Andrew Chaulk Department of Biology, Western University, London, ON, Canada

Tse-Yu Chen Florida Medical Entomology Laboratory, Institute of Food and Agricultural Sciences, University of Florida, Vero Beach, FL, USA

Zachary P. Cohen Department of Entomology, University of Wisconsin-Madison, Madison, WI, USA

Travis C. Collier Vector Genetics Laboratory, University of California, Davis, Davis, CA, USA

Michael S. Crossley Department of Entomology, University of Wisconsin-Madison, Madison, WI, USA
Department of Entomology, University of Georgia, Athens, GA, USA

Peter W. De Jong Laboratory of Entomology, Wageningen University, Wageningen, The Netherlands

Marcel Dicke Laboratory of Entomology, Wageningen University, Wageningen, The Netherlands

Josefa González Institute of Evolutionary Biology (CSIC-Universitat Pompeu Fabra), Barcelona, Spain

Nusha Keyghobadi Department of Biology, Western University, London, ON, Canada

Oscar D. Kirstein Emory University, Atlanta, GA, USA

Katharine L. Korunes Biology Department, Duke University, Durham, NC, USA

Yoosook Lee Florida Medical Entomology Laboratory, Institute of Food and Agricultural Sciences, University of Florida, Vero Beach, FL, USA

Boyd A. Mori Department of Agricultural, Food, and Nutrition Sciences, University of Alberta, Edmonton, AB, Canada

Mohamed A. F. Noor Biology Department, Duke University, Durham, NC, USA

Karla Saavedra-Rodriguez Colorado State University, Fort Collins, CO, USA

Kieran Samuk Biology Department, Duke University, Durham, NC, USA

Hanno Schmidt Institute of Organismic and Molecular Evolution, University of Mainz, Mainz, Germany

Sean D. Schoville Department of Entomology, University of Wisconsin-Madison, Madison, WI, USA

Kanishka Senevirathna Department of Agricultural, Food, and Nutrition Sciences, University of Alberta, Edmonton, AB, Canada

Chelsea T. Smartt Florida Medical Entomology Laboratory, Institute of Food and Agricultural Sciences, University of Florida, Vero Beach, FL, USA

Carlos Vargas-Chávez Institute of Evolutionary Biology (CSIC-Universitat Pompeu Fabra), Barcelona, Spain

Patrick Verbaarschot Laboratory of Entomology, Wageningen University, Wageningen, The Netherlands

Kim M. C. A. Vermeer Laboratory of Entomology, Wageningen University, Wageningen, The Netherlands

Part I
Approaches, and Evolutionary Forces and Dynamics

Insect Landscape Genomics

Andrew Chaulk and Nusha Keyghobadi

Abstract The composition and configuration of the landscape influence insect abundance, distribution, and diversity, with implications for the dynamics and spread of pests and disease vectors, as well as the persistence of threatened species. Here, we explore how genomic data have been used, within the spatially explicit frameworks provided by landscape genetics and genomics, to provide unique insights into how insects use and react to their surrounding landscapes. We review the literature on insect landscape genetics and genomics, identifying key trends and areas of application. We argue that the small size and short life span of most insects make them particularly well suited to exploring central issues of spatial and temporal scales in landscape genetics and genomics. We then present case studies demonstrating the power of landscape genomics approaches to provide novel insights into the ecological and evolutionary dynamics of insect pests, disease vectors, and pollinators of conservation concern.

Keywords Conservation · Disease vector · Genetics · Genomics · Insect · Landscape · Pest · Spatial scale · Temporal scale

1 Introduction

Insects interact with their surrounding landscapes in diverse and complex ways, and at a wide range of different spatial and temporal scales (Hunter 2002). The composition and configuration of the landscape in turn exert an important influence on insects, affecting abundance, distribution, and both within- and among-species diversity (Hunter 2002; Rossetti et al. 2017). Landscape structure influences the dynamics and spread of insect pest populations, as well as their exposure to and interactions with agents of control (Liebhold et al. 1993; Barron et al. 2020). For rare or threatened species, changes in landscape structure can be key drivers of decline,

A. Chaulk (✉) · N. Keyghobadi
Department of Biology, Western University, London, ON, Canada
e-mail: achaulk@uwo.ca

Julian R. Dupuis and Om P. Rajora (eds.), *Population Genomics: Insects*,
Population Genomics [Om P. Rajora (Editor-in-Chief)],
https://doi.org/10.1007/13836_2022_106,

and management of landscapes is recognized as being critical to maintaining such species and the ecosystem processes that they support (Samways et al. 2020). Genomic data provide unique insights into how insects use and react to their surrounding landscapes, and the tools of landscape genetics and genomics are increasingly used to better understand diverse aspects of the ecology and evolution of insects.

Population geneticists have long been interested in how geographic and landscape features influence genetic variation within and among populations (Mayr 1942; Dowdeswell et al. 1949; Dobzhansky et al. 1966; Kruckeberg 1969; Avise et al. 1979). Recognizing that the key processes of drift, selection, and particularly gene flow could be shaped by geographic context and landscape factors, many empirical studies have identified such effects. Since the early 2000s, much of this area of inquiry has been formally referred to as "landscape genetics" (Manel et al. 2003). Various definitions of landscape genetics have been proposed and considered (Manel et al. 2003; Storfer et al. 2007; Wang 2010; Dyer 2015). The broader of these definitions, which we adopt here, considers landscape genetics to comprise all studies that seek to understand how ecological processes and patterns occurring over heterogeneous landscapes influence genetic variation within species, by *quantifying and explicitly testing the relationship between landscape and genetic variables* (Storfer et al. 2007). Landscape genetics can provide insight into factors that impede or facilitate movement and intraspecific interactions across landscapes, that affect population size and growth rates, and that exert selection on populations (Manel and Holderegger 2013).

There are some notable and very early contributions to landscape genetics using insects as subjects. Perhaps one of the earliest studies in landscape genetics is that by Dowdeswell et al. (1949) on the meadow brown butterfly, *Maniola jurtina*. Here, the authors used wing spot number as a quantitative character to assess genetic differentiation among populations separated by putative ecological barriers (created by vegetation communities not habitable by the insect) at distances of <1 km. Although they found no evidence of genetic differentiation as a result of these putative barriers, this work nonetheless represents one of the earliest tests of a landscape genetic hypothesis (a hypothesis that is now termed "isolation by resistance" (McRae 2006)). Other examples of relatively early landscape genetic studies on insects include work by King (1987), which tested explicitly for the effects of rivers as barriers to gene flow between populations of a Melyrid beetle and Britten et al. (1995), which compared genetic differentiation and patterns of isolation by distance among populations of the butterfly, *Euphydryas editha*, in regions with very different landscape structure.

Several different conceptual and analytical frameworks for landscape genetic studies have been applied (Wagner and Fortin 2013; DiLeo and Wagner 2016; Balkenhol et al. 2019). These frameworks differ in how the spatially explicit genetic and landscape data are treated. These include distance-based, neighborhood-based, and patch-based approaches and, more recently, individual-based frameworks designed to handle continuous data (Moniem et al. 2016). These approaches are summarized and compared in other publications (e.g., Wagner and Fortin 2013;

Balkenhol et al. 2019). While landscape genetics initially placed a heavy emphasis on pairwise, distance-based analyses focused mainly on understanding patterns of gene flow across the landscape (Storfer et al. 2010), recently, in combination with increased use of genomic data, there is growing emphasis on neighborhood- and patch-based approaches for inference of selection and local adaptation (Cushman et al. 2018; Storfer et al. 2018; Balkenhol et al. 2019).

A distinction between "neutral" and "adaptive" landscape genetics has also been defined (Holderegger et al. 2006; Manel et al. 2010; Parisod and Holderegger 2012). Under this dichotomy, neutral landscape genetic studies are those that focus on patterns of genetic variation that are the products of processes other than selection, namely, genetic drift and gene flow. Specifically, neutral landscape genetic studies aim to understand how drift and gene flow (the latter being the more common process of interest) are shaped by landscape structure and landscape features. For example, many studies, including those on insects, examine the extent to which different intervening land covers may increase genetic differentiation at putatively neutral loci and, by inference, decrease gene flow (Keyghobadi et al. 1999; Keller et al. 2012; Schenau and Jha 2017). Inferences about the effects of habitat patch characteristics, such as size, shape, connectivity, and habitat quality on gene flow and drift, have also been achieved through studies of genetic differentiation and diversity at neutral markers (Habel et al. 2015; Marcus et al. 2015; Crawford and Keyghobadi 2018).

Adaptive landscape genetics on the other hand focuses on patterns of genetic variation that are a result of selection. Adaptive landscape genetic studies typically aim to identify genes or markers that underlie adaptation to environmental or landscape conditions, and at the same time assess which environmental and landscape variables are the key drivers of local adaptation (Manel et al. 2010; Rellstab et al. 2015). Such studies have identified adaptations to a variety of biotic and abiotic factors in insects, including climate and climatic variables (Waldvogel et al. 2018; Jackson et al. 2020), land cover and urbanization (Theodorou et al. 2018; Jaffé et al. 2019), host plant use (Manel et al. 2009; Silva et al. 2020), and insecticides (Crossley et al. 2017).

Over the past two decades, landscape genetics has become more rigorous in the quantitative assessment of relationships between genetic and landscape variables, borrowing concepts and methods from landscape ecology while making use of detailed landscape data obtained using remote sensing (Manel and Holderegger 2013; DiLeo and Wagner 2016; Leempoel et al. 2017). Most recently, the field has, like all of population genetics and molecular ecology, been marked by the rapidly increasing use of "genomic" data (Cushman et al. 2018; Storfer et al. 2018; Balkenhol et al. 2019).

The adaptive landscape genetics framework has sometimes been synonymized with "landscape genomics." For example, Holderegger et al. (2010) define "landscape genomics" as exploring "the interaction between environment and adaptive genetic variation in natural populations and individuals." Likewise, a recent review of landscape genomics by Storfer et al. (2018) focuses on methods for the detection of local adaptation and the identification of candidate loci under selection. We use a broader definition of landscape genomics, defining it as any study that explicitly examines relationships among landscape variables and genome-wide genetic variation – regardless of whether that variation, or the underlying processes driving it, are "neutral" or "adaptive." We will thus be consistent with other publications in this series, namely, Balkenhol et al. (2019), in our treatment of the concept of "landscape genomics." In practice, this definition means that landscape genomic studies are those that would fit the definition of landscape genetics studies and use a relatively large number (typically >100) of genetic markers, distributed throughout an organism's genome. These may include older studies using AFLPs (e.g., Honnay et al. 2006; Chaput-Bardy et al. 2008; Watanabe et al. 2014) but more recently are dominated by analyses of hundreds to thousands of single nucleotide polymorphisms (SNPs) using a variety of approaches (targeted SNP genotyping (Fountain et al. 2018), sequencing of reduced representation libraries (Hansen et al. 2018), and whole genome re-sequencing (Waldvogel et al. 2018)).

In this chapter, we provide an overview of landscape genetics and genomics as they have been applied to studies of insects. We begin with a comprehensive review of the published literature on insect landscape genetics and genomics and identify trends in taxonomic coverage, molecular markers used, aspects of study design including spatial scale, and research motivation and areas of application. We explore the traits of insects that make them particularly well suited to addressing certain landscape genetic and landscape genomic questions, particularly for questions of spatial and temporal scales. This is followed by a series of case studies that demonstrate the various ways landscape genomics have been applied to insect systems; these examples illustrate how landscape genomics can yield valuable insights into insect ecology and evolution and can be applied to solve problems in a number of important areas including agriculture, vector-borne illness, and conservation.

2 Review of the Insect Landscape Genetics and Genomics Literature

2.1 Search and Filtering Methods

We conducted a comprehensive review of the literature on insect landscape genetics and genomics, to assess the state of the field and how it has been changing over time. We began with a search for "Landscape Genetics/Genomics" articles that could then be refined to include insect-focused papers. For this initial search, we used a modified version of the search terms reported by Storfer et al. (2010). Briefly, we ran this initial query in the Web of Science (http://www.isiknowledge.com), limited to the years spanning 1998 to 2020, using the following search terms (within the title, keywords, or abstract): "landscape" and ("genetics" or "genetic variation" or "genetic structure" or **"genomics"**); "GIS" and ("genetics" or "gene flow" or "genetic structure" or **"genomics"**); "gene flow" and ("environmental factor" or "environmental variable"); "habitat fragmentation" and ("gene flow" or "genetics" or **"genomics"**); "connectivity" and ("gene flow" or "genetics" or **"genomics"**); "spatial epidemiology" and ("genetics" or "gene flow" or **"genomics"**); **"adaptive" and ("genetic variation" or "SNP*"); "circuit theory" and ("genetics" or "gene flow" or "genomics")**; and "geographical genetics" (bolded terms are those we added to the search terms of Storfer et al. (2010)).

Following this initial search, we performed two secondary refinements; the first included primarily names of insect orders: "insec*" or "Archaeognatha" or "Blattodea" or "Coleoptera" or "Dermaptera" or "Diptera" or "Embioptera" or "Ephemeroptera" or "Grylloblattodea" or "Hemiptera" or "Hymenoptera" or "Lepidoptera" or "Mantodea" or "Mecoptera" or "Megaloptera" or "Neuroptera" or "Notoptera" or "Odonata" or "Orthoptera" or "Phasmatodea" or "Phthiraptera" or "Plecoptera" or "Psocoptera" or "Raphidioptera" or "Siphonaptera" or "Thysanoptera" or "Trichoptera" or "Zoraptera" or "Zygentoma." This initial refinement, which we refer to as our "reduced" search, returned slightly less than 1,000 articles. A second more lenient search included terms from the first refinement in addition to common names for major insect groups: "bristletail" or "cockroach" or "termite" or "beetle" or "earwig" or "fly" or "*fly" or "*flies" or "mosquito*" or "*spinner" or "bug" or "true bug" or "cicada" or "wasp" or "bee" or "ant" or "moth" or "manti*" or "lacewing" or "*hopper" or "locust" or "cricket" or "katydid" or "weta" or "stick insect" or "*lice" or "louse" or "flea" or "thrips." We refer to this second, larger set of articles as our "expanded" search. We recognize that some articles have likely been missed; however, we are confident our queries properly represent the greater body of literature that encompasses insect-focused landscape genetic/genomic studies.

Our initial search retrieved a total of 13,412 articles, which were then reduced to 915 and 1,402 articles for the reduced and expanded secondary taxonomic searches, respectively. Results from each of these queries were then filtered to include only articles that focused primarily on insect taxa and that qualified as studies using landscape genetic/genomic techniques (i.e., tests for effects of landscape variables on genetic variation). Some articles included landscape genetic analyses on non-insect taxa in addition to a focal insect; in such cases, the article was retained only if a valid, landscape genetic/genomic analysis was performed on the insect (e.g., James et al. 2011). This final filtering resulted in 71 and 206 qualifying articles in the reduced and expanded taxonomic searches, respectively.

During our literature search we found that a significant proportion of retrieved records could be placed within the related field of phylogeography, either exclusively or with some degree of overlap with landscape genetic studies. While this is unsurprising given the complimentary nature of the two disciplines, the distinction between landscape genetics and phylogeography has sometimes been contentious with each having a broad and partially independent body of literature to draw from (Wang 2010; Bohonak and Vandergast 2011; Rissler 2016). Landscape genetic studies can perhaps be best distinguished from phylogeographic studies based on the emphasis of the former on the use of genetic diversity and differentiation metrics derived from allele frequency data, rather than the analysis of genealogies and application of phylogenetic methods, and on the explicit testing of relationships between these genetic metrics and landscape variables (Bohonak and Vandergast 2011; Epps and Keyghobadi 2015). There were of course records that either included analyses belonging to or straddling the boundary between the two disciplines (e.g., Margraf et al. 2007; Guzman et al. 2012). Additionally, some earlier records clearly fell into the purview of landscape genetics or some combination of landscape genetics and phylogeography, but the terminology associated with landscape genetics was not used, potentially due to the field having been relatively new at that time (e.g., Finn et al. 2006; Pedro and Sallum 2009). Thus, all records that explicitly labeled themselves as falling under the purview of phylogeography were reviewed to ensure no landscape genetic articles were missed.

After filtering the search results, we collected data from each article pertaining to the following: research motivation, relevant temporal and spatial scale(s), taxa studied, number of samples, number and type(s) of molecular markers used, whether analyses were individual or population based (or both), landscape and genetic variables used, and relevant bibliographic information (authors, year published, title). In rare cases where an article met conditions for multiple categories of a variable (e.g., two separate orders of insects were studied), the article was counted once for each category. Further details on the data collected are as follows:

Research Motivation We classified each article as belonging to one of three broad motivation classes, which were then divided into more specific subclasses. *Applied* studies included analyses that sought to answer questions related to a central species or system in the context of a specific societal, environmental, and/or economic problem. The *Applied* group was then further divided into four subclasses: *conservation* included landscape genetic or genomic studies whose focal species was formally classified as threatened or endangered, and three *pest* classes *(agricultural, disease, forest)* included studies whose focal species are classified as important agricultural, human/animal disease carrying, or forest pests, respectively. Studies in the *Investigative* class were defined as articles whose primary question is driven by a broader, often theoretical topic, and is not motivated by a focal application or species. This group was further divided into two subclasses: *conservation* included studies addressing fundamental questions motivated by conservation (e.g., population persistence related to N_e) but not focused directly on species of conservation concern, while *basic research* included any studies driven by more basic questions in landscape genetics or genomics (e.g., identifying driving factors shaping genetic differentiation/variability). A final *Methods* class included studies that used insect taxa to inform, test, or validate a novel landscape genetic or genomic method (e.g., sampling scheme, model). We collected these data from the expanded, "secondary" taxonomic filtering of the literature search.

Taxonomic Information Also using the expanded search, data relating to study organisms were collected from each article including species name(s), order(s), and whether single or multiple species were investigated.

Spatial Scale Information pertaining to spatial extent of the study area was collected from each article, based on linear distances separating the furthest sampled sites. These distances were collected directly from articles if reported. Otherwise, the distance between the two most distant sampling points was measured from maps provided by the authors. In some cases, where no figure was provided (e.g., authors report only GPS coordinates), sampling points were plotted, and linear distance was measured using Google maps (maps.google.com). As many studies did not directly report spatial extent, retrieving these data was very time consuming and was therefore collected only from our reduced literature search. We note that the reduced data set covered a broad range of spatial scales.

2.2 *Literature Search Results*

As expected, and in accordance with other reviews of landscape genetics, the number of insect landscape genetic and genomic studies has been increasing yearly since the early 2000s, although with a slight decrease in the number of published articles following 2016 (Fig. 1).

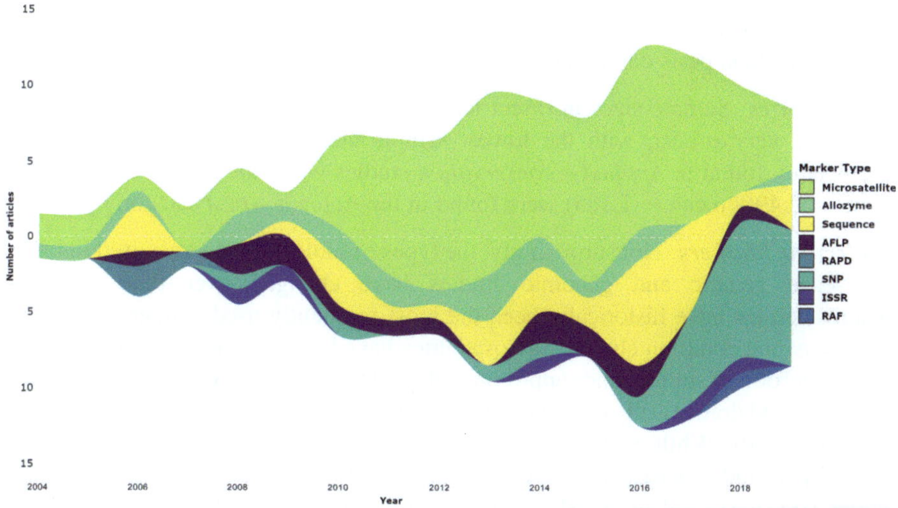

Fig. 1 Number of published insect landscape genetic and genomic studies per year, since 2004, by type of molecular marker used (AFLP amplified fragment length polymorphism; RAPD randomly amplified polymorphic DNA; SNP single nucleotide polymorphism; ISSR inter-simple sequence repeat; RAF randomly amplified DNA fingerprinting). Data are from our expanded literature search and represent 200 articles in total. The stream plot presented here shows two major revolutions in genetic marker technology, namely, the wide adoption of microsatellites, followed by breakthroughs allowing large numbers of single nucleotide polymorphisms (SNPs) to be characterized for non-model organisms. Our plot begins at 2004 as previous years had very few papers (six between 1995 and 2003), often with gaps of one or more years where no insect-focused landscape genetic studies were published

Research Motivation Most articles were investigative (57%) in nature, with the majority of those addressing basic research questions in landscape genetics or genomics (42% of total) and the remaining articles focusing more on fundamental questions motivated by issues in conservation (15% of total). Of applied studies, the majority were on species of conservation concern (17%), followed by studies involving agricultural pests (12%) and disease vectors (9%). While studies involving forest pests make up a minority of applied insect landscape genetic or genomic studies (3%), these include some very notable studies in forestry, specifically studies on the mountain pine beetle, *Dendroctonus ponderosae,* which we highlight later in this chapter. The smallest group of studies, those involving the development of novel methods in landscape genetics or genomics, represented 2% of articles.

Taxonomic Coverage A total of 12 insect orders were represented in our expanded literature survey, of which five orders make up the bulk of the articles (Lepidoptera, Coleoptera, Hymenoptera, Diptera, Orthoptera; 80%). We found that most studies focused on a single insect species (82%), while the remainder studied multiple species (including studies that included non-insect taxa). Investigative studies focused primarily on Lepidoptera (27%), with Coleoptera (15%), Hymenoptera (15%), and Orthoptera (14%) also being well-represented. Hymenoptera comprised the bulk (40%) of research in applied conservation studies. Hemiptera and Coleoptera made up the bulk (56%) of studies dealing with agricultural pests. Studies focusing on disease vectors comprised solely studies on Diptera, while all applied studies of forest pests comprised studies on Coleoptera or Blattodea.

Sample Sizes Sample sizes in insect landscape genetic and genomic studies were generally very robust, with the lowest sample sizes (min = 35, mean = 279) predictably found in *Applied – conservation* studies while the highest sample sizes (max = 2,490, mean = 1,185) were found in the *Applied – pest* categories.

Molecular Markers Not surprisingly, the types of molecular markers used in insect landscape genetic and genomic studies have changed over time (Fig. 1). Microsatellites have historically been the most frequently used marker. And while they remain a common choice, microsatellites have been used in a steadily decreasing proportion of studies since approximately 2015. The use of SNP markers, in turn, has increased dramatically over the same time period (Fig. 1), as have the numbers of SNPs per study. While SNPs first appear in our literature search in 2008, the number of loci per study (across all marker types) from 1995 to 2015 typically remained under 100 (max. 134 in 2014); beginning in 2016, however, the mean number of markers per study jumped dramatically to the thousands, due to studies using tens- to hundreds-of-thousands of SNPs. With the exception of RAPDs, which have fallen out of use since the late 2000s, a small proportion of studies continue to use alternative markers such as AFLPs, ISSRs, RAFs, and even allozymes.

Spatial Scale One of the most notable aspects of the insect landscape genetics and genomics literature is the tremendous range of spatial scales covered by different studies; articles in our reduced search reported spatial extents ranging from a few to thousands of kilometers. This is of note due to the importance of spatial scale in landscape genetics (and landscape ecology more generally). Interestingly, although insects systems provide opportunities to study processes across multiple spatial scales, we found that only ~10% of studies in our reduced literature search explicitly incorporated multiple spatial scales of analysis. In the following section, we highlight the potential of using insects to address questions of both spatial and temporal scales in landscape genetics and genomics.

3 Insects as Model Systems for Addressing Questions of Scale

Spatial and temporal scales have been identified as critical considerations in landscape genetics (Anderson et al. 2010; Cushman and Landguth 2010; Epps and Keyghobadi 2015; Fenderson et al. 2020). The spatial extent over which patterns of genetic variation are assessed, and the distances among sampling units (i.e., "grain"), determines which ecological and evolutionary processes dominate in the formation of the observed patterns. For example, among populations of the large marsh grasshopper in Switzerland, gene flow is an important determinant of genetic structure only up to distances of 3–4 km, while genetic drift is the driving factor determining genetic structure at broader spatial scales (Keller et al. 2013). Among three species of aquatic insects, different dispersal abilities result in a stronger or weaker influence of gene flow relative to drift at any particular spatial scale (Phillipsen et al. 2015). In the ant *Hypoponera opacior*, genetic structure at a very fine spatial scale (<10 m) is determined by the behaviors of wingless reproductives, while genetic structure at much broader scales (tens to hundreds of kilometers) is influenced by dispersal of winged reproductives (Foitzik et al. 2011). Examples such as these, demonstrating the ways in which different processes shape patterns of genetic variation at different scales, are quite common in the insect population and landscape genetics literature.

Different processes operating at different scales result, in turn, in different relationships between genetic and landscape variables at each scale. Spatial extent and grain of sampling therefore affect landscape genetic inferences (Cushman and Landguth 2010; Keller et al. 2013; Dupuis et al. 2019). For this reason, it is well established that the scale of sampling and analysis in landscape genetic and genomic studies should correspond to the scale of the phenomena of interest (e.g., gene flow or selection) (Anderson et al. 2010). However, researchers often do not know *a priori* what these appropriate scales are. Furthermore, there may be several phenomena of interest that operate at different spatial scales. Insects are arguably easier to sample and study at multiple scales compared to many other animals, and therefore can provide important insights into scale-related questions. Our literature search identified several insect studies that tested landscape genetic hypotheses across multiple spatial scales, and in doing so revealed the importance of different driving processes and factors at different scales. In their study of the grasshopper,

Mioscirtus wagneri, in central Spain, Ortego et al. (2012) found that the relative influence of geographic distance, elevation, and river basin differed across different spatial scales of analysis. When a much more distant and isolated population was included in their analyses, researchers found that only geographic distance was supported in structuring genetic information. However, after removing this population, the influence of all variables, including distance, elevation, and river basins could be detected. Multi-scale studies in the pitcher plant midge (*Metriocnemus knabi*) have found that the relative importance of the amount of habitat in the landscape versus its spatial configuration changes across spatial scales (Millette and Keyghobadi 2015), and that very fine scale patterns of genetic structure among larvae (i.e., patterns of differentiation among leaves within plants) are influenced by landscape variables measured at much broader spatial scales (amount of available host plants in the surrounding peatland; Rasic and Keyghobadi 2012). In a recent study, Dupuis et al. (2019) found that in a swallowtail butterfly, topographic variables best explain patterns of genetic variation at spatial scales of 250 to 1,000 meters, while climatic variables had higher explanatory power at a smaller scale (100 meters).

Given the smaller spatial scales at which many insects may operate, the prospect of replication across multiple landscapes is also more feasible. Although challenging in practice, replication across different landscapes is important because the specific characteristics and history of a given landscape can potentially affect inferences about the effects of landscape variables on gene flow (Anderson et al. 2010; Short Bull et al. 2011; Storfer et al. 2018). Replication across different landscapes also reduces the incidence of false positives and increases confidence in the results of studies of local adaptation and adaptive variation. A study by Haran et al. (2017) on a long-horned beetle, *Monochamus galloprovincialis*, provides an excellent example of testing landscape genetic hypotheses across multiple different landscapes. In this study, the researchers resampled over 30,000 study areas across the Iberian Peninsula at multiple different spatial scales ranging from ~200 to 1,000 kilometers in extent, and in over a hundred different locations. They demonstrated that the ability to detect effects of temperature, elevation, and density of pine cover on genetic differentiation varied with the location and characteristics of the study area. It was harder to detect significant effects in an area where two distinct and divergent lineages overlapped. Also, variance in the values of the landscape variable measured also mattered but showed complex interactions with spatial scale and with the specific variable being considered.

While the relative importance of various processes can differ depending on spatial scale, they also operate at different temporal scales. Spatial scale and temporal scale in population and landscape genetic studies are intimately linked with larger spatial scales typically encompassing processes operating at longer temporal scales (Balkenhol and Fortin 2015). Considerations of temporal scale in landscape genetics focus largely on rates of landscape change and corresponding rates of response in population genetic attributes (e.g., the rate of approach to drift-gene flow equilibrium or the rate of adaptation; Epps and Keyghobadi 2015). This topic is a significant concern in landscape genetics because the genetic attributes of populations (diversity, differentiation) do not change immediately in response to changes in the rates or strength of driving processes such as gene flow and drift (Latter 1973; Nei et al. 1977). Depending on the characteristics of the species and populations (e.g., generation time, dispersal rate), genetic attributes may take few to many generations to approach new equilibrium values (Varvio et al. 1986). Therefore, the genetic response of populations to landscape change may be marked by potentially significant "time lags" (Lloyd et al. 2013; Bolliger et al. 2014). As a result, contemporary landscape characteristics may not be correlated with contemporary patterns of genetic variation. A number of studies, including in insects, have shown that in some cases contemporary genetic variation may be better explained by past than current landscape structure (Keyghobadi et al. 2005; Holzhauer et al. 2006).

A key variable affecting time lags in landscape genetics is generation time, as longer generation times mean slower rates of change (per year) in the genetic attributes of populations (Landguth et al. 2010). Many insects have relatively short and often discrete generations when compared with other animals. Insects can therefore be expected to show more rapid responses to changes in landscape structure, making it easier for researchers to both elucidate meaningful relationships between genetic and landscape variables, and to determine how genetic variation tracks changes in environmental conditions and landscape structure (Garnas 2018). A number of papers on insects using both historic and contemporary landscape data find that contemporary landscape is the best predictor of genetic differentiation, suggesting rapid responses in many systems (Keyghobadi et al. 2005; Jha and Kremen 2013; Crossley et al. 2019a). On the other hand, effective population size also exerts an important influence on genetic time lags, with time lags generally being longer (i.e., genetic responses delayed) when effective population size is very large (Latter 1973; Varvio et al. 1986; Epps and Keyghobadi 2015). In those insects with very large population sizes therefore, including many pest or outbreaking species, this could counteract the effect of short generation time (Lloyd et al. 2013).

Overall, because of their relatively small size and short generation times, many insects arguably operate at spatial and temporal scales that are more tractable to study than in other organisms (Addicott et al. 1987; Woodward et al. 2005). For example, in an analysis of the effect of spatial scale on landscape genetic inference in the large marsh grasshopper, Keller et al. (2013) found that analyses conducted at relatively small spatial scales of up to 3 km had the highest explanatory power. In the Rocky Mountain apollo butterfly, a combination of relatively limited scale of dispersal (<~2 km) and short generation time, as well as relatively small local effective population sizes, allows the detection of rapid changes in genetic diversity and differentiation in response to demographic bottlenecks, including rapid shifts in spatial genetic structure as population sizes fluctuate (Jangjoo et al. 2016, 2020). These genetic responses to changes in population size would be difficult if not impossible to observe in species that operate at large spatial and temporal scales. Ultimately, studies on insects have contributed substantially to our understanding of the important influence of spatial and temporal scales in landscape genetics.

4 Case Studies

To illustrate the breadth of application of landscape genomics in insect systems, we have selected a group of studies to highlight in greater depth. These studies represent each of the applied motivations from our literature search (forestry, disease vectors, agriculture, and conservation of species at risk), utilize landscape genomic methods in interesting and sometimes novel ways, and reveal the insights that can be gained from landscape genomic studies.

4.1 The Mountain Pine Beetle: Range Expansion and Multi-Species Interactions

Periodic outbreaks of the mountain pine beetle (MPB), *Dendroctonus ponderosae* Hopkins (Fig. 2a), have occurred in North America approximately every 30–40 years (Alfaro et al. 2009). While these outbreaks have played an important role in forest rejuvenation, they have become increasingly severe, resulting in devastating damage to pine forests (Eng et al. 2005; Wulder et al. 2009; Fig. 2b). This beetle forms part of a complex ecological system, along with a genus of pathogenic fungi (*Ophiostoma* spp.), and their host trees (*Pinus* spp.; Mercado et al. 2014). The MPB and the fungi together form a symbiotic relationship in which beetles provide an efficient means of dispersal between tree hosts, while the fungus provides nutrition to the beetle (larval stages), prevents host trees from defending against beetle attack (using pitch), and aids in overcoming host tree defenses (Bleiker and Six 2007; Wang et al. 2012).

Historically, the Canadian MPB range was restricted to the southern interior of British Columbia (Bentz et al. 2010; Safranyik et al. 2010; Fig. 2c). During the most recent outbreak which began in the late 1990s, the MPB was reported east of the Canadian Rockies in western Alberta, Canada (Cullingham et al. 2011); here, the outbreak lead to unprecedented damage (on the order of millions of hectares) and concern that the MPB may spread into the vast tracts of pure jack pine of the North American boreal forest. As the Rocky Mountains had been previously assumed to be an absolute barrier, this led researchers to the obvious question: where did this unprecedented population expansion originate from and what factors may have facilitated its spread?

To answer this question, researchers conducted a series of landscape genetic and genomic studies, initially using available microsatellites and a mitochondrial locus before moving to a genomic approach. James et al. (2011) were interested in defining spatial genetic structure of both the beetle and its fungal symbiont, and introduced an analytical framework for determining how spatial patterns of genetic variation of multiple, interacting species are affected by their common environment ("multi-taxa integrated landscape genetics"). They found that isolation by distance best explained the observed genetic structure across the range for both species, although the total

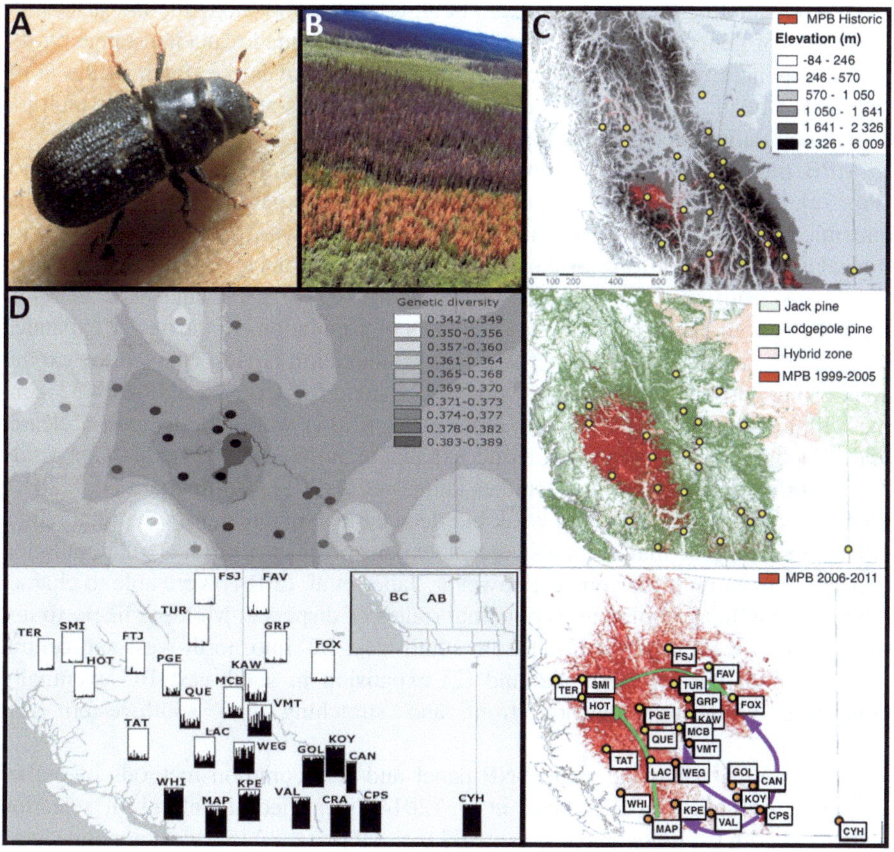

Fig. 2 The mountain pine beetle, *Dendroctonus ponderosae* (**a**), causes mass tree die-off (**b**), especially during outbreak periods. In its most recent outbreak, the MPB increased its range dramatically into much of north-central British Columbia (BC) and western Alberta (AB), Canada ((**c**) – *top panel* shows historic range before 1999; *central panel* shows distribution from 1999 to 2005). Janes et al. (2014) used landscape genomics methods to identify the route of expansion of MPB through BC and western AB ((**c**) – *bottom panel* modified to show likely routes of expansion of MPB (a) from southwestern range (MAP) north into HOT region, then east into AB (FAV/FOX) and (b) outwards from CPS region in all directions in southeastern BC and southwest AB). The authors identified two genetic clusters ((**d**) – *lower panel*), and a genetic diversity hotspot near Valemount (VMT) and Wells Gray (WEG), BC ((**d**) – *upper panel*), likely the result of the merging of two expansion fronts ((**c**) – *bottom panel*). (Panel (**a**) photo: Ron Long, Simon Fraser University, Bugwood.org, used with permission. Panel (**b**) photo: Dezene Huber, University of Northern British Columbia, used with permission. Panels (**c**, **d**) modified from Janes et al. 2014)

variation explained was low and primarily driven by structuring among major geographic regions. In the southern part of the MPB range, fungal genetic variation was a significant predictor of beetle genetic structure. In the northern, expanded region of the range, however, pine volume and climate variables were the best predictors of MPB spatial genetic structure. Using a much larger sample size and more microsatellite loci, Samarasekera et al. (2012) found moderate but detectable signatures of genetic structure across the beetle's expanded Canadian range; structure was generally weak but detectable in the south but broke down in the north (likely as a result of the more recent expansion), and the results suggested multiple source populations for the recent population expansion.

MPB populations had been previously restricted to the southern BC interior (Fig. 2c), leading researchers to postulate whether the expansion from the historical endemic range to northern BC and especially into AB was associated with physiological adaptations for long-distance dispersal, cold tolerance, or tolerance of potential toxins from novel host plants. Janes et al. (2014) used landscape genomic methods to both identify the potential sources of populations in the newly expanded range and to assess the evidence for genomic adaptation having enabled the expansion. Using a low-density genome-wide SNP panel (764 targeted SNPs in 532 samples; Illumina GoldenGate Tech.), Janes et al. (2014) were able to gain a clearer understanding of the recent epidemic expansion in Western Canada. This study reflected the results of previous studies (James et al. 2011; Cullingham et al. 2012; Samarasekera et al. 2012), with weak but detectable structure across the total range. All studies suggest that the most recent expansion was a product of population expansion from multiple sources; however, Janes et al. (2014) were able to characterize a more detailed picture of potential routes of dispersal. Multiple likely routes of expansion were described: (1) from south-west BC into north-west and central BC, then into northwestern AB and (2) expansion in south-east BC eventually contacting northern expansion front and stretching into southwestern AB (Fig. 2c, d).

Furthermore, using the same SNP panel and two common methods of outlier detection (both F_{ST} based), Janes et al. (2014) identified signatures of selection associated with the range expansion at 208 loci. Of these, 32 loci were annotated and mostly non-synonymous, and these were grouped by the authors into three broad functional categories: cholesterol/sterol association, ion transport, and actin contraction. While the exact roles these changes play, both physiologically and ecologically, remain poorly understood, they point to potential targets of selection related to phenology and shifts in developmental regimes, higher metabolic function relating to mobilizing energy for dispersal, or overcoming cold temperatures associated with overwintering in more northerly climates.

Genetic research on MPB has contributed to a better fundamental understanding of the dynamics of this complex system. This research has also enabled a more accurate assessment of the risk posed to Canadian forests by this insect; work using microsatellite markers confirmed jack pine as a suitable host for MPB, necessitating a range map revision and reinforcing importance of control efforts on the eastern front of the infestation (Cullingham et al. 2011). However, while recent research using SNPs has provided a more detailed picture of likely routes of invasion and spread, as well as evidence of local adaptation, this work has not directly influenced established MPB control programs. Management priorities and a discrepancy between the scale at which the MPB has been studied and the broad scale at which MPB control strategies are implemented means that outcomes from landscape genetic studies have not directly impacted "boots on the ground" control work. Furthermore, while identifying putative genes for local adaptation is a necessary first step, this information is of limited use to management practitioners until researchers can overcome the challenges in linking genetic differences to variation in behavior or physiology (pers. comm. Whitehouse, C., April 2021).

4.2 *Insect Disease Vectors: Human-Mediated Spread and Efficacy of Control Strategies*

Vector-borne diseases present a particularly challenging context for landscape genetic studies. Similar to the MPB system, the dynamics of disease-causing agents (e.g., malaria-inducing *Plasmodium* spp.) rely on interactions between the vector (disease-carrying agent) and the host, but also on the dispersal characteristics of the agent and the vector within the intervening landscape (Leo et al. 2016). Studies of these systems can be incredibly complex, often involving effects of human activity on vector range, as well as long-distance travel, which can influence both the vector and the disease (Hemming-Schroeder et al. 2018). Recently, several review articles have detailed ways in which landscape genetics and genomics can aid in understanding the dynamics of vectors and the diseases they transmit (Schwabl et al. 2017; Hemming-Schroeder et al. 2018; Kozakiewicz et al. 2018). These reviews highlight the complexity involved in applying landscape genetic and genomic methods to host-vector-pathogen systems. Additionally, Leo et al. (2016) have adapted the multi-taxa integrated landscape genetic approach of James et al. (2011) for zoonotic infectious disease emergence.

Researchers have used landscape genetic techniques to investigate the spread of important disease vectors, including the invasive mosquito species *Aedes albopictus* (Medley et al. 2015) and *Aedes japonicus japonicus* (Egizi et al. 2016). While these studies did not use genomic techniques (seven and nine microsatellites, respectively), they both identified a significant impact of human-mediated transportation on population connectivity, with human-aided dispersal being important for range expansions in both species. However, to date only a relatively small number of landscape genomic studies, per se, have investigated mosquito vectors. Schmidt et al. (2017) investigated spatial genetic structure and drivers of gene flow among populations of *Aedes albopictus* in a city within the species' native range. Using 3084 unlinked SNPs the authors found evidence for both IBD and human-mediated passive dispersal, again associated with transportation networks. Importantly, as highlighted by the authors, the use of high-density genome-wide markers in *Ae. albopictus* allowed the characterization of genetic structure and gene flow at a much finer scale than was possible with traditional markers. This enabled the authors to make much more detailed suggestions for controlling pest populations in the region. These studies also demonstrate the potential to use landscape genomic techniques to investigate loci under selection at range boundaries or in newly colonized regions in vector-borne disease systems (e.g., northern portions of range for *Ae. albopictus*), as well as identifying heterogeneity in insecticide resistance, akin to the Colorado potato beetle (Crossley et al. 2017).

In another study, Schmidt et al. (2018) used landscape genomic techniques to investigate a slower-than-predicted spread of the disease-controlling bacterium *Wolbachia pipientis* (wPip) through populations of *Aedes aegypti* in Cairns, Australia (Fig. 3a). *Wolbachia* is a genus of gram-negative endosymbiotic bacterium that has been used to control insect populations due to its ability to induce cytoplasmic incompatibility in some insect species (Bourtzis 2008). Importantly, infection of

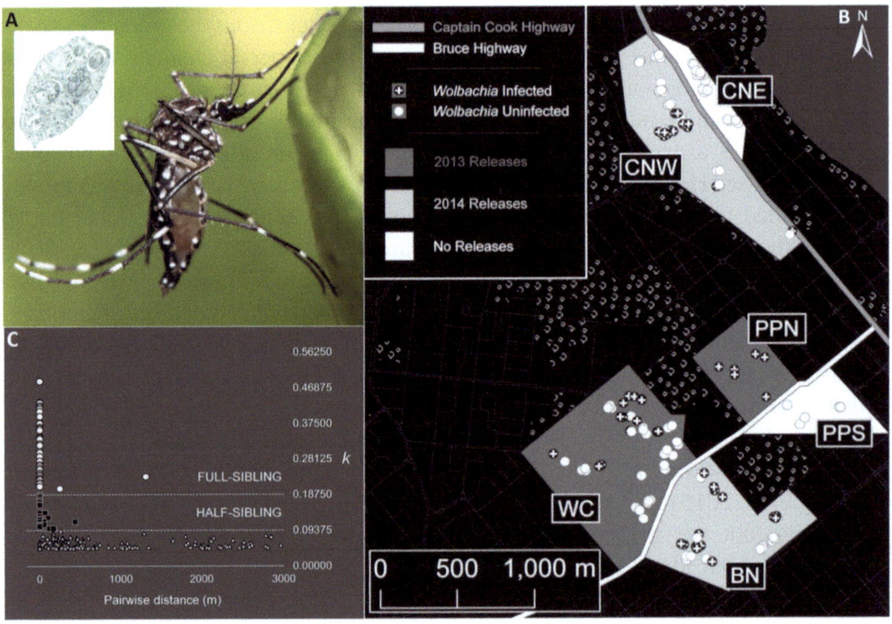

Fig. 3 The yellow fever mosquito, *Aedes aegypti* (**a**), is the primary vector for several devastating arbovirses (e.g., dengue, Zika viruses). *Ae. aegypti* mosquitoes infected with *Wolbachia,* an endosymbiotic bacterium ((**a**) – *inset*), display reduced viral transmission rates (Ferguson et al. 2015). Schmidt et al. (2018) sampled *Ae. aegypti* populations from six geographic regions (**b**); in some of these regions *Wolbachia*-infected mosquitoes had been released in previous years (light and dark gray shaded areas), while others had no releases (white areas). They found that while the relatively small *Ae. aegypti* mosquitoes could travel surprisingly far distances, as inferred from the spatial distributions of full- and half-sibs (**c**), highways had a significant influence on their genetic structure, acting as barriers to dispersal and gene flow, and therefore the transmission of *Wolbachia.* (Panel (**a**) photos: Muhammad Mahdi Karim and Scott O'Neill (inset), used under GFDL License 1.2 and CC BY 2.5, respectively. Panels (**b**, **c**) adapted from Schmidt et al. (2018); the underlying road network is derived from "Australia Oceania Continent Roads" made available by MapCruzin.com and OpenStreetMap.org under the Open Database License)

Aedes aegypti with wMel, a *Drosophila melanogaster*-derived *Wolbachia* strain, limits both the infection of mosquitoes with several important arboviruses (Dengue, Yellow fever, Chikungunya), and the transmission potential of infected mosquitoes to uninfected hosts (Ryan et al. 2020 and references therein). wMel-infected *Ae. aegypti* were released in multiple sites in Cairns (Hoffmann et al. 2011; Schmidt et al. 2018; Fig. 3b); however, the bacteria's spread throughout the population was proceeding at a much slower rate than expected (Turelli and Barton 2017). Interestingly, Schmidt et al. (2018) found that while there was some evidence for passive, human-mediated dispersal of *Ae. aegypti,* highways were found to act primarily as a moderate barrier to dispersal, likely slowing the spread of wMel throughout the Cairns populations (Fig. 3b, c).

While Schmidt et al. (2017, 2018) applied landscape genomics to identify factors limiting the efficacy of *Wolbachia*-infected mosquito release in established control programs, other spatially explicit population genomic studies have directly informed novel efforts at mosquito control. In a recent study, Schmidt et al. (2021) used spatial genomics to identify which islands act as sources of incursion of the invasive *Ae. albopictus* in the central Indo-Pacific. They were successfully able to identify different dispersal pathways, and this approach has now been adopted by management authorities working with Queensland Health; new incursions are being genotyped and traced via this method in an ongoing effort to stop *Ae. albopictus* from spreading further south (pers. comm. Schmidt T.L., April 2021).

Spatial genomic data were also used recently to characterize dispersal of *Ae. aegypti* at a very fine level of detail, for the purposes of population management using *Wolbachia*-infected mosquitoes. Filipović et al. (2020) developed a novel method to estimate the dispersal kernel (a probability density function which encompasses individual dispersal events, including those that lead to successful establishment and reproduction post-arrival) using spatial analyses of relatedness metrics generated from genome-wide SNP markers. As effective population management using *Wolbachia*-infected mosquitoes relies heavily on the ability to predict the dispersal of infected individuals, having accurate, local predictions of dispersal characteristics is important. Using ddRADseq, Filipović et al. (2020) generated ~60,000 SNPs from which they estimated genetic relatedness between all pairs of 215 female mosquitoes collected from two districts in Singapore. Combining these estimates with pairwise geographic (Euclidean) distances, the authors were able to parameterize the effective dispersal distance kernel. Using their close-kin method, Filipović et al. (2020) estimated that a buffer zone of ~250 meters would be required to achieve complete elimination of *Ae. aegypti* within the core treatment (i.e., mosquito release) area, a recommendation that is now being implemented in the ongoing control strategy in Singapore (pers. comm. Rašić G., April 2021; https://www.nea.gov.sg/corporate-functions/resources/research/wolbachia-aedes-mosquito-suppression-strategy).

4.3 Colorado Potato Beetle: Effects of Changes in Agricultural Land Use and the Emergence of Insecticide Resistance

The Colorado potato beetle (CPB), *Leptinotarsa decemlineata* Say (Fig. 4a), has been an important agricultural pest of solanaceous crops in North America since the mid-1800s (Walsh 1865; Riley 1871). In addition to its proliferation within North America and in many other countries, the CPB is notorious for its ability to rapidly develop resistance to chemical insecticides; to date it is documented to have resistance to 56 pesticides (Whalon and Mota-Sanchez 2020). In a series of studies, researchers used landscape genomic techniques to investigate the impact of historic and contemporary land cover on population structure of the CPB, the effect of agricultural practices on genetic variation in CPB populations, and, spatial patterns of variation at loci associated with insecticide resistance (Fig. 4).

Several studies in landscape genetics and genomics, including some on insect systems, have used historical land cover data to distinguish past versus current drivers of contemporary spatial genetic structure (Keyghobadi et al. 2005; Vandergast et al. 2007; Jha and Kremen 2013). This is particularly relevant in landscapes that are subject to frequent flux in their composition or configuration (e.g., agricultural landscapes). Crossley et al. (2019a) were interested in determining whether historical potato land cover had any detectable effect on contemporary patterns of genetic differentiation in two major potato-producing regions within the USA. The authors selected multiple time points spanning a period of over one and a half centuries within which the CPB was first reported as associated with the spread of potato crops in the regions (1860), through the peak period of potato production (1910), to more aggregated and intensified contemporary potato agriculture (2012). Using contemporary genetic distances estimated with a GBS-derived panel of ~7,400 SNPs, Crossley et al. (2019a) modeled landscape resistance to CPB transmission, due to absence of the host plant, during each time period. Compared to historic land cover, they found that contemporary land cover had a stronger, albeit still only weakly limiting, effect on contemporary genetic differentiation. While the authors suggest that short genetic time lags may be characteristic of agricultural insect pest systems in general, they also introduced an interesting caveat: the spatial locations of populations have changed significantly over time due to changes in agricultural intensity and land use (Crossley et al. 2019a). The authors point out that in such highly dynamic systems where spatial locations of populations shift significantly over time, our ability to detect lagged effects of historic land cover on contemporary populations will be reduced. These results were supported in a second study investigating the effects of crop rotation practices on genetic differentiation and diversity in the CPB system, where Crossley et al. (2019b) found that populations separated by a higher proportion of non-potato land cover were more genetically differentiated.

In a related study, Crossley et al. (2017) used landscape genomics techniques to investigate local adaptation to insecticide use in CPB (Fig. 4b, c). Specifically, the

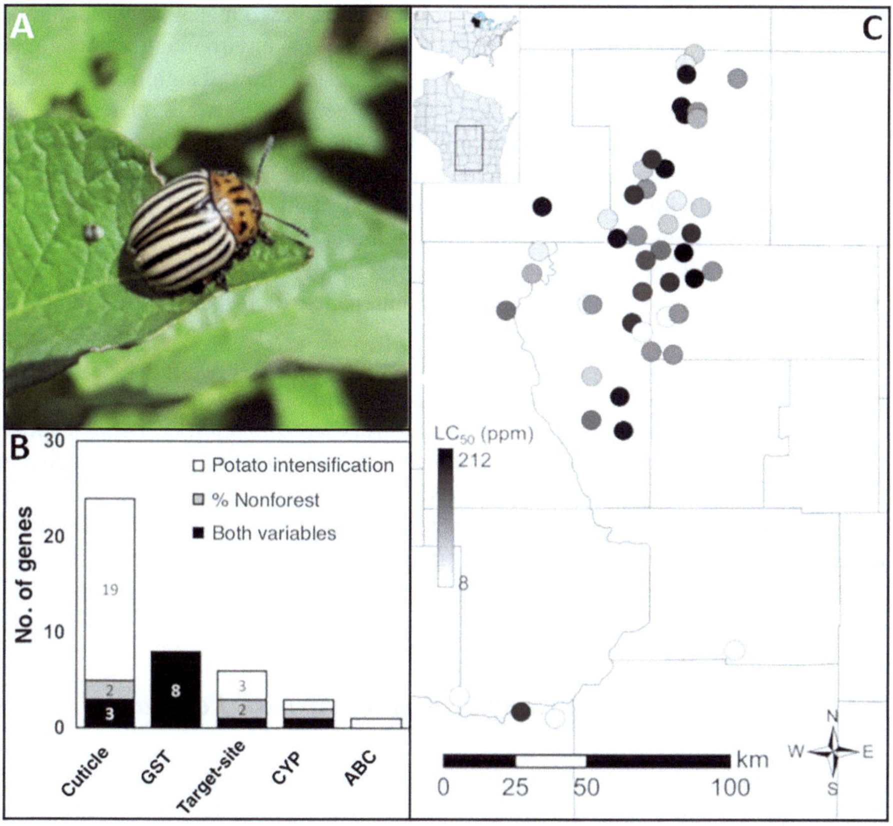

Fig. 4 The Colorado potato beetle, *Leptinotarsa decemlineata* (**a**), is the subject of a series of landscape genomic studies focused on identifying the impact of historic or contemporary land use, agricultural practices, and insecticides on neutral and adaptive genetic variation (Crossley et al. 2017, 2019a, b). Here we illustrate results on adaptive variation in relation to insecticide use. Crossley et al. (2017) identified 42 candidate genes likely involved in insecticide resistance, acting through various possible mechanisms ((**b**) numbers within bars represent counts of genes, unlabeled bars = 1 gene; Cuticle, outermost protective layer of an insect's exoskeleton; GST, glutathione S-transferase; Target-site, the physical site of action of an insecticide in the insect; CYP, cytochrome p450; ABC, ATP-binding cassette transporter). Using the spatial pattern of insecticide (imidacloprid) resistance ((**c**) LC_{50} is the expected insecticide concentration required to kill 50% of the population) combined with genetic data, the authors concluded that insecticide resistance in this species is most likely due to both de novo and standing genetic variation in many genes involved in a variety of insecticide resistance mechanisms. (Figures adapted from Crossley et al. (2017, 2019b))

authors were interested in determining whether insecticide resistance could be attributed to single genes of large effect, or multiple, small-effect genes, and whether resistance originated from standing or de novo genetic variation (via a spatial signature of an "origin and spread"). The authors used two environmental variables related to insecticide use: the first, "potato intensification," accounted for the intensity of insecticide use in potato crops across the landscape, while the second, "percent non-forest habitat," accounted for insecticide-based management including but not limited to potato crops. Using tests for genome–environment associations, Crossley et al. (2017) identified 462 putative genes that showed significant correlation with these environmental variables. Of these genes, ~9% were related to three modes of known insecticide resistance (Fig. 4b). Additionally, the authors found that insecticide resistance in the CPB in the examined regions likely arose from standing, genetic variation, with high spatial heterogeneity at small scales and high nucleotide diversity surrounding resistance-related SNPs among populations (Fig. 4c). Using landscape genomic methods, these researchers were able to provide substantial evidence for the evolutionary basis and maintenance of insecticide resistance in this important pest.

While the results of Crossley et al. (2019b) suggest that rotating potato cover over time and space can not only reduce CPB abundance, but also gene flow and genetic diversity, the factors governing US potato growers' crop rotation decisions include many more factors than CPB pressure alone. This, compounded by the high degree of coordination among farmers that would be required to control CPB with crop rotations, has meant that although landscape genomic studies in this system have produced actionable outcomes, to date farmers have not reaped the potential benefits of implementing the results of this work (pers. comm. Crossley M.S., April 2021).

4.4 The Jandaíra Bee: Conservation of At-Risk Pollinators

The global loss of insect biodiversity and biomass has received considerable attention recently (Wagner 2020; Harvey et al. 2020). Among both the general public and scientists, conservation of insect pollinators is a particularly pressing issue (Potts et al. 2016; Rhodes 2018). Pollinators provide a key ecosystem service and underlie billions of dollars of agricultural production, yet they are subject to many direct and indirect anthropogenic stresses (Losey and Vaughan 2006; Vanbergen and Insect Pollinators Initiative 2013). Landscape genetics and genomics can provide a better understanding of critical landscape and environmental factors affecting declining populations of insect pollinators (Jha and Kremen 2013; Moniem et al. 2016; Theodorou et al. 2018).

The Jandaíra bee, *Melipona subnitida* Ducke, is a tropical stingless bee native to the tropical dry forests of northeastern Brazil (Fig. 5a). It is a pollinator of native plants in threatened habitats, and is also economically important, producing honey and providing pollination services for local crops (Jaffé et al. 2015; Giannini et al. 2017). It is threatened by deforestation and relocation of colonies for agricultural use (Rego and Albuquerque 2006). Species distribution models forecasting suitable habitat under climate change identified potential refugia at higher elevations for both *M. subnitida* and its native plant resources (Giannini et al. 2017); however, the bee's ability to colonize these areas will depend on its ability to navigate a landscape intensely disrupted by human activity. To understand effects of habitat loss on this important pollinator, as well as its potential to colonize high elevation areas under

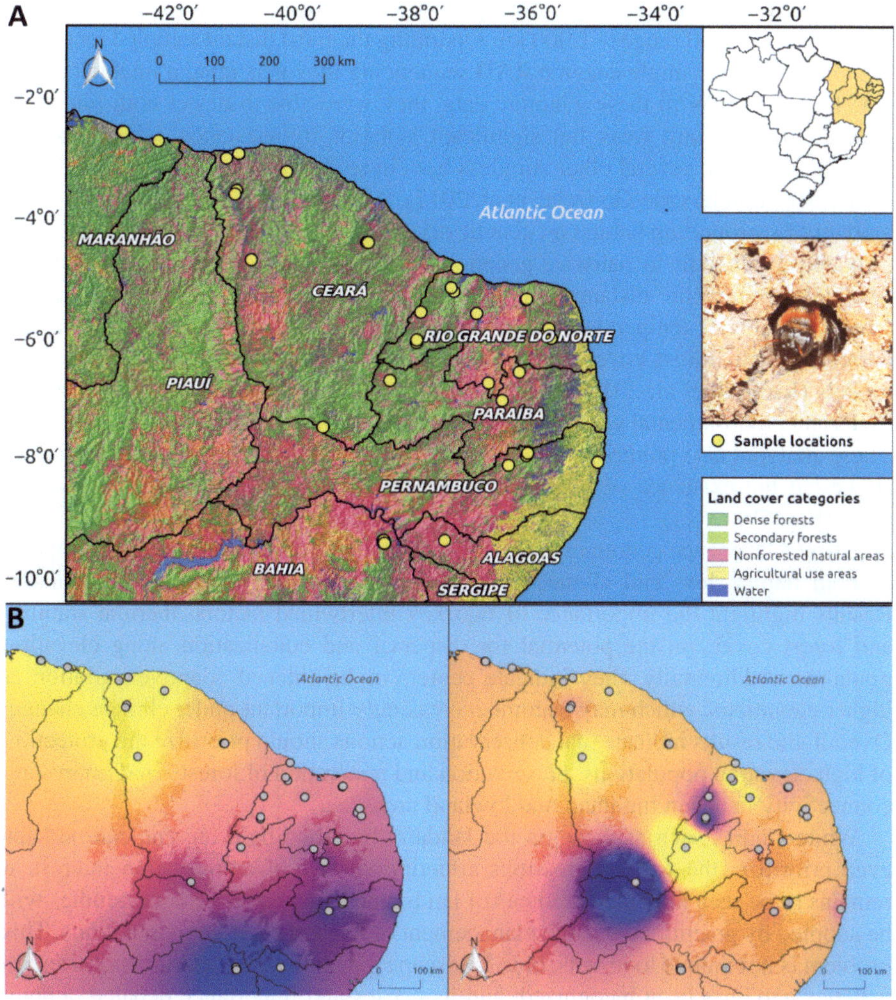

Fig. 5 The stingless bee, *Melipona subnitida* ((**a**) – inset), is an important pollinator of conservation concern restricted to northeastern Brazil. Sampling locations (yellow points) from the study of Jaffé et al. (2019) span the entire range of the species and are overlaid here on a land cover map (1; map source: http://mapbiomas.org/). Heat maps (**b**) show the spatial distribution of adaptive genetic variation identified in *M. subnitida* using latent factor mixed models and are represented here as interpolated spatial principal components (sPCA). Jaffé et al. (2019) interpret patterns for sPC1 (left) and sPC2 (right) as adaptive latitudinal and altitudinal clines, respectively. Shaded areas have elevations of at least 500 masl. (Figures adapted from Jaffé et al. (2019))

climate change, Jaffé et al. (2019) examined whether genomic diversity and gene flow were significantly affected by the amount of remaining native habitat and other environmental variables. To identify loci that could be important for adaptation to future conditions, they also tested for signatures of contemporary adaptation along climatic gradients driven by latitude and altitude.

The authors sampled a single freshly emerged worker bee from each of 160 nests across the bee's full range (~1,000 km separating the most distant sampled locations; Fig. 5a) and using single enzyme RAD sequencing they generated a final dataset of 3,454 SNP loci. With these genomic data, they were able to detect both fine scale genetic structure and weak but significant isolation-by-resistance, which had not been detectable in several other stingless bees in studies using microsatellites (Jaffé et al. 2016; Landaverde-Gonzalez et al. 2017). Although they found no effect of the amount of surrounding habitat on genetic diversity or gene flow, resistance distances provided a better fit to pairwise genetic relatedness between individuals than did Euclidean geographic distance. These analyses indicated that gene flow increased where the annual temperature range was lower; this factor, in turn, was highly correlated with forest cover. The researchers also employed latent factor mixed models (Rellstab et al. 2015) to identify associations between potentially adaptive SNPs and environmental variables, while accounting for spatial population structure. These analyses also pointed to an important role of ambient temperature and further suggested the existence of populations locally adapted to high and low elevation areas (Fig. 5b).

Using a landscape genomics approach, this study uncovered the potential impacts of both habitat loss and climate change on this regionally important pollinator. Results highlight the importance of two key intertwined factors, thermal stability and forest cover, on the potential for dispersal and colonization along elevation gradients. Additionally, they indicate centers of heightened adaptive variation at higher elevations, which may become increasingly important under climate change. Overall, the results indicate that conservation actions should prioritize the protection of high elevation populations, preservation and restoration of forests, and improving connectivity between highland and lowland areas.

While studies such as those on the Jandaíra bee shed light on, and can perhaps even prioritize, the various pressures affecting species of conservation interest, it remains to be seen how the outcomes of landscape genetic and genomic studies will be adopted by practitioners in the management of threatened insect populations. This problem is not unique to insects, or to landscape genetics per se, as the "conservation genetics gap" has been recognized, and its causes and importance debated, among conservation geneticists and practitioners (Hoban et al. 2013; Shafer et al. 2015); while genetic data from non-model organisms have never been so easy to generate, they are often deemed unnecessary or unattainable in management settings (Shafer et al. 2015).

Unlike efforts linked to pest control, conservation of at-risk species has a much more established bank of applied grey literature to draw from. To investigate the degree to which landscape genetics and genomics have informed the management of at-risk insect species, we queried conservation documents (e.g., recovery strategies, action plans, species assessments, multi-year reviews) for insect taxa from both the USA and Canada. American documents were accessed via the Environmental Conservation Online System's species search tool (https://ecos.fws.gov/ecp/species-reports), while Canadian reports were accessed via the species at risk public registry search tool (https://www.canada.ca/en/environment-climate-change/services/species-risk-public-registry.html). In total, 374 documents were collected representing 90 American and 81 Canadian insect species (179 documents for US listed species and 188 documents for Canadian species at risk). While most documents were species-specific, some that covered multiple species (e.g., annual reports) were also included as they were linked to species profiles. Documents were downloaded and queried for search terms using the R packages rvest (v0.3.6; Wickham 2020) and pdfsearch (v0.3.0; LeBeau 2018), respectively. Each document was queried using a set of search terms modified from the terms used in our original literature search as the search function in pdfsearch cannot handle logical operators. These terms were as follows: "landscape genetic," "landscape genomic," "genetic*," "genomic*," "genetic structure," "adaptive variation," "SNP," "gene flow," "genetic connectivity," and "spatial genetic."

Of 374 documents surveyed, only 12 referenced published landscape genetic or genomic studies. These covered six species of butterfly, two species of bumble bee, and one document provided amendments to recovery criteria for seven species of endangered karst invertebrates. While many conservation documents discussed risks associated with genetic factors (e.g., drift, inbreeding), much fewer directly referenced or included the results of genetic analyses (e.g., barcoding, genetic variation) in their reports, primarily due to a lack of available genetic resources already established for those species. In many cases, genetic data and analyses at the most basic level (i.e., quantifying genetic diversity, population structure) are still relegated to the "research gaps" portion of many species reports, though some reports do include early-stage plans for gathering genetic data. While Jaffé et al. (2019) illustrate very well the insight landscape genomics can bring to managing species at risk, our review of conservation documents from the USA and Canada suggests limited reference to, or application of, the results of such studies in the management of species at risk. Indeed, a lack of even basic genetic data on most species highlights how far we are from applying more involved landscape genetic approaches to the wholesale management of at-risk insects.

We note also one current challenge for the application of modern genomic methods to at-risk insect species: the ability to collect tissue samples of sufficient size and quality. Lethal sampling of entire individuals is often inadvisable, or simply not permitted, for at-risk insect species. Non-lethal tissue samples, such as legs, wing clips or exuviae (Holehouse et al. 2003; Vila et al. 2009; Ožana et al. 2020), are adequate for genotyping using microsatellites or AFLPs (Châline et al. 2004; Keyghobadi et al. 2009; Oi et al. 2013), but these do not typically provide sufficient DNA quantity or quality for commonly used reduced representation NGS-based genotyping methods such as RADSeq (Graham et al. 2015; Schmidt et al. 2020). Effective approaches to the genomic analysis of low quantity and quality DNA samples have been developed in forensics and ancient DNA studies, but these depend on the availability of genomic resources (e.g., a sequenced genome or high quality genomic libraries), interrogate a relatively small number of markers, or may be prohibitively expensive for large numbers of samples (Carpenter et al. 2013; Enk et al. 2014; Zhou et al. 2016). Wide-scale application of landscape genomics approaches in threatened insects will require further development and refinement of genome-wide genotyping techniques that can be consistently and readily applied to very small quantities of DNA available from non-lethal tissue samples (Shafer et al. 2015; Schmidt et al. 2020).

Landscape genetics and genomics have the potential to provide valuable insights into the management of insect populations, be they pests, species of conservation concern, or anything in between. In our initial literature survey, studies focused on insects of conservation concern and those aimed at insect pests (in forestry, agriculture, or disease) were roughly equally represented (17% and 22%, respectively), indicating the adoption of landscape genetic and genomic approaches in both applied settings. The case studies presented here highlight in further detail the value of landscape genomics in both contexts. Although research motivations varied, the approaches highlighted in the case studies enabled researchers to identify specific populations, habitats, and landscape factors that represent critical foci for management and revealed the potential for dispersal, spread, or adaptation under rapidly changing environmental conditions.

5 Conclusions and Future Prospects

Many major advances have occurred over the nearly two decades of formal landscape genetics research. Reduced sequencing costs and novel genetic techniques have allowed researchers to generate far more data from non-model organisms. In combination with advances in statistical frameworks and study design, the breadth of questions that can be addressed, and the power to tackle those questions, have expanded massively. As we look to the future, we believe that enduring questions relating to scale, or that otherwise require more tractable or discrete generation times, may encourage researchers to look to insects as model organisms in landscape genetics and genomics. Insects and landscape genomics have a lot to offer each other. On the one hand, the small size and short life span of most insects make them well suited to exploring and overcoming issues of temporal and spatial scales. Insects therefore offer the potential to reveal important insights in landscape genomics that could be difficult to achieve in organisms that operate at much larger scales. On the other hand, landscape genomics approaches offer important tools for addressing questions about insect population biology. The inclusion of insects as important model organisms in landscape genetics was recommended by Balkenhol et al. (2016). In their publication, the authors reflected on the state of landscape genetics research at that time. One of their recommendations for the field, to expand the taxonomic breadth of the discipline, highlighted the potential power of using insects to address topics requiring a higher degree of control and/or replication. Some studies have indeed taken advantage of the unique benefits afforded to researchers by insect systems, such as short and often discrete generation times, extreme breadth in dispersal ability and behavior, and ease of handling. Nonetheless, our literature search indicates that there is considerable scope to expand research that explicitly takes advantage of these traits of insects to characterize changing patterns and processes across spatial and temporal scales, and to assess time lags, perform comparative multi-species studies, and incorporate replication at the landscape level. To approach the sometimes very fine spatial scales at which many insects operate, advancements and increases in the availability of very high-resolution, fine-scale remote sensing data will be key. We anticipate that the increasing availability of LIDAR and drone technology will allow for more accurate characterization of key landscape variables at such fine scales and allow researchers to better match the scale of landscape data with the scales at which insects operate. For example, unmanned aerial vehicles (i.e., drones) equipped with high-resolution stereo photogrammetric sensors are increasingly being deployed for the generation of small-scale maps for the purposes of species identification and distribution modeling (Guillaume et al. 2021). We also note opportunities to extend recent advances in the analysis of spatial genomic data from insects by incorporating landscape data in novel ways. For example, the approach of Filipović et al. (2020) to generate a dispersal kernel based on sibling relatedness could be extended to include landscape resistance rather than geographic distance alone, as in Compton et al. (2007). In the specific context of a disease vector, such an approach could lead to even more detailed management recommendations that consider differential dispersal rates through a variety of land cover types and configurations.

Finally, we emphasize that a concerted effort is required by researchers and practitioners to address the gap between basic landscape genetic research and the application of that work by management practitioners. Landscape genetics and genomics can provide valuable insights that could be implemented by practitioners in a wide variety of applied contexts. However the actionable recommendations arising from the basic research often do not take into account factors outside of the system itself, as is evident from the mismatch in scales used in research compared to the scales at which populations are managed, for example in the Colorado potato beetle or mountain pine beetle systems. It remains clear that in many cases more communication and engagement is necessary, at all stages of study design and execution, for landscape genetics research to provide truly actionable outcomes for species management.

Acknowledgments We thank Jessica Jeong for assistance with the literature review, Dezene Huber for help finding mountain pine beetle images, and Julian Dupuis and one anonymous reviewer for their comments on an earlier version.

References

Addicott JF, Aho JM, Antolin MF, Padilla DK, Richardson JS, Soluk DA. Ecological neighborhoods: scaling environmental patterns. Oikos. 1987;49:340–6. https://doi.org/10.2307/3565770.

Alfaro R, Campbell E, Hawkes B. Historical frequency, intensity and extent of Mountain Pine Beetle Disturbance in British Columbia. Mt Pine Beetle Initiat Nat Resour Can, 2009.

Anderson CD, Epperson BK, Fortin M-J, Holderegger R, James PMA, Rosenberg MS, et al. Considering spatial and temporal scale in landscape-genetic studies of gene flow. Mol Ecol. 2010;19:3565–75. https://doi.org/10.1111/j.1365-294X.2010.04757.x.

Avise JC, Giblin-Davidson C, Laerm J, Patton JC, Lansman RA. Mitochondrial DNA clones and matriarchal phylogeny within and among geographic populations of the pocket gopher, Geomys pinetis. Proc Natl Acad Sci. 1979;76:6694–8. https://doi.org/10.1073/pnas.76.12.6694.

Balkenhol N, Fortin M-J. Basics of study design: sampling landscape heterogeneity and genetic variation for landscape genetic studies. In: Landscape genetics. Wiley; 2015. p. 58–76.

Balkenhol N, Cushman SA, Waits LP, Storfer A. Current status, future opportunities, and remaining challenges in landscape genetics. In: Balkenhol N, Cushman SA, Storfer AT, Waits LP, editors. Landscape genetics. Chichester: Wiley; 2016. p. 247–56.

Balkenhol N, Dudaniec RY, Krutovsky KV, Johnson JS, Cairns DM, Segelbacher G, et al. Landscape genomics: understanding relationships between environmental heterogeneity and genomic characteristics of populations. In: Rajora OP, editor. Population genomics: concepts, approaches and applications. Cham: Springer International Publishing; 2019. p. 261–322.

Barron MC, Liebhold AM, Kean JM, Richardson B, Brockerhoff EG. Habitat fragmentation and eradication of invading insect herbivores. J Appl Ecol. 2020;57:590–8. https://doi.org/10.1111/1365-2664.13554.

Bentz BJ, Régnière J, Fettig CJ, Hansen EM, Hayes JL, Hicke JA, et al. Climate change and Bark Beetles of the Western United States and Canada: direct and indirect effects. Bioscience. 2010;60:602–13. https://doi.org/10.1525/bio.2010.60.8.6.

Bleiker KP, Six DL. Dietary benefits of fungal associates to an eruptive Herbivore: potential implications of multiple associates on host population dynamics. Environ Entomol. 2007;36: 1384–96. https://doi.org/10.1093/ee/36.6.1384.

Bohonak AJ, Vandergast AG. The value of DNA sequence data for studying landscape genetics. Mol Ecol. 2011;20:2477–9. https://doi.org/10.1111/j.1365-294X.2011.05122.x.

Bolliger J, Lander T, Balkenhol N. Landscape genetics since 2003: status, challenges and future directions. Landsc Ecol. 2014;29:361–6. https://doi.org/10.1007/s10980-013-9982-x.

Bourtzis K. Wolbachia- based technologies for insect pest population control. In: Aksoy S, editor. Transgenesis and the management of Vector-Borne disease. New York: Springer; 2008. p. 104–13.

Britten HB, Brussard PF, Murphy DD, Ehrlich PR. A test for isolation-by-distance in Central Rocky Mountain and Great Basin populations of Edith's Checkerspot Butterfly (Euphydryas editha). J Hered. 1995;86:204–10. https://doi.org/10.1093/oxfordjournals.jhered.a111563.

Carpenter ML, Buenrostro JD, Valdiosera C, Schroeder H, Allentoft ME, Sikora M, et al. Pulling out the 1%: whole-genome capture for the targeted enrichment of ancient DNA sequencing libraries. Am J Hum Genet. 2013;93:852–64. https://doi.org/10.1016/j.ajhg.2013.10.002.

Châline N, Ratnieks FLW, Raine NE, Badcock NS, Burke T. Non-lethal sampling of honey bee, Apis mellifera, DNA using wing tips. Apidologie. 2004;35:311–8. https://doi.org/10.1051/apido:2004015.

Chaput-Bardy A, Lemaire C, Picard D, Secondi J. In-stream and overland dispersal across a river network influences gene flow in a freshwater insect, Calopteryx splendens. Mol Ecol. 2008;17: 3496–505. https://doi.org/10.1111/j.1365-294X.2008.03856.x.

Compton BW, McGarigal K, Cushman SA, Gamble LR. A Resistant-Kernel Model of connectivity for amphibians that breed in Vernal Pools. Conserv Biol. 2007;21:788–99. https://doi.org/10.1111/j.1523-1739.2007.00674.x.

Crawford LA, Keyghobadi N. Analysis of genetic diversity in a peatland specialist butterfly suggests an important role for habitat quality and small habitat patches. Conserv Genet. 2018;19:1109–21. https://doi.org/10.1007/s10592-018-1082-7.

Crossley MS, Chen YH, Groves RL, Schoville SD. Landscape genomics of Colorado potato beetle provides evidence of polygenic adaptation to insecticides. Mol Ecol. 2017;26:6284–300. https://doi.org/10.1111/mec.14339.

Crossley MS, Rondon S, Schoville SD. Patterns of genetic differentiation in Colorado potato beetle correlate with contemporary, not historic, potato land cover. Evol Appl. 2019a;12:804–14. https://doi.org/10.1111/eva.12757.

Crossley MS, Rondon SI, Schoville SD. Effects of contemporary agricultural land cover on Colorado potato beetle genetic differentiation in the Columbia Basin and Central Sands. Ecol Evol. 2019b;9:9385–94. https://doi.org/10.1002/ece3.5489.

Cullingham CI, Cooke JEK, Dang S, Davis CS, Cooke BJ, Coltman DW. Mountain pine beetle host-range expansion threatens the boreal forest. Mol Ecol. 2011;20:2157–71. https://doi.org/10.1111/j.1365-294X.2011.05086.x.

Cullingham CI, James PMA, Cooke JEK, Coltman DW. Characterizing the physical and genetic structure of the lodgepole pine × jack pine hybrid zone: mosaic structure and differential introgression. Evol Appl. 2012;5:879–91. https://doi.org/10.1111/j.1752-4571.2012.00266.x.

Cushman SA, Landguth EL. Scale dependent inference in landscape genetics. Landsc Ecol. 2010;25:967–79.

Cushman SA, Shirk AJ, Howe GT, Murphy MA, Dyer RJ, Joost S. The least cost path from landscape genetics to landscape genomics: challenges and opportunities to explore NGS data in a spatially explicit context. Front Genet. 2018;9 https://doi.org/10.3389/fgene.2018.00215.

DiLeo MF, Wagner HH. A landscape ecologist's agenda for landscape genetics. Curr Landsc Ecol Rep. 2016;1:115–26. https://doi.org/10.1007/s40823-016-0013-x.

Dobzhansky T, Anderson WW, Pavlovsky O. Genetics of natural populations. XXXVIII. Continuity and change in populations of Drosophila pseudoobscura in Western United States. Evolution. 1966;20:418–27. https://doi.org/10.2307/2406640.

Dowdeswell WH, Fisher RA, Ford EB. The quantitative study of populations in the Lepidoptera II. Maniola jurtina L. Heredity. 1949;3:67–84. https://doi.org/10.1038/hdy.1949.3.

Dupuis JR, Cullingham CI, Nielsen SE, Sperling FAH. Environmental effects on gene flow in a species complex of vagile, hilltopping butterflies. Biol J Linn Soc. 2019;127:417–28. https://doi.org/10.1093/biolinnean/blz043.

Dyer RJ. Is there such a thing as landscape genetics? Mol Ecol. 2015;24:3518–28. https://doi.org/10.1111/mec.13249.

Egizi A, Kiser J, Abadam C, Fonseca DM. The hitchhiker's guide to becoming invasive: exotic mosquitoes spread across a US state by human transport not autonomous flight. Mol Ecol. 2016;25:3033–47. https://doi.org/10.1111/mec.13653.

Eng M, Fall A, Hughes J, Shore TL, Riel WG, Hall P, et al. Provincial-level projection of the current mountain Pine Beetle outbreak: an overview of the model (BCMPB v2) and results of year 2 of the project. 2005–20, 2005.

Enk JM, Devault AM, Kuch M, Murgha YE, Rouillard J-M, Poinar HN. Ancient whole genome enrichment using Baits built from modern DNA. Mol Biol Evol. 2014;31:1292–4. https://doi.org/10.1093/molbev/msu074.

Epps CW, Keyghobadi N. Landscape genetics in a changing world: disentangling historical and contemporary influences and inferring change. Mol Ecol. 2015;24:6021–40. https://doi.org/10.1111/mec.13454.

Fenderson LE, Kovach AI, Llamas B. Spatiotemporal landscape genetics: investigating ecology and evolution through space and time. Mol Ecol. 2020;29:218–46. https://doi.org/10.1111/mec.15315.

Ferguson NM, Kien DTH, Clapham H, Aguas R, Trung VT, Chau TNB, et al. Modeling the impact on virus transmission of Wolbachia-mediated blocking of dengue virus infection of Aedes aegypti. Sci Transl Med. 2015;7:279ra37. https://doi.org/10.1126/scitranslmed.3010370.

Filipović I, Hapuarachchi HC, Tien W-P, Razak MAB, Lee C, Tan CH, et al. Using spatial genetics to quantify mosquito dispersal for control programs. BMC Biol. 2020;18:104. https://doi.org/10.1186/s12915-020-00841-0.

Finn DS, Theobald DM, Black WC, Poff NL. Spatial population genetic structure and limited dispersal in a Rocky Mountain alpine stream insect. Mol Ecol. 2006;15:3553–66. https://doi.org/10.1111/j.1365-294X.2006.03034.x.

Foitzik S, Rueger MH, Kureck IM, Metzler D. Macro- and microgeographic genetic structure in an ant species with alternative reproductive tactics in sexuals. J Evol Biol. 2011;24:2721–30. https://doi.org/10.1111/j.1420-9101.2011.02397.x.

Fountain T, Husby A, Nonaka E, DiLeo MF, Korhonen JH, Rastas P, et al. Inferring dispersal across a fragmented landscape using reconstructed families in the Glanville fritillary butterfly. Evol Appl. 2018;11:287–97. https://doi.org/10.1111/eva.12552.

Garnas JR. Rapid evolution of insects to global environmental change: conceptual issues and empirical gaps. Curr Opin Insect Sci. 2018;29:93–101. https://doi.org/10.1016/j.cois.2018.07.013.

Giannini TC, Costa WF, Cordeiro GD, Imperatriz-Fonseca VL, Saraiva AM, Biesmeijer J, et al. Projected climate change threatens pollinators and crop production in Brazil. PLoS One. 2017;12:e0182274. https://doi.org/10.1371/journal.pone.0182274.

Graham CF, Glenn TC, McArthur AG, Boreham DR, Kieran T, Lance S, et al. Impacts of degraded DNA on restriction enzyme associated DNA sequencing (RADSeq). Mol Ecol Resour. 2015;15:1304–15. https://doi.org/10.1111/1755-0998.12404.

Guillaume AS, Leempoel K, Rochat E, Rogivue A, Kasser M, Gugerli F, et al. Multiscale very high resolution topographic models in Alpine ecology: Pros and Cons of Airborne LiDAR and Drone-based Stereo-Photogrammetry Technologies. Remote Sens (Basel). 2021;13:1588. https://doi.org/10.3390/rs13081588.

Guzman NV, Lanteri AA, Confalonieri VA. Colonization ability of two invasive weevils with different reproductive modes. Evol Ecol. 2012;26:1371–90. https://doi.org/10.1007/s10682-012-9564-4.

Habel JC, Brueckmann SV, Krauss J, Schwarzer J, Weig A, Husemann M, et al. Fragmentation genetics of the grassland butterfly Polyommatus coridon: stable genetic diversity or extinction debt? Conserv Genet. 2015;16:549–58. https://doi.org/10.1007/s10592-014-0679-8.

Hansen AK, Justesen MJ, Olsen MT, Solodovnikov A. Genomic population structure and conservation of the red listed Carabus arcensis (Coleoptera: Carabidae) in island–mainland habitats of Northern Europe. Insect Conserv Divers. 2018;11:255–66. https://doi.org/10.1111/icad.12268.

Haran JM, Rossi J-P, Pajares J, Bonifacio L, Naves P, Roques A, et al. Multi-scale and multi-site resampling of a study area in spatial genetics: implications for flying insect species. PeerJ. 2017;5:e4135. https://doi.org/10.7717/peerj.4135.

Harvey JA, Heinen R, Armbrecht I, Basset Y, Baxter-Gilbert JH, Bezemer TM, et al. International scientists formulate a roadmap for insect conservation and recovery. Nat Ecol Evol. 2020;4:174–6. https://doi.org/10.1038/s41559-019-1079-8.

Hemming-Schroeder E, Lo E, Salazar C, Puente S, Yan G. Landscape genetics: a toolbox for studying Vector-Borne diseases. Front Ecol Evol. 2018;6 https://doi.org/10.3389/fevo.2018.00021.

Hoban SM, Hauffe HC, Pérez-Espona S, Arntzen JW, Bertorelle G, Bryja J, et al. Bringing genetic diversity to the forefront of conservation policy and management. Conserv Genet Resour. 2013;5:593–8. https://doi.org/10.1007/s12686-013-9859-y.

Hoffmann AA, Montgomery BL, Popovici J, Iturbe-Ormaetxe I, Johnson PH, Muzzi F, et al. Successful establishment of Wolbachia in Aedes populations to suppress dengue transmission. Nature. 2011;476:454–7. https://doi.org/10.1038/nature10356.

Holderegger R, Kamm U, Gugerli F. Adaptive vs. neutral genetic diversity: implications for landscape genetics. Landsc Ecol. 2006;21:797–807. https://doi.org/10.1007/s10980-005-5245-9.

Holderegger R, Buehler D, Gugerli F, Manel S. Landscape genetics of plants. Trends Plant Sci. 2010;15:675–83. https://doi.org/10.1016/j.tplants.2010.09.002.

Holehouse KA, Hammond RL, Bourke AFG. Non-lethal sampling of DNA from bumble bees for conservation genetics. Insectes Sociaux. 2003;50:277–85. https://doi.org/10.1007/s00040-003-0672-6.

Holzhauer SIJ, Ekschmitt K, Sander A-C, Dauber J, Wolters V. Effect of historic landscape change on the genetic structure of the bush-cricket Metrioptera roeseli. Landsc Ecol. 2006;21:891–9. https://doi.org/10.1007/s10980-005-0438-9.

Honnay O, Coart E, Butaye J, Adriaens D, Van Glabeke S, Roldán-Ruiz I. Low impact of present and historical landscape configuration on the genetics of fragmented Anthyllis vulneraria populations. Biol Conserv. 2006;127:411–9. https://doi.org/10.1016/j.biocon.2005.09.006.

Hunter MD. Landscape structure, habitat fragmentation, and the ecology of insects. Agric Forest Entomol. 2002;4:159–66. https://doi.org/10.1046/j.1461-9563.2002.00152.x.

Jackson JM, Pimsler ML, Oyen KJ, Strange JP, Dillon ME, Lozier JD. Local adaptation across a complex bioclimatic landscape in two montane bumble bee species. Mol Ecol. 2020;29:920–39. https://doi.org/10.1111/mec.15376.

Jaffé R, Pope N, Carvalho AT, Maia UM, Blochtein B, de Carvalho CA, et al. Bees for development: Brazilian Survey reveals how to optimize stingless beekeeping. PLoS One. 2015;10: e0121157. https://doi.org/10.1371/journal.pone.0121157.

Jaffé R, Castilla A, Pope N, Imperatriz-Fonseca VL, Metzger JP, Arias MC, et al. Landscape genetics of a tropical rescue pollinator. Conserv Genet. 2016;17:267–78. https://doi.org/10.1007/s10592-015-0779-0.

Jaffé R, Veiga JC, Pope NS, Lanes ECM, Carvalho CS, Alves R, et al. Landscape genomics to the rescue of a tropical bee threatened by habitat loss and climate change. Evol Appl. 2019;12: 1164–77. https://doi.org/10.1111/eva.12794.

James PMA, Coltman DW, Murray BW, Hamelin RC, Sperling FAH. Spatial genetic structure of a Symbiotic Beetle-Fungal System: toward multi-taxa integrated landscape genetics. PLoS One. 2011;6 https://doi.org/10.1371/journal.pone.0025359.

Janes JK, Li Y, Keeling CI, Yuen MMS, Boone CK, Cooke JEK, et al. How the Mountain Pine Beetle (Dendroctonus ponderosae) Breached the Canadian Rocky Mountains. Mol Biol Evol. 2014;31:1803–15. https://doi.org/10.1093/molbev/msu135.

Jangjoo M, Matter SF, Roland J, Keyghobadi N. Connectivity rescues genetic diversity after a demographic bottleneck in a butterfly population network. Proc Natl Acad Sci. 2016;113: 10914–9.

Jangjoo M, Matter SF, Roland J, Keyghobadi N. Demographic fluctuations lead to rapid and cyclic shifts in genetic structure among populations of an alpine butterfly, Parnassius smintheus. J Evol Biol. 2020;33:668–81. https://doi.org/10.1111/jeb.13603.

Jha S, Kremen C. Urban land use limits regional bumble bee gene flow. Mol Ecol. 2013;22:2483–95. https://doi.org/10.1111/mec.12275.

Keller D, van Strien MJ, Holderegger R. Do landscape barriers affect functional connectivity of populations of an endangered damselfly? Freshw Biol. 2012;57:1373–84. https://doi.org/10.1111/j.1365-2427.2012.02797.x.

Keller D, Holderegger R, van Strien MJ. Spatial scale affects landscape genetic analysis of a wetland grasshopper. Mol Ecol. 2013;22:2467–82. https://doi.org/10.1111/mec.12265.

Keyghobadi N, Roland J, Strobeck C. Influence of landscape on the population genetic structure of the alpine butterfly Parnassius smintheus (Papilionidae). Mol Ecol. 1999;8:1481–95. https://doi.org/10.1046/j.1365-294x.1999.00726.x.

Keyghobadi N, Roland J, Matter SF, Strobeck C. Among- and within-patch components of genetic diversity respond at different rates to habitat fragmentation: an empirical demonstration. Proc R Soc B Biol Sci. 2005;272:553–60. https://doi.org/10.1098/rspb.2004.2976.

Keyghobadi N, Crawford LA, Maxwell SA. Successful analysis of AFLPs from non-lethally sampled wing tissues in butterflies. Conserv Genet. 2009;10:2021. https://doi.org/10.1007/s10592-009-9884-2.

King PS. Macro- and microgeographic structure of a spatially subdivided beetle species in nature. Evolution. 1987;41:401–16. https://doi.org/10.1111/j.1558-5646.1987.tb05806.x.

Kozakiewicz CP, Burridge CP, Funk WC, VandeWoude S, Craft ME, Crooks KR, et al. Pathogens in space: advancing understanding of pathogen dynamics and disease ecology through landscape genetics. Evol Appl. 2018;11:1763–78. https://doi.org/10.1111/eva.12678.

Kruckeberg AR. The implications of ecology for plant systematics. TAXON. 1969;18:92–120. https://doi.org/10.2307/1218592.

Landaverde-Gonzalez P, Enriquez E, Ariza MA, Murray T, Paxton RJ, Husemann M. Fragmentation in the clouds? The population genetics of the native bee Partamona bilineata (Hymenoptera: Apidae: Meliponini) in the cloud forests of Guatemala. Conserv Genet. 2017;18:631–43. https://doi.org/10.1007/s10592-017-0950-x.

Landguth EL, Cushman SA, Schwartz MK, McKelvey KS, Murphy M, Luikart G. Quantifying the lag time to detect barriers in landscape genetics. Mol Ecol. 2010;19:4179–91. https://doi.org/10.1111/j.1365-294X.2010.04808.x.

Latter BDH. The Island Model of population differentiation: a general solution. Genetics. 1973;73: 147–57. https://doi.org/10.1093/genetics/73.1.147.

LeBeau B. {pdfsearch}: search tools for PDF files. J Open Source Softw. 2018;3:668.

Leempoel K, Duruz S, Rochat E, Widmer I, Orozco-terWengel P, Joost S. Simple rules for an efficient use of geographic information systems in molecular ecology. Front Ecol Evol. 2017;5 https://doi.org/10.3389/fevo.2017.00033.

Leo SST, Gonzalez A, Millien V, Cristescu ME. Multi-taxa integrated landscape genetics for zoonotic infectious diseases: deciphering variables influencing disease emergence. Genome. 2016;59:349–61. https://doi.org/10.1139/gen-2016-0039.

Liebhold AM, Rossi RE, Kemp WP. Geostatistics and geographic information systems in applied insect ecology. Annu Rev Entomol. 1993;38:303–27. https://doi.org/10.1146/annurev.en.38.010193.001511.

Lloyd MW, Campbell L, Neel MC. The power to detect recent fragmentation events using genetic differentiation methods. PLoS One. 2013;8:e63981. https://doi.org/10.1371/journal.pone.0063981.

Losey JE, Vaughan M. The economic value of ecological services provided by insects. Bioscience. 2006;56:311–23. https://doi.org/10.1641/0006-3568(2006)56[311:TEVOES]2.0.CO;2.

Manel S, Holderegger R. Ten years of landscape genetics. Trends Ecol Evol. 2013;28:614–21. https://doi.org/10.1016/j.tree.2013.05.012.

Manel S, Schwartz MK, Luikart G, Taberlet P. Landscape genetics: combining landscape ecology and population genetics. Trends Ecol Evol. 2003;18:189–97. https://doi.org/10.1016/S0169-5347(03)00008-9.

Manel S, Conord C, Despres L. Genome scan to assess the respective role of host-plant and environmental constraints on the adaptation of a widespread insect. BMC Evol Biol. 2009;9 https://doi.org/10.1186/1471-2148-9-288.

Manel S, Joost S, Epperson BK, Holderegger R, Storfer A, Rosenberg MS, et al. Perspectives on the use of landscape genetics to detect genetic adaptive variation in the field. Mol Ecol. 2010;19: 3760–72. https://doi.org/10.1111/j.1365-294X.2010.04717.x.

Marcus T, Boch S, Durka W, Fischer M, Gossner MM, Mueller J, et al. Living in Heterogeneous Woodlands - are habitat continuity or quality drivers of genetic variability in a flightless ground Beetle? Plos One. 2015;10:e0144217. https://doi.org/10.1371/journal.pone.0144217.

Margraf N, Verdon A, Rahier M, Naisbit RE. Glacial survival and local adaptation in an alpine leaf beetle. Mol Ecol. 2007;16:2333–43. https://doi.org/10.1111/j.1365-294X.2007.03318.x.

Mayr E. Systematics and the origin of species. New York: Columbia University Press; 1942.

McRae BH. Isolation by resistance. Evolution. 2006;60:1551–61. https://doi.org/10.1111/j.0014-3820.2006.tb00500.x.

Medley KA, Jenkins DG, Hoffman EA. Human-aided and natural dispersal drive gene flow across the range of an invasive mosquito. Mol Ecol. 2015;24:284–95. https://doi.org/10.1111/mec.12925.

Mercado JE, Hofstetter RW, Reboletti DM, Negrón JF. Phoretic symbionts of the Mountain Pine Beetle (*Dendroctonus ponderosae* Hopkins). Forest Sci. 2014;60:512–26. https://doi.org/10.5849/forsci.13-045.

Millette KL, Keyghobadi N. The relative influence of habitat amount and configuration on genetic structure across multiple spatial scales. Ecol Evol. 2015;5:73–86. https://doi.org/10.1002/ece3.1325.

Moniem HEMA, Schemerhorn BJ, Dewoody JA, Holland JD. Landscape genetics of a pollinator longhorn beetle [Typocerus v. velutinus (Olivier)] on a continuous habitat surface. Mol Ecol. 2016;25:5015–28. https://doi.org/10.1111/mec.13819.

Nei M, Chakravarti A, Tateno Y. Mean and variance of FST in a finite number of incompletely isolated populations. Theor Popul Biol. 1977;11:291–306. https://doi.org/10.1016/0040-5809(77)90013-2.

Oi CA, López-Uribe MM, Cervini M, Del Lama MA. Non-lethal method of DNA sampling in euglossine bees supported by mark–recapture experiments and microsatellite genotyping. J Insect Conserv. 2013;17:1071–9. https://doi.org/10.1007/s10841-013-9582-8.

Ortego J, Aguirre MP, Cordero PJ. Landscape genetics of a specialized grasshopper inhabiting highly fragmented habitats: a role for spatial scale. Divers Distrib. 2012;18:481–92. https://doi.org/10.1111/j.1472-4642.2011.00840.x.

Ožana S, Pyszko P, Dolný A. Determination of suitable insect part for non-lethal DNA sampling: case study of DNA quality and regeneration capability of dragonflies. Insect Conserv Divers. 2020;13:319–27. https://doi.org/10.1111/icad.12400.

Parisod C, Holderegger R. Adaptive landscape genetics: pitfalls and benefits. Mol Ecol. 2012;21:3644–6. https://doi.org/10.1111/j.1365-294X.2012.05675.x.

Pedro PM, Sallum MAM. Spatial expansion and population structure of the neotropical malaria vector, Anopheles darlingi (Diptera: Culicidae). Biol J Linn Soc. 2009;97:854–66. https://doi.org/10.1111/j.1095-8312.2009.01226.x.

Phillipsen IC, Kirk EH, Bogan MT, Mims MC, Olden JD, Lytle DA. Dispersal ability and habitat requirements determine landscape-level genetic patterns in desert aquatic insects. Mol Ecol. 2015;24:54–69. https://doi.org/10.1111/mec.13003.

Potts SG, Imperatriz-Fonseca V, Ngo HT, Aizen MA, Biesmeijer JC, Breeze TD, et al. Safeguarding pollinators and their values to human well-being. Nature. 2016;540:220–9. https://doi.org/10.1038/nature20588.

Rasic G, Keyghobadi N. From broadscale patterns to fine-scale processes: habitat structure influences genetic differentiation in the pitcher plant midge across multiple spatial scales. Mol Ecol. 2012;21:223–36. https://doi.org/10.1111/j.1365-294X.2011.05280.x.

Rego M, Albuquerque P. Rediscovery of Melipona subnitida Ducke (Hymenoptera: Apidae) in the "Restinga" in the Nacional Park Lençóis Maranhenses, Barreirinhas, MA, Brazil. Neotrop Entomol, 2006

Rellstab C, Gugerli F, Eckert AJ, Hancock AM, Holderegger R. A practical guide to environmental association analysis in landscape genomics. Mol Ecol. 2015;24:4348–70. https://doi.org/10.1111/mec.13322.

Rhodes CJ. Pollinator decline – an ecological calamity in the making? Sci Prog. 2018;101:121–60. https://doi.org/10.3184/003685018X15202512854527.

Riley CV. First[-ninth] annual report on the noxious, beneficial and other insects, of the state of Missouri, made to the State board of agriculture, pursuant to an appropriation for this purpose from the Legislature of the state. Jefferson City; 1871.

Rissler LJ. Union of phylogeography and landscape genetics. Proc Natl Acad Sci. 2016;113:8079–86. https://doi.org/10.1073/pnas.1601073113.

Rossetti MR, Tscharntke T, Aguilar R, Batáry P. Responses of insect herbivores and herbivory to habitat fragmentation: a hierarchical meta-analysis. Ecol Lett. 2017;20:264–72. https://doi.org/10.1111/ele.12723.

Ryan PA, Turley AP, Wilson G, Hurst TP, Retzki K, Brown-Kenyon J, et al. Establishment of wMel Wolbachia in Aedes aegypti mosquitoes and reduction of local dengue transmission in Cairns and surrounding locations in northern Queensland, Australia. Gates Open Res. 2020;3 https://doi.org/10.12688/gatesopenres.13061.2.

Safranyik L, Carroll AL, Régnière J, Langor DW, Riel WG, Shore TL, et al. Potential for range expansion of Mountain Pine Beetle into the Boreal Forest of North America. Can Entomol. 2010;142:415–42. https://doi.org/10.4039/n08-CPA01.

Samarasekera GDNG, Bartell NV, Lindgren BS, Cooke JEK, Davis CS, James PMA, et al. Spatial genetic structure of the mountain pine beetle (Dendroctonus ponderosae) outbreak in western Canada: historical patterns and contemporary dispersal. Mol Ecol. 2012;21:2931–48. https://doi.org/10.1111/j.1365-294X.2012.05587.x.

Samways MJ, Barton PS, Birkhofer K, Chichorro F, Deacon C, Fartmann T, et al. Solutions for humanity on how to conserve insects. Biol Conserv. 2020;242:108427. https://doi.org/10.1016/j.biocon.2020.108427.

Schenau E, Jha S. High levels of male diploidy but low levels of genetic structure characterize Bombus vosnesenskii populations across the Western US. Conserv Genet. 2017;18:597–605. https://doi.org/10.1007/s10592-016-0900-z.

Schmidt TL, Rašić G, Zhang D, Zheng X, Xi Z, Hoffmann AA. Genome-wide SNPs reveal the drivers of gene flow in an urban population of the Asian Tiger Mosquito, Aedes albopictus. PLoS Negl Trop Dis. 2017;11:e0006009. https://doi.org/10.1371/journal.pntd.0006009.

Schmidt TL, Filipović I, Hoffmann AA, Rašić G. Fine-scale landscape genomics helps explain the slow spatial spread of Wolbachia through the Aedes aegypti population in Cairns, Australia. Heredity. 2018;120:386–95. https://doi.org/10.1038/s41437-017-0039-9.

Schmidt DA, Campbell NR, Govindarajulu P, Larsen KW, Russello MA. Genotyping-in-Thousands by sequencing (GT-seq) panel development and application to minimally invasive DNA samples to support studies in molecular ecology. Mol Ecol Resour. 2020;20:114–24. https://doi.org/10.1111/1755-0998.13090.

Schmidt TL, Swan T, Chung J, Karl S, Demok S, Yang Q, et al. Spatial population genomics of a recent mosquito invasion. Mol Ecol. 2021;30:1174–89. https://doi.org/10.1111/mec.15792.

Schwabl P, Llewellyn MS, Landguth EL, Andersson B, Kitron U, Costales JA, et al. Prediction and prevention of parasitic diseases using a landscape genomics framework. Trends Parasitol. 2017;33:264–75. https://doi.org/10.1016/j.pt.2016.10.008.

Shafer ABA, Wolf JBW, Alves PC, Bergström L, Bruford MW, Brännström I, et al. Genomics and the challenging translation into conservation practice. Trends Ecol Evol. 2015;30:78–87. https://doi.org/10.1016/j.tree.2014.11.009.

Short Bull RA, Cushman SA, Mace R, Chilton T, Kendall KC, Landguth EL, et al. Why replication is important in landscape genetics: American black bear in the Rocky Mountains. Mol Ecol. 2011;20:1092–107. https://doi.org/10.1111/j.1365-294X.2010.04944.x.

Silva CS, Cordeiro EMG, de Paiva JB, Dourado PM, Carvalho RA, Head G, et al. Population expansion and genomic adaptation to agricultural environments of the soybean looper, Chrysodeixis includens. Evol Appl. 2020;13:2071–85. https://doi.org/10.1111/eva.12966.

Storfer A, Murphy MA, Evans JS, Goldberg CS, Robinson S, Spear SF, et al. Putting the "landscape" in landscape genetics. Heredity. 2007;98:128–42. https://doi.org/10.1038/sj.hdy.6800917.

Storfer A, Murphy MA, Spear SF, Holderegger R, Waits LP. Landscape genetics: where are we now? Mol Ecol. 2010;19:3496–514. https://doi.org/10.1111/j.1365-294X.2010.04691.x.

Storfer A, Patton A, Fraik AK. Navigating the interface between landscape genetics and landscape genomics. Front Genet. 2018;9 https://doi.org/10.3389/fgene.2018.00068.

Theodorou P, Radzeviciute R, Kahnt B, Soro A, Grosse I, Paxton RJ. Genome-wide single nucleotide polymorphism scan suggests adaptation to urbanization in an important pollinator, the red-tailed bumblebee (Bombus lapidarius L.). Proc R Soc B-Biol Sci. 2018;285:20172806. https://doi.org/10.1098/rspb.2017.2806.

Turelli M, Barton NH. Deploying dengue-suppressing Wolbachia: robust models predict slow but effective spatial spread in Aedes aegypti. Theor Popul Biol. 2017;115:45–60. https://doi.org/10.1016/j.tpb.2017.03.003.

Vanbergen AJ, the Insect Pollinators Initiative. Threats to an ecosystem service: pressures on pollinators. Front Ecol Environ. 2013;11:251–9. https://doi.org/10.1890/120126.

Vandergast AG, Bohonak AJ, Weissman DB, Fisher RN. Understanding the genetic effects of recent habitat fragmentation in the context of evolutionary history: phylogeography and landscape genetics of a southern California endemic Jerusalem cricket (Orthoptera: Stenopelmatidae: Stenopelmatus). Mol Ecol. 2007;16:977–92. https://doi.org/10.1111/j.1365-294X.2006.03216.x.

Varvio S-L, Chakraborty R, Nei M. Genetic variation in subdivided populations and conservation genetics. Heredity. 1986;57:189–98. https://doi.org/10.1038/hdy.1986.109.

Vila M, Auger-Rozenberg MA, Goussard F, Lopez-Vaamonde C. Effect of non-lethal sampling on life-history traits of the protected moth Graellsia isabelae (Lepidoptera: Saturniidae). Ecol Entomol. 2009;34:356–62. https://doi.org/10.1111/j.1365-2311.2008.01084.x.

Wagner DL. Insect declines in the anthropocene. In: Douglas AE, editor. Annual review of entomology, vol. 65. Annual Reviews. Palo Alto; 2020. p. 457–80.

Wagner HH, Fortin M-J. A conceptual framework for the spatial analysis of landscape genetic data. Conserv Genet. 2013;14:253–61. https://doi.org/10.1007/s10592-012-0391-5.

Waldvogel A-M, Wieser A, Schell T, Patel S, Schmidt H, Hankeln T, et al. The genomic footprint of climate adaptation in *Chironomus riparius*. Mol Ecol. 2018;27:1439–56. https://doi.org/10.1111/mec.14543.

Walsh BD. The new potato bug, and its natural history. Pract Entamol. 1865;2:13–6.

Wang IJ. Recognizing the temporal distinctions between landscape genetics and phylogeography. Mol Ecol. 2010;19:2605–8. https://doi.org/10.1111/j.1365-294X.2010.04715.x.

Wang B, Salcedo C, Lu M, Sun J. Mutual interactions between an invasive bark beetle and its associated fungi. Bull Entomol Res. 2012;102:71–7. https://doi.org/10.1017/S000748531100037X.

Watanabe K, Kazama S, Omura T, Monaghan MT. Adaptive genetic divergence along narrow environmental gradients in four stream insects. PLoS One. 2014;9 https://doi.org/10.1371/journal.pone.0093055.

Whalon ME, Mota-Sanchez D. Arthropod pesticide resistance database | Michigan State University. https://www.pesticideresistance.org/. Accessed 12 Jun 2020

Wickham H. rvest: easily harvest (Scrape) Web Pages, 2020.

Woodward G, Ebenman B, Emmerson M, Montoya JM, Olesen JM, Valido A, et al. Body size in ecological networks. Trends Ecol Evol. 2005;20:402–9. https://doi.org/10.1016/j.tree.2005.04.005.

Wulder MA, White JC, Grills D, Nelson T, Coops NC, Ebata T. Aerial overview survey of the mountain pine beetle epidemic in British Columbia: communication of impacts. J Ecosyst Manag. 2009;10:45.

Zhou Y, Guo F, Yu J, Liu F, Zhao J, Shen H, et al. Strategies for complete mitochondrial genome sequencing on Ion Torrent PGM^{TM} platform in forensic sciences. Forensic Sci Int Genet. 2016;22:11–21. https://doi.org/10.1016/j.fsigen.2016.01.004.

Transposable Elements in Anopheles Species: Refining Annotation Strategies Towards Population Genomics Analyses

Carlos Vargas-Chávez and Josefa González

Abstract Mosquitoes from the *Anopheles* genus have been extensively studied given their role as vectors of several human infectious diseases. While the genomes of 27 species are available, only a few of them have been studied from a population genomics perspective. Additionally, most of these studies have relied on single nucleotide polymorphisms while other genetic variants, including transposable element (TE) insertions have been neglected. TEs are important players in genome function, structure, and evolution. Indeed, given their capacity to generate mutations and to respond to environmental changes, TEs have been shown to play an important role in adaptation. However, given their repetitive nature they remain challenging to annotate. In this work, we aim at summarizing the current knowledge on TE content and TE diversity in anopheline genomes as the necessary first step to include these relevant genetic variants in population genomics studies. The comparison of the TE content obtained using different annotation strategies in three anopheline species suggests that combining strategies is essential to fully characterize the TE repertoire. As a first approach to the characterization of the TE repertoire in all 27 available anopheline genomes, we used *TEfam*, a database specialized in mosquitoes TEs, to annotate them. We found that the TE content, and its composition, varied among anopheline species, with some TE families restricted to specific phylogenetic groups. Furthermore, we showed that using all the available libraries allowed us to uncover the TE diversity to a larger extent. The ongoing efforts to improve the already available genome sequences together with the generation of new population genomic resources should allow us to assess in the near future the role of TE insertions in the evolution and function of anopheline genomes.

Keywords Bioinformatics · Comparative genomics · De novo annotation · Homology-based annotation · Malaria · Population dynamics · Structure-based annotation · Vector species

C. Vargas-Chávez · J. González (✉)
Institute of Evolutionary Biology (CSIC-Universitat Pompeu Fabra), Barcelona, Spain
e-mail: josefa.gonzalez@csic.es

Julian R. Dupuis and Om P. Rajora (eds.), *Population Genomics: Insects*,
Population Genomics [Om P. Rajora (Editor-in-Chief)],
https://doi.org/10.1007/13836_2020_86,

1 Introduction

Anopheles mosquitoes have been extensively studied given their role as vectors of several human infectious diseases including malaria, a disease that caused 228 million cases and 405,000 deaths worldwide in 2018 alone (World Health Organization 2019), and lymphatic filariasis which infected 120 million people in 2000, leaving around 40 million incapacitated (World Health Organization 2020). The *Anopheles* genus includes at least 465 species which are organized into seven subgenera, and further divided into informal taxonomic categories such as series and complexes (Harbach 2013). Among anopheline species, the *Anopheles gambiae* complex is of particular interest given that some of its members have the highest vectorial capacity. This complex of species has an intricate evolutionary history, with extensive introgression (Fontaine et al. 2015). While members of this complex are hard to differentiate morphologically (Krzywinski and Besansky 2003), they have several behavioral and ecological differences (Derua et al. 2012). Population genomics studies also aid in the identification of new species and taxa of the *Anopheles* genus as exemplified by the recent discovery of *An. fontenillei* and *An.* TENGRELA (Barrón et al. 2019; Tennessen et al. 2020).

Discovering new species is not the only application of population genomic studies in anopheline species. As an example, the *Anopheles gambiae* 1000 Genomes (Ag1000G) Project, which has analyzed individuals from populations of *An. gambiae* and *An. coluzzii* from several African countries, has confirmed that these species are among the most genetically diverse eukaryotic species (Miles et al. 2017; The Anopheles gambiae Genomes Consortium 2020). This population genomics project has also allowed us to identify genes that are relevant for conferring insecticide resistance to these species (The Anopheles gambiae Genomes Consortium 2020; Miles et al. 2017; Grau-Bové et al. 2020). This is highly relevant as insecticide resistance is one of the major threats for malaria control strategies. To date, most population genomics studies in *Anopheles* used single-nucleotide polymorphisms (SNPs) taking advantage of the relative ease to obtain large numbers of these genome-wide markers (Fouet et al. 2017; Kamdem et al. 2017; Miles et al. 2017; Ogola et al. 2019; The Anopheles gambiae Genomes Consortium 2020; Grau-Bové et al. 2020). However, other often neglected genetic variants, such as transposable elements (TEs), have also been shown to be relevant for the study of the adaptive evolution of mosquito species such as the Asian tiger mosquito *Aedes albopictus* (Goubert et al. 2017). TEs are especially informative to perform population genomic studies as they often have phenotypic effects that are more complex than the ones produced by point mutations (Guio and González 2019). However, they remain understudied as they are repetitive sequences and thus challenging to annotate in genomes.

TEs are mobile genetic sequences that can replicate and change their relative position within the genome. When they insert into a new genomic location, they can cause frameshift mutations or disrupt the gene structure rendering the protein non-functional (Bourque et al. 2018). Additionally, TE insertions can affect the structure of genes via the addition of alternative transcription start sites, splicing sites, or terminators. TEs can also disperse regulatory sequences such as enhancers, insulators, and repressive elements, thus affecting gene expression (Chuong et al. 2017). Furthermore, TEs can affect gene expression by disrupting the epigenetic landscape of nearby genes (Rebollo et al. 2012; Elbarbary et al. 2016; Bourque et al. 2018; Guio et al. 2018; Lerat et al. 2019). Besides affecting gene structure and gene expression through their transposition, TE insertions can also lead to structural mutations such as inversions and deletions when they act as substrates for ectopic recombination (Gray 2000). TE-induced mutations frequently have neutral or detrimental consequences for the host, thus TEs have traditionally been considered junk parasites (Chuong et al. 2017). However, it has now been demonstrated that TEs can play a major role in adaptation given their capacity to generate a great variety of mutations and to respond to environmental changes (Casacuberta and González 2013; Schrader and Schmitz 2019).

TEs are divided into two main categories: Class I and Class II. Additionally, both classes can be further divided into at least nine orders, and 29 superfamilies taking several mechanistic and enzymatic criteria into consideration (Wicker et al. 2007; Kapitonov and Jurka 2008, Arkhipova 2017). Superfamilies are further subdivided in families based on DNA sequence similarity. Class I elements use RNA intermediates in their transposition process, and three orders have been identified in anopheline species: long terminal repeat (LTR) retrotransposons (Tubío et al. 2005), long interspersed elements (LINEs) (Biedler and Tu 2003), and short interspersed elements (SINEs; Fig. 1) (Neafsey et al. 2015). Class II TEs use DNA intermediates to transpose and two orders have been identified in members of the *Anopheles* genus: terminal inverted repeats (TIRs) and Helitrons (Fig. 1) (Neafsey et al. 2015). Additionally, both classes have non-autonomous elements, such as MITEs and TRIMs, derived from autonomous elements, that transpose in trans by using the machinery from other related autonomous elements (Fig. 1) (Tu 2001; Zhou and Cahan 2012).

In this chapter, we aim at providing a broad overview on the current knowledge on transposable elements in *Anopheles* species. To this end, we first briefly summarized the discovery of the first individual TE families in the *Anopheles* genus, followed by a description of the strategies that have been applied to perform genome-wide TE identification, highlighting the advantages and disadvantages of each strategy. We then performed a TE annotation in the 27 anopheline genomes available by using the largest mosquito-specific TE database. Finally, we review the few available TE population genomic studies, that to date have been focused on a few individual families, and discuss what could be accomplished by incorporating TEs in population genomics studies.

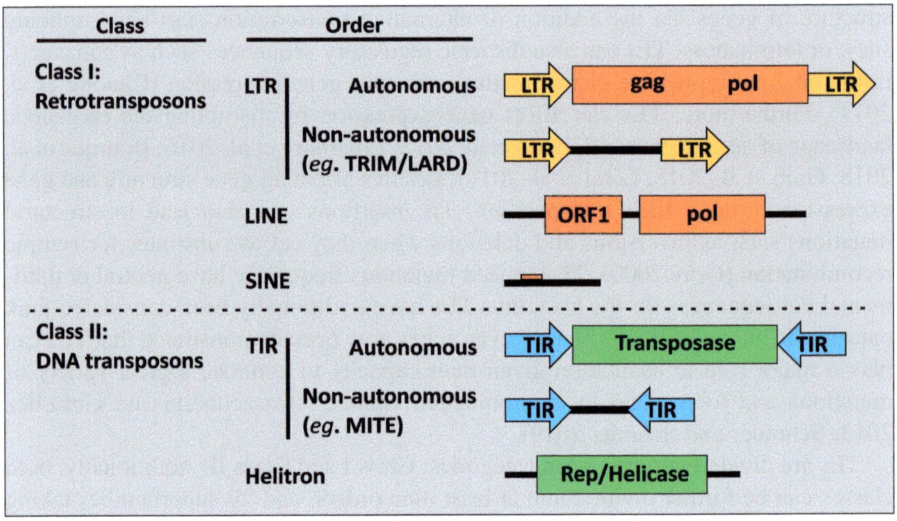

Fig. 1 TE orders identified so far in anopheline genomes. Class I: yellow arrows represent the long terminal repeats (LTR), which contain the promoter needed to produce a template RNA, and the black line represents non-coding features. *Gag* encodes a capsid protein and *pol* encodes a reverse transcriptase with different domains according to the superfamily it belongs to. While not an order *per se*, there are two types of non-autonomous degenerated LTR retrotransposons without coding potential: Terminal-repeat Retrotransposons in Miniature (TRIM) when shorter than 4 kb and Large Retrotransposon Derivative (LARD) when larger than 4 kb. Long interspersed elements (LINEs) retrotranspose without the need of LTRs. *ORF1* is an open reading frame with unknown function that is present in some of these elements. The small box in the SINE element represents the internal RNA polymerase III promoter region. Members of this order are non-autonomous and use the machinery of LINEs to transpose. Class II: blue arrows represent the terminal inverted repeats (TIR) which are recognized by the transposase. Miniature Inverted–Repeat Transposable Elements (MITEs) are a type of non-autonomous TIR transposon. Helitrons lack TIRs and transpose via a rolling-circle mechanism

2 Diverse Strategies Have Been Used to Characterize the TE Repertoire in Anopheline Genomes

Before the availability of full genome sequences, numerous TE families were identified in *An. gambiae* (Boulesteix and Biémont 2005): several families from the non-LTR retrotransposons order (Besansky 1990; Besansky et al. 1992), and several families mostly from the Tc1-mariner superfamily from the TIR order (Robertson 1993; Shao and Tu 2001). In 2002, with the publication of the *Anopheles gambiae* genome (Holt et al. 2002), the full TE repertoire of this species became available, allowing us to study the impact of TEs in this species and to perform comparisons with other well-studied models, such as *Drosophila melanogaster* (Quesneville et al. 2003). Through time, diverse strategies have been used to characterize the full TE repertoire in both *An. gambiae* and other anopheline species.

These diverse strategies can be divided into three main categories: homology-based, structure-based, and de novo-based (Bergman and Quesneville 2007).

Homology-based approaches rely on the identification of TEs through their sequence similarity to previously known TE families. Several tools that allow the identification of TEs based on homology have been used in anopheline genomes: *TEpost* and *TEcombine* (Biedler and Tu 2003), *RepeatMasker* (http://www.repeatmasker.org), *Censor* (Kohany et al. 2006), and *TESeeker* (Kennedy et al. 2011). Homology-based strategies are particularly reliable for identifying TEs with a low number of copies that other methods could miss. This approach is generally quick and highly sensitive; however, it depends on the quality of the database used to identify the TEs. In *Anopheles*, the main database used for the homology-based annotations is *Repbase* (Bao et al. 2015). Additionally, there are two manually-curated databases specialized in TEs in mosquitoes: *TEfam* (http://tefam.biochem.vt.edu/tefam/) and *AnoTExcel* (Fernández-Medina et al. 2011).

Structure-based approaches focus on the identification of canonical features that are particular to each TE superfamily. *FINDMITE* (Tu 2001), *LTR_STRUC* (McCarthy and McDonald 2003), *LTR_FINDER* (Xu and Wang 2007), and *MUSTv2* (Ge et al. 2017) have been used for the identification of TEs in the *Anopheles* genus. While these methods commonly have a high false discovery rate, they are especially relevant for the identification of elements with a conserved structure and that lack coding capacity such as TRIMs, LARDs, and MITEs (Han and Wessler 2010). These approaches do not depend on finding similarities with the sequences of known elements and thus allow the detection of understudied elements, such as DIRS with low representation in databases (Malicki et al. 2020).

Three de novo-based methods have been used in anopheline genomes: *RepeatModeler* (http://www.repeatmasker.org; Flynn et al. 2020), *PILER* (Edgar and Myers 2005), and *RepeatScout* (Price et al. 2005). These approaches take advantage of the repetitive nature of TEs and identify them by locating repetitive features in the genome. While this strategy can identify potentially novel and unknown TEs, most methods are unable to accurately classify all the TEs identified into specific superfamilies. They are also prone to mistakenly classify unrelated features in the genome, such as multicopy gene families, tandem repeats, segmental duplications, and satellites, as TEs.

In summary, each approach has its strengths and its weaknesses, and different success rates at identifying elements from each TE superfamily. Therefore, using a combination of these approaches is the best strategy to characterize the full TE repertoire with a high accuracy (Ou et al. 2019). Furthermore, pipelines comprising several approaches that have been successfully used in other species, such as *REPET* and *McClintock*, could also be used for a comprehensive TE annotation in members of the *Anopheles* genus (Flutre et al. 2011; Nelson et al. 2017).

3 Hybrid Strategies Are Essential to Fully Characterize the TE Repertoire in Anopheline Genomes

To evaluate the effect of using different TE annotation strategies in anopheline genomes, we reviewed the works that have assessed the TE content using different strategies in three species: *An. gambiae*, *An. darlingi,* and *An. funestus*. In *An. gambiae*, while the percentage of the genome spanned by TEs has not changed much, the percentages at the order level have changed across studies (Table 1). Using a homology-based strategy, it was determined that around 17% of the *An. gambiae* genome was spanned by TEs (Table 1) (Holt et al. 2002). A later work using homology, structure, and de novo approaches found this percentage to be 17.78%, with 3.94% of them corresponding to LTRs (Table 1) (Neafsey et al. 2015). Finally, based only on a de novo strategy, it was determined that 17.45% of the genome was covered by TEs and 2.61% was assigned to LTRs (Table 1) (Petersen et al. 2019). The high accuracy of the homology-based annotation performed by Holt et al. (2002) could be explained by the high quality of the TE library used, given that several TE families were already characterized in *An. gambiae* (Besansky 1990; Besansky et al. 1992; 1994, 1996; Robertson 1993; Romans et al. 1998; Grossman et al. 1999; Hill et al. 2001; Shao and Tu 2001; Tu 2001).

While the fraction of the genome covered by TEs has not changed, the number of families described in the *An. gambiae* genome has changed across studies. One of the first analyses focused on the detection of non-LTRs and identified 104 families using a hybrid homology/structure-based approach (Table 1) (Biedler and Tu 2003). Most of the follow-up studies searching for TE families were not restricted to a single order. Using a homology-based approach, 405 families were identified by Quesneville et al. (2003). Afterwards, using a de novo strategy, 38 new families were identified, thus illustrating the usefulness of de novo approaches (Fernández-Medina et al. 2011). A more recent work, also following a de novo strategy, identified the largest number of TE families to date: 655 families (Table 1) (Petersen et al. 2019). Nevertheless, the authors were unable to accurately classify more than two thirds of these families. Although some of these unclassified families might represent real TE families, others could correspond to non-TE repeated sequences. Lastly, the most comprehensive analysis used a combination of homology, structure, and de novo approaches and identified 392 TE families (Table 1) (Neafsey et al. 2015). The differences in the number of families identified in the aforementioned studies are at least partly explained by the different methodologies used, and when the same methodology is applied, by the different thresholds used to define a family. For example, while most works considered two TE families different if they are at least 20% divergent, Fernández-Medina et al. (2011) used a 10% divergence threshold.

In contrast to what was observed in *An. gambiae*, the percentage of the genome covered by TEs in *An. darlingi* has doubled across studies. Along with the publication of the genome, and following homology and structure-based approaches, it was determined that 2.29% of the genome was spanned by TEs (Table 1) (Marinotti et al. 2013). However, upon a reassessment of the genome following homology and

Table 1 Summary of the TE annotation results for the three *Anopheles* species for which annotations were performed with more than one approach. Percentages refer to the fraction of the whole genome spanned by TEs from each category. Families refers to the number of families or consensus sequences identified on each genome. The analysis performed by Fernández-Medina et al. (2017) using the transcriptome is marked with an asterisk

Species	Method	TE content (%)	DNA (%)	LINE (%)	SINE (%)	LTR (%)	Others (%)	Families	Reference
An. gambiae	Homology	17.54	–	–	–	–	–	–	Holt et al. (2002)
	Homology	–	–	–	–	–	–	405	Quesneville et al. (2003)
	Homology and structure	–	–	3.00	–	–	–	104	Biedler and Tu (2003)
	De novo	–	–	–	–	–	–	183	Fernández-Medina et al. (2011)
	Homology, structure and de novo	17.78	4.42	3.48	0.43	3.94	5.51	392	Neafsey et al. (2015)
	De novo	17.45	5.49	3.18	0.90	2.61	5.27	655	Petersen et al. (2019)
An. darlingi	Homology and structure	2.29	0.7	0.9	0.51	0.19	–	–	Marinotti et al. (2013)
	Homology and de novo	5.60	2.44	1.25	–	1.93	–	–	Diesel et al. (2019)
An. funestus	Homology, structure and de novo	4.03	0.40	0.64	0.36	0.50	2.13	334	Neafsey et al. (2015)
	Homology	–	–	–	–	–	–	211	Fernández-Medina et al. (2017)*

de novo-based approaches, 5.60% of the genome was reported to be covered by TEs (Diesel et al. 2019). This highlights the importance of applying de novo-based methodologies in newly sequenced species despite having reliable TE databases from a member of the same genus.

Although most TE annotation studies are based only on the analysis of the genome sequence, the addition of transcriptomic data allows the detection of active TE copies and might also lead to the identification of new families. This is the case for *An. funestus* TE annotation, where both approaches were used. Based on the genome assembly analysis, 334 TE families were identified using homology, structure, and de novo approaches (Neafsey et al. 2015). A subsequent work, which searched for active TE families in the transcriptome using a homology-based approach, identified 211 putatively active TE families, including 62 new TE families (Table 1) (Fernández-Medina et al. 2017). These families were validated by aligning the TE RNA sequence against the DNA genome assembly, and by mapping the DNA raw reads to the TE RNA sequence. This suggests that besides genomics, transcriptomics data should also be taken into account to fully identify the TE repertoire of a given species.

4 TE Content and TE Diversity in Members of the *Anopheles* Genus

To the best of our knowledge, there are two large studies comparing the TE repertoire among anopheline species, one comprising 11 genomes (Neafsey et al. 2015) and the other 21 (De Melo and Wallau 2020). However, the genomes of 27 anopheline species, belonging to three subgenera and four series, and including eight species from the *An. gambiae* complex, are available in either *VectorBase* (Giraldo-Calderón et al. 2015) or NCBI (http://www.ncbi.nlm.nih.gov/genome/) (Fig. 2a; Appendix 1) (Holt et al. 2002; Marinotti et al. 2013; Jiang et al. 2014; Zhou et al. 2014; Logue et al. 2015; Neafsey et al. 2015; Lau et al. 2016; Barrón et al. 2019; Kingan et al. 2019; Waterhouse et al. 2020; Chakraborty et al. 2020; Chida et al. 2020; Compton et al. 2020). Thus, to extend the analysis of the TE repertoire to all 27 currently available *Anopheles* genomes, we annotated their TE insertions using the full *TEfam* database (http://tefam.biochem.vt.edu/tefam/, December 2019, date last accessed) with *RepeatMasker* version open-4.0.9 using default parameters (http://www.repeatmasker.org). Note that the *TEfam* database contains 3,192 consensus sequences from 13 TE libraries: about half of the consensus sequences belong to 11 anopheline species libraries, and the remaining sequences belong to *Aedes aegypti* and *Culex pipiens* libraries.

Because our annotation is based only on a homology-based approach, while those performed by Neafsey et al. (2015) and De Melo and Wallau (2020) are based on a

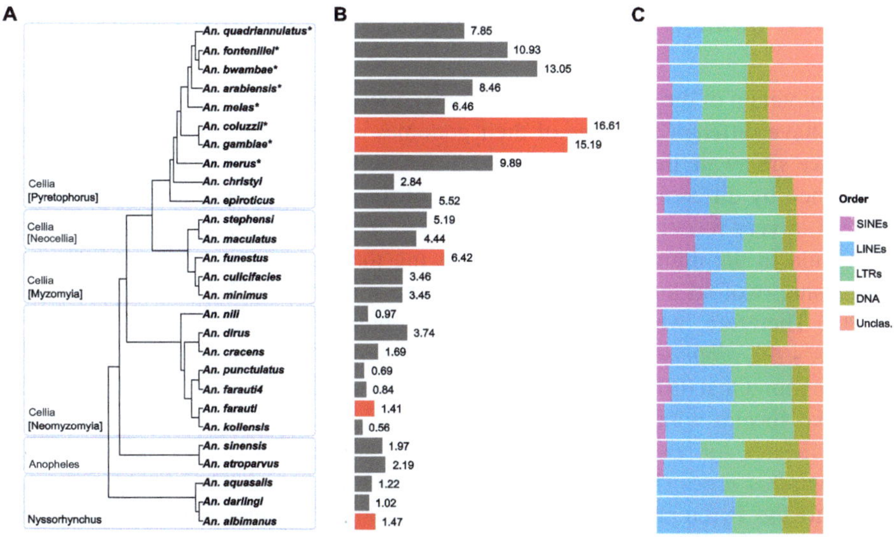

Fig. 2 Dendrogram and TE content for the 27 anopheline species analyzed in this work. (**a**) The dendrogram was reconstructed using information from (Beebe et al. 2012, 2015; Kamali et al. 2014; Fontaine et al. 2015; Freitas et al. 2015; Neafsey et al. 2015; Lau et al. 2016; Barrón et al. 2019). Subgenus and series are grouped in boxes of different colors. Members of the *An. gambiae* complex are marked with an asterisk. (**b**) The bar plot shows the percentage of the genome masked in each species using *RepeatMaske*r with the full *TEfam* database (http://tefam.biochem.vt.edu/tefam/). Bars in red mark highly contiguous genome assemblies with less than 350 contigs. (**c**) Barplot with the TE composition of each species classified at the order level. Unclas. stands for unclassified elements

combination of approaches, before analyzing our results, we first compared the annotations. We obtained similar total TE content in all the species compared with Neafsey et al. (2015), with the differences ranging from 0.03 to 2.59% (Appendix 1). When comparing with the results from De Melo and Wallau (2020), we also observed similar results, but with slightly larger differences which ranged from 0 to 8.03%. However, these larger differences can be mostly explained by the different reference genomes used in the two studies. For example, the largest difference in TE content was observed in *An. sinensis* where De Melo and Wallau (2020) used the *AsinC2* reference while we used the *AsinS2* reference. Thus, the similarity of our results with those from previous works suggests that our annotation using only a homology-based approach reproduces the previously observed results. This is most likely because *TEfam* contains the TEs that Neafsey et al. (2015) had previously identified and De Melo and Wallau (2020) also used this database in one of their TE identification strategies.

4.1 TE Content and Phylogenomics in Anopheline Species

The annotation of the 27 available genomes showed that species from the *Anopheles* genus have different fractions of their genomes composed of TEs (Fig. 2b). However, we are aware that these observed differences in TE content could partially be explained by differences in the contiguity of the genomes analyzed (Appendix 1). Indeed, a very recent reassessment of the *An. stephensi* genome led to a 15.8-fold reduction in the number of contigs, allowing the identification of novel TEs and increasing twofold the fraction of the genome spanned by TEs (Chakraborty et al. 2020). Therefore, we focused on the five genomes with the smallest number of contigs – *An. albimanus*, *An. coluzzii*, *An. farauti*, *An. funestus,* and *An. gambiae* – and we found that differences in TE content were still present (Fig. 2b; Appendix 1). The two species with the largest fraction of their genome covered by TEs are the sister species *An. coluzzii* (16.61%) and *An. gambiae* (15.19%), both from the Pyretophorus series (Fig. 2b). *An. funestus* which belongs to the Myzomyia series, sister to the Pyretophorus series, has a smaller TE content: 6.42%. Finally, *An. farauti* (1.41%) and *An. albimanus* (1.47%) have the lowest TE content and are phylogenetically distant from the aforementioned species. Therefore, we observed a strong phylogenetic signal on the evolution of TE content in anopheline species, similar to what has been reported in the genus *Drosophila* (Sessegolo et al. 2016). However, further de novo TE analysis is required to confirm this observation. Additionally, we also observed that larger genomes had a larger fraction covered by TEs, which has also been previously reported in other species (Canapa et al. 2015; Neafsey et al. 2015; Sessegolo et al. 2016).

4.2 TE Families Are Distributed Non-homogenously Between Members of the Anopheles Genus

The differences between the members of the genus involved not only the fraction of the genome covered by TEs, but also its composition. SINEs represented a large fraction of the TE repertoire in the Myzomyia and Neocellia series (Fig. 2c). Notably, while almost 40% of the identified TEs in *An. stephensi* are SINEs, almost no SINEs were identified in the members of the Nyssorhynchus subgenus (Fig. 2c). On the other hand, this subgenus seems enriched for LINE elements and both this and the *Anopheles* subgenus contain a large fraction of DNA elements. Members of the *An. gambiae* complex were enriched for unclassified elements. Upon closer inspection, we observed that they corresponded to MITEs, thus suggesting that while these non-autonomous elements are shared between members of this complex, they are not widely shared by other species in the genus.

Next, we assessed whether TE families from each library were uniquely found in each species, series, or subgenus. We observed that most of the families from the libraries of *An. arabiensis*, *An. melas*, *An. merus,* and *An. quadriannulatus* (all members of the *An. gambiae* complex) were identified in all eight members of the complex (Appendix 2). Similarly, we identified most members of the libraries of *An. gambiae*, *An. christyi*, *An. albimanus,* and *An. funestus* in their corresponding series or even subgenus. Comparably, most of the elements in the library of *An. stephensi* were identified in both the Myzomyia and the Neocellia series. This suggests that these families originated after these series separated from the Pyretophorus series and thus represent elements unique to these lineages. In the case of the *An. dirus* library, most of its members were identified only in its own genome and in that of its sister clade *An. cracens* (Appendix 2). Given that these sequences are not shared among all members of the series, it seems that these families might have a recent origin. An intriguing case is the library of *An. epiroticus*, where most of its members were identified in its genome and in that of *An. funestus*, which does not belong to the same series (Appendix 2). The TE families shared between these two species and not found in other members of their respective series might represent horizontally transferred TEs (HTTs). However, a recent study assessing HTTs between anopheline species found that the number of HTT events between *An. epiroticus* and *An. funestus* was similar to the one found between other species (De Melo and Wallau 2020). Thus, HTT events between *An. epiroticus* and *An. funestus* do not fully explain that 82% of the *An. epiroticus* families are also present in *An. funestus*.

Finally, we determined the fraction of the TEs in each species that was identified using each of the 13 TE libraries. We observed that in five of the anopheline species with a corresponding TE library – *An. dirus*, *An. funestus*, *An. stephensi*, *An. epiroticus* and *An. gambiae* – their own library identified the largest fraction of TEs (Appendix 3). Additionally, the libraries of *An. dirus*, *An. stephensi,* and *An. gambiae* also identified a large fraction of the TEs in the genomes of other members of the same clade. In the remaining six species (four of which belong to the *An. gambiae* complex), the largest fraction of their TE repertoire was identified using the TE library of *An. gambiae* (Appendix 3). Furthermore, members of the *An. gambiae* complex had a large fraction of their TEs identified using libraries from other members of the complex besides their own. This suggests that members of the *An. gambiae* complex share most, if not all, of their TEs as previously suggested (Neafsey et al. 2015). Therefore, while using the highly studied library of *An. gambiae* allowed capturing a large fraction of the TE diversity in most genomes, the use of the remaining libraries allowed the identification of TEs unique to a given species or clade. These results suggest that generating TE libraries for the species still lacking one is very important to prevent biases in TE annotation.

In summary, we found that although the genome contiguity might have an impact in the estimated fraction of the genome covered by TEs, focusing on the species with the most contiguous genomes, TE content varies among anopheline species. These differences were not restricted to the TE content, they also included the composition of the TE repertoire. We found that the presence of some families of each TE library is restricted to specific phylogenetic groups and that performing the TE annotation using the full set of libraries allows uncovering the TE diversity to a larger extent. Moreover, the availability of genome sequences for a given species from different geographic populations along with the annotation of their full TE repertoire should allow improving the knowledge on population dynamics of TEs in the genus *Anopheles*.

5 Population Genomics of Transposable Elements in *Anopheles*

The population dynamics of TEs in anopheline species have been studied for many years before the advent of high-throughput sequencing technologies. One of the first studies characterized TE insertion polymorphisms of four TE families on 10 individuals from the same inbred strain of *An. gambiae* (Mukabayire and Besansky 1996). They found similar levels of polymorphism compared to *D. melanogaster* laboratory stocks, and also similar to this species some of the TE families were not randomly distributed but concentrated in centromeric and pericentromeric regions. Later works determined that some families from the *Tc1-mariner* superfamily also had polymorphic insertions and further showed that they allowed differentiating between anopheline species and different strains of *An. gambiae* (Grossman et al. 1999). A TE family that has been used in numerous works both as a marker (see below) and to study its dynamics is *Herves* (Box 1). This family was identified in members of the *An. gambiae* complex and it was demonstrated to be active (Arensburger et al. 2005). Additionally, it has been proposed as a valuable tool for transforming insect cells.

Box 1 *Herves*: A Versatile TE to Study TE Dynamics, Anopheline Population Genomics and a Potential Transformation Tool

Herves is an active *hAT* element first discovered in *An. gambiae* in 2005 and found to be present in other anopheline genomes as well (Arensburger et al. 2005). After the initial discovery, insertion polymorphisms of this TE in members of the *An. gambiae* complex from populations sampled over a period of 2 years in Mozambique were described, and it was suggested that *Herves* was present before the divergence of the complex (O'Brochta et al. 2006). The following year, analysis of the *Herves* TE insertions in populations of *An. gambiae* from four African countries suggested that *Herves* had been horizontally transferred into the ancestor of the *An. gambiae* complex from an unknown source (Subramanian et al. 2007). They proposed that the transfer did not occur recently as they observed a low number of insertions with internal deletions and a low copy number. Both facts suggest that the copy-

(continued)

Box 1 (continued)
number equilibrium has not been reached, which is unexpected if the TE was introduced in the distant past.

Besides studying its TE dynamics, *Herves* insertions have also been successfully used as markers to study the population dynamics of populations of *An. gambiae* and *An. coluzzii* from different islands in the Gulf of Guinea (Salgueiro et al. 2013). They observed that the mosquito's populations mirrored biogeography patterns, suggesting multiple, yet sporadic, colonizations of this vector species into the Gulf of Guinea islands.

Finally, *Herves* was demonstrated to be capable of transforming insect cells (Arensburger et al. 2005). However, this TE has not been routinely used to transform insects, perhaps given the great success as transformation tools of other TEs such as the PiggyBac element (Laptev et al. 2017; Häcker and Schetelig 2018).

TE dynamics have also been studied for non-autonomous elements in *An. gambiae*, particularly the *Mighty* MITE family (Struchiner et al. 2009). Struchiner et al. (2009) evaluated whether the population dynamics of *Mighty* could be reproduced using two different mathematical models: one following the constraints of the master gene hypothesis and the other following similar constraints but with different deletion rates for active and inactive elements. They observed that both models poorly fit the observed trajectory for the *Mighty* family, thus highlighting the need for new models to better reproduce the dynamics of non-autonomous TEs in anopheline genomes.

Besides studying the TE population dynamics, TEs can also be used as markers to study the population dynamics of their hosts. Using TE display, the TE insertion polymorphisms for MITE families in several individuals from different colonies of *An. gambiae* were compared against each other and with individuals of *An. arabiensis* as a proof of principle of the usage of TEs as markers (Biedler and Tu 2003). While the authors highlight the value of MITEs as markers given that many families are interspersed throughout the genome, and because of their relatively high levels of insertion polymorphisms, they also warn of the limitation of using Class II TEs given that their excision could be confused with a non-insertion state. A type of marker without this limitation are SINEs which are considered homoplasy-free given that the chance of independent insertions in the same position is remote and that there are no known mechanisms for SINE excision. This coupled with their small size makes them ideal markers for studying populations or even as markers to distinguish between cryptic species (Barnes et al. 2005).

Finally, insertion polymorphisms have been used to study the population dynamics from anopheline species from different geographical origins. Esnault et al. (2008) characterized the insertion polymorphisms from samples of *An. gambiae* and *An. coluzzii* from different sites throughout their geographical range. While they were able to clearly differentiate between both species, they did not observe structure between populations of the same species, thus suggesting that most insertion sites were widespread and highly polymorphic between populations. However, other works using different TE polymorphisms have been able to observe population structure in individuals of *An. gambiae* from Cameroon (Boulesteix et al. 2007) and in individuals from different islands in the Gulf of Guinea (Salgueiro et al. 2013) (Box 1).

Overall, while there are several studies devoted to the study of population dynamics of TEs and their hosts in anopheline species, most are focused on a single or a few TE families. However, the availability of several whole genome sequences from the same species will allow performing genome-wide studies taking all TE families into consideration. This will allow studying population dynamics using TEs at a higher resolution than ever before.

6 Future Perspectives

While the available anopheline genomes have different degrees of completeness and contiguity, there are ongoing efforts to improve them using long-reads and genome scaffolding (Waterhouse et al. 2020). These improved genomes along with high-quality TE annotations based on hybrid TE detection strategies, which allow to exploit the strengths and mitigate the weaknesses of individual strategies, will undoubtedly allow us to grasp the full extent of the TE diversity in the genus (Flutre et al. 2011; Nelson et al. 2017; Goerner-Potvin and Bourque 2018; Flynn et al. 2020). Creating a comprehensive database with TEs libraries from all the available anopheline species would allow us to detect TEs that might have gone undetected in previous annotation efforts. Using this database, homologous TE families could be identified and used to build probabilistic models of nucleotide sequences using profile hidden Markov models to allow the identification of highly diverged TE insertions in other species (Andrieu et al. 2004). As such, we would be able to evaluate the evolutionary origin of given TE families and detect horizontal transfer events with more precision (Tu and Li 2009).

Until now, most TE annotation works have focused on a single genome per species. Characterizing the full TE repertoire in more than one genome of the same species would allow us to uncover hidden TE diversity as has been done in *D. melanogaster* (Ellison and Cao 2020). Additionally, large datasets such as those generated by the *Anopheles gambiae* 1000 Genomes Project (Miles et al. 2017; The Anopheles gambiae Genomes Consortium 2020) could be used to determine the presence/absence of polymorphic TEs (Mukabayire and Besansky 1996; Biedler and Tu 2003). This would allow us to study the population dynamics of TEs that might have a greater impact on the host's fitness than the traditionally studied SNPs (Guio and González 2019).

Furthermore, the role of TEs in the adaptation of anopheline mosquitoes to changing environmental conditions, including adaptation to urban environments, could also be assessed. Adaptation driven by several stresses including saline, desiccation, and exposure to xenobiotic agents such as insecticides, has increased the prevalence of mosquitoes in urban environments (Kamdem et al. 2017; Carrasco et al. 2019; Holmes and Benoit 2019; Surendran et al. 2019). Given that TEs can modulate gene expression under stress conditions, it would be interesting to determine the functional effects of polymorphic TEs under different stress conditions (Chénais et al. 2012; Horváth et al. 2017; Fouché et al. 2020). Furthermore, besides affecting gene expression, TE insertions can have a high impact in their host's biology through the generation of chromosomal inversions. These inversions can lead to reproductive isolation, thus allowing the host to rapidly speciate and adapt to novel ecological niches. In *An. arabiensis*, the *Odysseus* TE was found in the proximity of the recent and polymorphic 2Rd' inversion (Mathiopoulos et al. 1998). Additionally, TEs have also been identified in the vicinity of an older and widely shared inversion (Sharakhov et al. 2006). Chromosomal inversion breakpoints are challenging to detect because they are often embedded in repeat regions, more so in fragmented genomes (Pettersson et al. 2019). Therefore, a larger number of highly contiguous assemblies would allow a better understanding of the role of TEs in chromosomal inversions.

Finally, identifying TEs in the transcriptome will allow us to accurately identify active copies and recent insertions which are likely to be involved in contemporary adaptation. Also, the identification of active TE copies in anopheline genomes could provide novel tools for the genetic manipulation of mosquitoes, like *piggyBac*, a TE from the TIR order which has already been successfully used for the stable germline transformation of *An. gambiae* (Tu and Coates 2004; Volohonsky et al. 2015; Laptev et al. 2017; Häcker and Schetelig 2018).

Acknowledgments We thank all members of González Lab for comments on the manuscript. This work was supported by the Ministerio de Ciencia, Innovación y Universidades/AEI/FEDER, EU (BFU2017-82937-P) and by the 'Secretaria d'Universitats i Recerca. Departament d'Economia i Coneixement. Generalitat de Catalunya' (GRC 2017 SGR 880).

Appendix 1

Statistics of the 27 available anopheline genomes. In bold the species members of the *An. gambiae* complex. Genomes that were downloaded from NCBI are marked with an asterisk; all other genomes were downloaded from *VectorBase*

Species	Subgenus [series]	Contigs	Genome size (Mb)	Gapless genome size (Mb)	Total number of TEs	Mb covered by TEs	% of genome covered by TEs	% of genome using Neafsey et al. (2015)	% of genome De Melo and Wallau (2020)
An. quadriannulatus	Cellia [Pyretophorus]	2,823	283.83	208.97	86,215	16.41	7.85%	7.69%	7.85%
An. fontenillei*	Cellia [Pyretophorus]	69,201	232.63	232.57	131,114	25.42	10.93%		
An. bwambae*	Cellia [Pyretophorus]	30,458	288.96	288.84	148,482	37.69	13.05%		
An. arabiensis	Cellia [Pyretophorus]	1,214	246.57	211.45	85,938	17.88	8.46%	9.38%	9.79%
An. melas	Cellia [Pyretophorus]	20,229	224.16	203.43	77,969	13.14	6.46%	7.29%	7.77%
An. coluzzii	Cellia [Pyretophorus]	206	251.41	251.41	123,959	41.75	16.61%		11.17%
An. gambiae	Cellia [Pyretophorus]	8	273.11	252.46	127,817	38.34	15.19%	17.78%	19.55%
An. merus	Cellia [Pyretophorus]	2,027	288.05	217.32	95,170	21.5	9.89%	11.43%	9.78%
An. christyi	Cellia [Pyretophorus]	30,369	172.66	169.99	31,807	4.82	2.84%	2.81%	2.23%
An. epiroticus	Cellia [Pyretophorus]	2,673	223.49	202.64	55,402	11.19	5.52%	6.27%	6.11%
An. stephensi	Cellia [Neocellia]	8,946	196.18	196.18	63,780	10.19	5.19%	5.04%	5.73%
An. maculatus	Cellia [Neocellia]	47,797	141.89	131.83	33,994	5.85	4.44%		4.40%
An. funestus	Cellia [Myzomyia]	4	210.99	210.94	41,414	13.54	6.42%	4.03%	5.54%

An. culicifacies	Cellia [Myzomyia]	16,162	203	187.16	42,644	6.47	3.46%		3.80%
An. minimus	Cellia [Myzomyia]	678	201.79	186.4	38,419	6.44	3.45%		4.70%
An. nili	Cellia [Neomyzomyia]	51,048	98.32	98.17	5,611	0.95	0.97%		
An. dirus	Cellia [Neomyzomyia]	1,266	216.31	197.74	35,867	7.5	3.79%	5.09%	6.40%
*An. cracens**	Cellia [Neomyzomyia]	34,365	326.47	325.76	39,195	5.5	1.69%		
An. punctulatus	Cellia [Neomyzomyia]	20,774	146.16	146.16	9,215	1.01	0.69%		0.38%
*An. farauti no. 4**	Cellia [Neomyzomyia]	14,407	146.38	146.38	9,788	1.23	0.84%		
An. farauti	Cellia [Neomyzomyia]	310	183.1	175.82	13,052	2.48	1.41%		2.17%
An. koliensis	Cellia [Neomyzomyia]	41,925	151.11	151.11	7,879	0.85	0.56%		0.13%
An. sinensis	Anopheles	10,448	375.76	190.28	20,850	3.75	1.97%		10.00%
An. atroparvus	Anopheles	1,320	225.3	188.96	15,742	4.14	2.19%		4.90%
*An. aquasalis**	Nyssorhynchus	16,504	162.94	162.35	15,046	1.98	1.22%		
An. darlingi	Nyssorhynchus	2,221	136.95	136.92	12,951	1.4	1.02%		2.63%
An. albimanus	Nyssorhynchus	201	173.34	163.5	15,317	2.4	1.47%	1.98%	2.70%

Appendix 2

Number of identified TE families using each of the 13 libraries in *TEfam*. Aalb stands for *An. albimanus*, Aara for *An. arabiensis*, Achr for *An. christyi*, Adir for *An. dirus*, Aepi for *An. epiroticus*, Afun for *An. funestus*, Agam for *An. gambiae*, Amel for *An. melas*, Amer for *An. merus*, Aqua for *An. quadriannulatus*, Aste for *An. stephensi*, Aaeg for *Aedes aegypti* and Cpip for *Culex pipiens*. Shaded in red values larger than 80% of total sequences in the library. In parenthesis the number of families in each library

		Anopheles						
		Cellia [Pyretophorus]						
		An. gambiae complex						
Species	Subgenus [Series]	Aqua (99)	Aara (213)	Amel (81)	Agam (392)	Amer (244)	Achr (30)	Aepi (168)
An. quadriannulatus	Cellia [Pyretophorus]	93 (94%)	194 (91%)	75 (93%)	373 (95%)	223 (90%)	25 (83%)	119 (71%)
An. fontenillei	Cellia [Pyretophorus]	94 (95%)	203 (95%)	79 (98%)	379 (97%)	232 (90%)	28 (93%)	126 (75%)
An. bwambae	Cellia [Pyretophorus]	90 (91%)	198 (93%)	77 (95%)	376 (96%)	218 (90%)	26 (87%)	131 (78%)
An. arabiensis	Cellia [Pyretophorus]	88 (89%)	197 (92%)	74 (91%)	368 (94%)	217 (90%)	27 (90%)	123 (73%)
An. melas	Cellia [Pyretophorus]	86 (87%)	195 (92%)	75 (93%)	367 (94%)	212 (90%)	24 (80%)	120 (71%)
An. coluzzii	Cellia [Pyretophorus]	86 (87%)	197 (92%)	76 (94%)	376 (96%)	227 (90%)	26 (87%)	123 (73%)
An. gambiae	Cellia [Pyretophorus]	88 (89%)	202 (95%)	75 (93%)	379 (97%)	222 (90%)	28 (93%)	127 (76%)
An. merus	Cellia [Pyretophorus]	86 (87%)	193 (91%)	75 (93%)	369 (94%)	234 (90%)	25 (83%)	125 (74%)
An. christyi	Cellia [Pyretophorus]	61 (62%)	157 (74%)	52 (64%)	318 (81%)	192 (79%)	29 (97%)	124 (74%)
An. epiroticus	Cellia [Pyretophorus]	54 (55%)	149 (70%)	50 (62%)	325 (83%)	175 (72%)	26 (87%)	165 (98%)
An. stephensi	Cellia [Neocellia]	44 (44%)	115 (54%)	37 (46%)	264 (67%)	147 (60%)	19 (63%)	117 (70%)
An. maculatus	Cellia [Neocellia]	38 (38%)	135 (63%)	36 (44%)	273 (70%)	148 (61%)	21 (70%)	126 (75%)
An. funestus	Cellia [Myzomyia]	47 (47%)	134 (63%)	39 (48%)	300 (77%)	157 (64%)	24 (80%)	137 (82%)
An. culicifacies	Cellia [Myzomyia]	51 (52%)	126 (59%)	37 (46%)	280 (71%)	156 (64%)	22 (73%)	120 (71%)
An. minimus	Cellia [Myzomyia]	52 (53%)	139 (65%)	35 (43%)	301 (77%)	153 (63%)	22 (73%)	128 (76%)
An. nili	Cellia [Neomyzomyia]	17 (17%)	43 (20%)	8 (10%)	144 (37%)	77 (32%)	7 (23%)	65 (39%)
An. dirus	Cellia [Neomyzomyia]	44 (44%)	117 (55%)	31 (38%)	284 (72%)	147 (60%)	18 (60%)	126 (75%)
An. cracens	Cellia [Neomyzomyia]	34 (34%)	100 (47%)	29 (36%)	249 (64%)	134 (55%)	17 (57%)	111 (66%)
An. punctulatus	Cellia [Neomyzomyia]	22 (22%)	64 (30%)	13 (16%)	154 (39%)	87 (36%)	11 (37%)	80 (48%)
An. farauti no. 4	Cellia [Neomyzomyia]	19 (19%)	78 (37%)	12 (15%)	198 (51%)	99 (41%)	13 (43%)	93 (55%)
An. farauti	Cellia [Neomyzomyia]	23 (23%)	55 (26%)	10 (12%)	168 (43%)	94 (39%)	13 (43%)	88 (52%)
An. koliensis	Cellia [Neomyzomyia]	15 (15%)	69 (32%)	10 (12%)	161 (41%)	84 (34%)	10 (33%)	73 (43%)
An. sinensis	Anopheles	37 (37%)	98 (46%)	22 (27%)	249 (64%)	126 (52%)	12 (40%)	112 (67%)
An. atroparvus	Anopheles	38 (38%)	95 (45%)	18 (22%)	275 (70%)	145 (59%)	12 (40%)	115 (68%)
An. aquasalis	Nyssorhynchus	18 (18%)	54 (25%)	4 (5%)	177 (45%)	75 (31%)	8 (27%)	76 (45%)
An. darlingi	Nyssorhynchus	8 (8%)	43 (20%)	2 (2%)	108 (28%)	47 (19%)	6 (20%)	46 (27%)
An. albimanus	Nyssorhynchus	16 (16%)	44 (21%)	8 (10%)	139 (35%)	61 (25%)	7 (23%)	55 (33%)

(continued)

		Anopheles				*Aedes*	*Culex*
		Cellia [Neocellia]	Cellia [Myzomyia]	Cellia [Neomyzomyia]	Nyssorhynchus		
Species	Subgenus [Series]	Aste (67)	Afun (105)	Adir (162)	Aalb (54)	Aaeg (1089)	Cpip (488)
An. quadriannulatus	Cellia [Pyretophorus]	43 (64%)	73 (70%)	108 (67%)	19 (35%)	486 (45%)	177 (36%)
An. fontenillei	Cellia [Pyretophorus]	49 (73%)	77 (73%)	107 (66%)	22 (41%)	603 (55%)	221 (45%)
An. bwambae	Cellia [Pyretophorus]	49 (73%)	79 (75%)	111 (69%)	28 (52%)	600 (55%)	202 (41%)
An. arabiensis	Cellia [Pyretophorus]	44 (66%)	71 (68%)	110 (68%)	19 (35%)	504 (46%)	187 (38%)
An. melas	Cellia [Pyretophorus]	42 (63%)	62 (59%)	98 (60%)	19 (35%)	507 (47%)	188 (39%)
An. coluzzii	Cellia [Pyretophorus]	48 (72%)	74 (70%)	113 (70%)	23 (43%)	539 (49%)	196 (40%)
An. gambiae	Cellia [Pyretophorus]	52 (78%)	73 (70%)	108 (67%)	21 (39%)	542 (50%)	200 (41%)
An. merus	Cellia [Pyretophorus]	44 (66%)	65 (62%)	102 (63%)	19 (35%)	523 (48%)	181 (37%)
An. christyi	Cellia [Pyretophorus]	51 (76%)	71 (68%)	106 (65%)	22 (41%)	525 (48%)	187 (38%)
An. epiroticus	Cellia [Pyretophorus]	55 (82%)	77 (73%)	112 (69%)	23 (43%)	554 (51%)	179 (37%)
An. stephensi	Cellia [Neocellia]	67 (100%)	81 (77%)	123 (76%)	22 (41%)	502 (46%)	189 (39%)
An. maculatus	Cellia [Neocellia]	63 (94%)	83 (79%)	125 (77%)	26 (48%)	486 (45%)	160 (33%)
An. funestus	Cellia [Myzomyia]	55 (82%)	103 (98%)	115 (71%)	24 (44%)	531 (49%)	161 (33%)
An. culicifacies	Cellia [Myzomyia]	58 (87%)	85 (81%)	115 (71%)	21 (39%)	518 (48%)	160 (33%)
An. minimus	Cellia [Myzomyia]	62 (93%)	93 (89%)	125 (77%)	28 (52%)	545 (50%)	160 (33%)
An. nili	Cellia [Neomyzomyia]	24 (36%)	36 (34%)	53 (33%)	12 (22%)	293 (27%)	99 (20%)
An. dirus	Cellia [Neomyzomyia]	55 (82%)	74 (70%)	162 (100%)	20 (37%)	471 (43%)	168 (34%)
An. cracens	Cellia [Neomyzomyia]	52 (78%)	67 (64%)	158 (98%)	17 (31%)	416 (38%)	141 (29%)
An. punctulatus	Cellia [Neomyzomyia]	31 (46%)	35 (33%)	95 (59%)	10 (19%)	398 (37%)	145 (30%)
An. farauti no. 4	Cellia [Neomyzomyia]	37 (55%)	55 (52%)	110 (68%)	17 (31%)	441 (40%)	160 (33%)
An. farauti	Cellia [Neomyzomyia]	33 (49%)	41 (39%)	92 (57%)	8 (15%)	374 (34%)	132 (27%)
An. koliensis	Cellia [Neomyzomyia]	30 (45%)	41 (39%)	92 (57%)	8 (15%)	378 (35%)	142 (29%)
An. sinensis	Anopheles	44 (66%)	61 (58%)	107 (66%)	29 (54%)	554 (51%)	181 (37%)
An. atroparvus	Anopheles	47 (70%)	67 (64%)	114 (70%)	34 (63%)	522 (48%)	182 (37%)
An. aquasalis	Nyssorhynchus	28 (42%)	36 (34%)	58 (36%)	52 (96%)	432 (40%)	148 (30%)
An. darlingi	Nyssorhynchus	16 (24%)	23 (22%)	30 (19%)	49 (91%)	309 (28%)	116 (24%)
An. albimanus	Nyssorhynchus	28 (42%)	35 (33%)	50 (31%)	53 (98%)	337 (31%)	129 (26%)

Appendix 3

Percentage of the identified TEs using each of the 13 libraries in TEfam. In parenthesis the number of families in each library. Aalb stands for *An. albimanus*, Aara for *An. arabiensis*, Achr for *An. christyi*, Adir for *An. dirus*, Aepi for *An. epiroticus*, Afun for *An. funestus*, Agam for *An. gambiae*, Amel for *An. melas*, Amer for *An. merus*, Aqua for *An. quadriannulatus*, Aste for *An. stephensi*, Aaeg for *Aedes aegypti* and Cpip for *Culex pipiens*. Shaded in red values larger than 20%.

		Anopheles						
		Cellia [Pyretophorus]						
		An. gambiae complex						
Species	Subgenus [Series]	Aqua (99)	Aara (213)	Amel (81)	Agam (392)	Amer (244)	Achr (30)	Aepi (168)
An. quadriannulatus	Cellia [Pyretophorus]	10.91%	15.51%	9.83%	33%	19.4%	1.13%	2.61%
An. fontenillei	Cellia [Pyretophorus]	9.6%	17.73%	8.46%	34.48%	18.65%	0.93%	2.68%
An. bwambae	Cellia [Pyretophorus]	10.1%	14.97%	8.54%	36.68%	17.78%	0.95%	2.94%
An. arabiensis	Cellia [Pyretophorus]	10.75%	15.35%	9.27%	34.12%	19.11%	1.16%	2.62%
An. melas	Cellia [Pyretophorus]	10.5%	15.56%	10.68%	31.44%	19.82%	1.11%	2.77%
An. coluzzii	Cellia [Pyretophorus]	9.8%	15.31%	8.32%	36.89%	18.08%	1.07%	2.94%
An. gambiae	Cellia [Pyretophorus]	9.87%	15.02%	8.56%	37.06%	18.16%	1.02%	2.89%
An. merus	Cellia [Pyretophorus]	8.97%	14.32%	9.17%	33.26%	22.79%	0.98%	2.87%
An. christyi	Cellia [Pyretophorus]	2.84%	8.85%	4.37%	25.89%	12.36%	21.76%	7.96%
An. epiroticus	Cellia [Pyretophorus]	1.86%	4.81%	3.03%	22.21%	7.37%	1.98%	44.05%
An. stephensi	Cellia [Neocellia]	0.56%	2.06%	0.65%	14.53%	3.69%	0.35%	5.64%
An. maculatus	Cellia [Neocellia]	1.44%	3.77%	0.73%	20.78%	5.08%	0.44%	9.3%
An. funestus	Cellia [Myzomyia]	2.06%	4.96%	0.83%	23.96%	4.68%	1.2%	6.16%
An. culicifacies	Cellia [Myzomyia]	1.3%	3.92%	0.8%	16.97%	4.48%	1.05%	7.23%
An. minimus	Cellia [Myzomyia]	1.91%	4.03%	0.89%	21.33%	4.36%	1.04%	8.76%
An. nili	Cellia [Neomyzomyia]	1.07%	7.08%	0.49%	38.75%	8.6%	0.3%	9.36%
An. dirus	Cellia [Neomyzomyia]	0.99%	4.03%	0.85%	20.23%	5.46%	0.56%	6.89%
An. cracens	Cellia [Neomyzomyia]	1.34%	2.75%	0.62%	8.09%	3.91%	0.9%	7.02%
An. punctulatus	Cellia [Neomyzomyia]	0.62%	7.52%	0.37%	38.83%	8.42%	2.07%	4.03%
An. farauti no. 4	Cellia [Neomyzomyia]	0.76%	6.08%	0.42%	37.39%	8.95%	1.24%	4.22%
An. farauti	Cellia [Neomyzomyia]	0.5%	5.6%	0.61%	39.4%	8.04%	1.15%	4.88%
An. koliensis	Cellia [Neomyzomyia]	0.51%	7.3%	0.49%	41.03%	7.2%	1.31%	4.08%
An. sinensis	Anopheles	1.22%	12.83%	1.96%	21.99%	7.33%	0.61%	8.45%
An. atroparvus	Anopheles	2.4%	7.96%	0.71%	34.14%	8.46%	0.34%	8.93%
An. aquasalis	Nyssorhynchus	0.6%	4.79%	0.05%	40.21%	6.77%	1.11%	3.25%
An. darlingi	Nyssorhynchus	0.2%	7.37%	0.02%	50.27%	8.22%	0.31%	3.25%
An. albimanus	Nyssorhynchus	0.29%	5.76%	0.06%	43.58%	7.93%	0.19%	2.94%

		Anopheles				*Aedes*	*Culex*
		Cellia [Neocellia]	Cellia [Myzomyia]	Cellia [Neomyzomyia]	Nyssorhynchus		
Species	Subgenus [Series]	Aste (67)	Afun (105)	Adir (162)	Aalb (54)	Aaeg (1089)	Cpip (488)
An. quadriannulatus	Cellia [Pyretophorus]	0.83%	0.84%	1.47%	0.12%	2.58%	1.78%
An. fontenillei	Cellia [Pyretophorus]	0.98%	0.86%	1.49%	0.11%	2.47%	1.56%
An. bwambae	Cellia [Pyretophorus]	1.08%	0.98%	1.6%	0.12%	2.61%	1.65%
An. arabiensis	Cellia [Pyretophorus]	0.89%	0.87%	1.44%	0.11%	2.51%	1.79%
An. melas	Cellia [Pyretophorus]	0.93%	0.76%	1.51%	0.09%	2.73%	2.09%
An. coluzzii	Cellia [Pyretophorus]	0.98%	0.92%	1.51%	0.1%	2.59%	1.48%
An. gambiae	Cellia [Pyretophorus]	0.99%	0.9%	1.61%	0.08%	2.37%	1.46%
An. merus	Cellia [Pyretophorus]	0.99%	0.85%	1.49%	0.14%	2.51%	1.66%
An. christyi	Cellia [Pyretophorus]	2.14%	3.34%	3.18%	0.2%	4.66%	2.45%
An. epiroticus	Cellia [Pyretophorus]	3.73%	2.92%	3.15%	0.14%	3.43%	1.32%
An. stephensi	Cellia [Neocellia]	57.15%	7.39%	4.59%	0.12%	2.17%	1.1%
An. maculatus	Cellia [Neocellia]	37.28%	7.33%	8%	0.2%	3.99%	1.67%
An. funestus	Cellia [Myzomyia]	8.89%	34.22%	6.26%	0.3%	4.83%	1.65%
An. culicifacies	Cellia [Myzomyia]	38.37%	13.88%	5.88%	0.2%	4.59%	1.33%
An. minimus	Cellia [Myzomyia]	31.78%	13.08%	6.33%	0.22%	4.89%	1.38%
An. nili	Cellia [Neomyzomyia]	6.07%	2.42%	11.27%	0.63%	9.23%	4.72%
An. dirus	Cellia [Neomyzomyia]	4.02%	4.29%	45.69%	0.27%	4.33%	2.4%
An. cracens	Cellia [Neomyzomyia]	4.45%	5.83%	59.89%	0.17%	3.67%	1.37%
An. punctulatus	Cellia [Neomyzomyia]	4.81%	6.12%	10.01%	0.27%	9.45%	7.48%
An. farauti no. 4	Cellia [Neomyzomyia]	5.34%	8.5%	12.2%	0.32%	8.74%	5.83%
An. farauti	Cellia [Neomyzomyia]	4.29%	8.27%	12.89%	0.35%	9.07%	4.95%
An. koliensis	Cellia [Neomyzomyia]	4.05%	5.64%	11.71%	0.33%	9.9%	6.45%
An. sinensis	Anopheles	8.82%	4.06%	13.59%	1.02%	15.55%	2.57%
An. atroparvus	Anopheles	5.09%	3.7%	10.84%	1.96%	11.41%	4.08%
An. aquasalis	Nyssorhynchus	1.17%	0.98%	3.43%	23.39%	10.55%	3.7%
An. darlingi	Nyssorhynchus	1.21%	0.51%	3.75%	11.92%	7.65%	5.32%
An. albimanus	Nyssorhynchus	1.02%	0.73%	2%	22.88%	7.92%	4.69%

References

Andrieu O, Fiston A-S, Anxolabéhère D, Quesneville H. Detection of transposable elements by their compositional bias. BMC Bioinformatics. 2004;5:94.

Arensburger P, Kim YJ, Orsetti J, Aluvihare C, O'brochta DA, Atkinson PW. An active transposable element, Herves, from the African malaria mosquito *Anopheles gambiae*. Genetics. 2005;169:697–708.

Arkhipova IR. Using bioinformatic and phylogenetic approaches to classify transposable elements and understand their complex evolutionary histories. Mob DNA. 2017;8:1–14.

Bao W, Kojima KK, Kohany O. Repbase update, a database of repetitive elements in eukaryotic genomes. Mob DNA. 2015;6:4–9.

Barnes MJ, Lobo NF, Coulibaly MB, Sagnon NF, Costantini C, Besansky NJ. SINE insertion polymorphism on the X chromosome differentiates *Anopheles gambiae* molecular forms. Insect Mol Biol. 2005;14:353–63.

Barrón MG, Paupy C, Rahola N, Akone-Ella O, Ngangue MF, Wilson-Bahun TA, et al. A new species in the major malaria vector complex sheds light on reticulated species evolution. Sci Rep. 2019;9:1–13.

Beebe NW, Russell TL, Burkot TR, Lobo NF, Cooper RD. The systematics and bionomics of malaria vectors in the Southwest Pacific. In: Manguin S, editor. Anopheles mosquitoes – new insights into malaria vectors. Rijeka: IntechOpen; 2012. https://doi.org/10.5772/55999.

Beebe NW, Russell T, Burkot TR, Cooper RD. *Anopheles punctulatus* group: evolution, distribution, and control. Annu Rev Entomol. 2015;60:335–50.

Bergman CM, Quesneville H. Discovering and detecting transposable elements in genome sequences. Brief Bioinform. 2007;8:382–92.

Besansky NJ. A retrotransposable element from the mosquito *Anopheles gambiae*. Mol Cell Biol. 1990;10:863–71.

Besansky NJ, Paskewitz SM, Hamm DM, Collins FH. Distinct families of site-specific retrotransposons occupy identical positions in the rRNA genes of *Anopheles gambiae*. Mol Cell Biol. 1992;12:5102–10.

Besansky NJ, Bedell JA, Mukabayire O. Q: a new retrotransposon from the mosquito *Anopheles gambiae*. Insect Mol Biol. 1994;3:49–56.

Besansky NJ, Mukabayire O, Bedell JA, Lusz H. Pegasus, a small terminal inverted repeat transposable element found in the white gene of *Anopheles gambiae*. Genetica. 1996;98:119–29.

Biedler J, Tu Z. Non-LTR retrotransposons in the African malaria mosquito, *Anopheles gambiae*: unprecedented diversity and evidence of recent activity. Mol Biol Evol. 2003;20:1811–25.

Boulesteix M, Biémont C. Transposable elements in mosquitoes. Cytogenet Genome Res. 2005;110:500–9.

Boulesteix M, Simard F, Antonio-Nkondjio C, Awono-Ambene HP, Fontenille D, Biémont C. Insertion polymorphism of transposable elements and population structure of *Anopheles gambiae* M and S molecular forms in Cameroon. Mol Ecol. 2007;16:441–52.

Bourque G, Burns KH, Gehring M, Gorbunova V, Seluanov A, Hammell M, et al. Ten things you should know about transposable elements. Genome Biol. 2018;19:199.

Canapa A, Barucca M, Biscotti MA, Forconi M, Olmo E. Transposons, genome size, and evolutionary insights in animals. Cytogenet Genome Res. 2015;147:217–39.

Carrasco D, Lefèvre T, Moiroux N, Pennetier C, Chandre F, Cohuet A. Behavioural adaptations of mosquito vectors to insecticide control. Curr Opin Insect Sci. 2019;34:48–54.

Casacuberta E, González J. The impact of transposable elements in environmental adaptation. Mol Ecol. 2013;22:1503–17.

Chakraborty M, Ramaiah A, Adolfi A, Halas P, Kaduskar B, Ngo LT, et al. Hidden features of the malaria vector mosquito, Anopheles stephensi, revealed by a high-quality reference genome. bioRxiv. 2020; https://doi.org/10.1101/2020.05.24.113019.

Chénais B, Caruso A, Hiard S, Casse N. The impact of transposable elements on eukaryotic genomes: from genome size increase to genetic adaptation to stressful environments. Gene. 2012;509:7–15.

Chida AR, Ravi S, Jayaprasad S, Paul K, Saha J, Suresh C, et al. A near-chromosome level genome assembly of Anopheles stephensi. bioRxiv. 2020; https://doi.org/10.1101/2020.04.27.063040.

Chuong EB, Elde NC, Feschotte C. Regulatory activities of transposable elements: from conflicts to benefits. Nat Rev Genet. 2017;18:71–86.

Compton A, Liang J, Chen C, Lukyanchikova V, Qi Y, Potters M, et al. The beginning of the end: a chromosomal assembly of the New World malaria mosquito ends with a novel telomere. G3 (Bethesda). 2020;10(10):3811–9.

De Melo ES, Wallau GDL. Transposable elements are constantly exchanged by horizontal transfer reshaping mosquito genomes. bioRxiv. 2020; https://doi.org/10.1101/2020.06.23.166744.

Derua YA, Alifrangis M, Hosea KM, Meyrowitsch DW, Magesa SM, Pedersen EM, et al. Change in composition of the *Anopheles gambiae* complex and its possible implications for the transmission of malaria and lymphatic filariasis in north-eastern Tanzania. Malar J. 2012;11:1–9.

Diesel JF, Ortiz MF, Marinotti O, Vasconcelos ATR, Loreto ELS. A re-annotation of the *Anopheles darlingi* mobilome. Genet Mol Biol. 2019;42:125–31.

Edgar RC, Myers EW. PILER: identification and classification of genomic repeats. Bioinformatics. 2005;21:152–8.

Elbarbary RA, Lucas BA, Maquat LE. Retrotransposons as regulators of gene expression. Science. 2016;351(6274):aac7247.

Ellison CE, Cao W. Nanopore sequencing and Hi-C scaffolding provide insight into the evolutionary dynamics of transposable elements and piRNA production in wild strains of *Drosophila melanogaster*. Nucleic Acids Res. 2020;48:290–303.

Esnault C, Boulesteix M, Duchemin JB, Koffi AA, Chandre F, Dabiré R, et al. High genetic differentiation between the M and S molecular forms of *Anopheles gambiae* in Africa. PLoS One. 2008;3:e1968.

Fernández-Medina RD, Struchiner CJ, Ribeiro JMC. Novel transposable elements from *Anopheles gambiae*. BMC Genomics. 2011;12:260.

Fernández-Medina RD, Carareto CMA, Struchiner CJ, Ribeiro JMC. Transposable elements in the Anopheles funestus transcriptome. Genetica. 2017;145:275–93.

Flutre T, Duprat E, Feuillet C, Quesneville H. Considering transposable element diversification in de novo annotation approaches. PLoS One. 2011;6:e16526.

Flynn JM, Hubley R, Goubert C, Rosen J, Clark AG, Feschotte C, et al. Repeat Modeler 2 for automated genomic discovery of transposable element families. Proc Natl Acad Sci. 2020;117:9451.

Fontaine MC, Pease JB, Steele A, Waterhouse RM, Neafsey DE, Sharakhov IV, et al. Extensive introgression in a malaria vector species complex revealed by phylogenomics. Science. 2015;347(6217):1258524.

Fouché S, Badet T, Oggenfuss U, Plissonneau C, Francisco CS, Croll D. Stress-driven transposable element de-repression dynamics and virulence evolution in a fungal pathogen. Mol Biol Evol. 2020;37:221–39.

Fouet C, Kamdem C, Gamez S, White BJ. Extensive genetic diversity among populations of the malaria mosquito Anopheles moucheti revealed by population genomics. Infect Genet Evol. 2017;48:27–33.

Freitas LA, Russo CAM, Voloch CM, Mutaquiha OCF, Marques LP, Schrago CG. Diversification of the genus Anopheles and a neotropical clade from the late cretaceous. PLoS One. 2015;10:1–12.

Ge R, Mai G, Zhang R, Wu X, Wu Q, Zhou F. MUSTv2: an improved de novo detection program for recently active miniature inverted repeat transposable elements (MITEs). J Integr Bioinform. 2017;14:1–7.

Giraldo-Calderón GI, Emrich SJ, MacCallum RM, Maslen G, Emrich S, Collins F, et al. VectorBase: An updated Bioinformatics Resource for invertebrate vectors and other organisms related with human diseases. Nucleic Acids Res. 2015;43(D1):D707–D13.

Goerner-Potvin P, Bourque G. Computational tools to unmask transposable elements. Nat Rev Genet. 2018;19:688–704.

Goubert C, Henri H, Minard G, Valiente Moro C, Mavingui P, Vieira C, et al. High-throughput sequencing of transposable element insertions suggests adaptive evolution of the invasive Asian tiger mosquito towards temperate environments. Mol Ecol. 2017;26:3968–81.

Grau-Bové X, Tomlinson S, O'reilly AO, Harding NJ, Miles A, Kwiatkowski D, et al. Evolution of the insecticide target Rdl in African Anopheles is driven by interspecific and interkaryotypic introgression. Mol Biol Evol. 2020;37:2900–17.

Gray YH. It takes two transposons to tango: transposable-element-mediated chromosomal rearrangements. Trends Genet. 2000;16:461–8.

Grossman GL, Cornel AJ, Rafferty CS, Robertson HM, Collins FH. Tsessebe, Topi and Tiang: three distinct Tc1-like transposable elements in the malaria vector, *Anopheles gambiae*. Genetica. 1999;105:69–80.

Guio L, González J. New insights on the evolution of genome content: population dynamics of transposable elements in flies and humans. In: Anisimova M, editor. Evolutionary genomics: statistical and computational methods. New York: Springer; 2019. https://doi.org/10.1007/978-1-4939-9074-0.

Guio L, Vieira C, González J. Stress affects the epigenetic marks added by natural transposable element insertions in *Drosophila melanogaster*. Sci Rep. 2018;8:12197.

Häcker I, Schetelig MF. Molecular tools to create new strains for mosquito sexing and vector control. Parasit Vectors. 2018;11:645.

Han Y, Wessler SR. MITE-hunter: a program for discovering miniature inverted-repeat transposable elements from genomic sequences. Nucleic Acids Res. 2010;38:e199.

Harbach RE. The phylogeny and classification of Anopheles. In: Manguin S, editor. Anopheles mosquitoes – new insights into malaria vectors. Rijeka: IntechOpen; 2013. https://doi.org/10.5772/54695.

Hill SR, Leung SS, Quercia NL, Vasiliauskas D, Yu J, Pasic I, et al. Ikirara insertions reveal five new *Anopheles gambiae* transposable elements in islands of repetitious sequence. J Mol Evol. 2001;52:215–31.

Holmes CJ, Benoit JB. Biological adaptations associated with dehydration in mosquitoes. Insects. 2019;10(11):375.

Holt RA, Mani Subramanian G, Halpern A, Sutton GG, Charlab R, Nusskern DR, et al. The genome sequence of the malaria mosquito *Anopheles gambiae*. Science. 2002;298:129–49.

Horváth V, Merenciano M, González J. Revisiting the relationship between transposable elements and the eukaryotic stress response. Trends Genet. 2017;33:832–41.

Jiang X, Peery A, Hall AB, Sharma A, Chen XG, Waterhouse RM, et al. Genome analysis of a major urban malaria vector mosquito, Anopheles stephensi. Genome Biol. 2014;15:459.

Kamali M, Marek PE, Peery A, Antonio-Nkondjio C, Ndo C, Tu Z, et al. Multigene phylogenetics reveals temporal diversification of major African malaria vectors. PLoS One. 2014;9:1–9.

Kamdem C, Fouet C, Gamez S, White BJ. Pollutants and insecticides drive local adaptation in African malaria mosquitoes. Mol Biol Evol. 2017;34:1261–75.

Kapitonov VV, Jurka J. A universal classification of eukaryotic transposable elements implemented in Repbase. Nat Rev Genet. 2008;9:411–2.

Kennedy RC, Unger MF, Christley S, Collins FH, Madey GR. An automated homology-based approach for identifying transposable elements. BMC Bioinformatics. 2011;12:130.

Kingan SB, Heaton H, Cudini J, Lambert CC, Baybayan P, Galvin BD, et al. A high-quality de novo genome assembly from a single mosquito using pacbio sequencing. Genes. 2019;10:62.

Kohany O, Gentles AJ, Hankus L, Jurka J. Annotation, submission and screening of repetitive elements in Repbase: Repbase submitter and censor. BMC Bioinformatics. 2006;7:1–7.

Krzywinski J, Besansky NJ. Molecular systematics of Anopheles: from subgenera to subpopulations. Annu Rev Entomol. 2003;48:111–39.

Laptev IA, Raevskaya NM, Filimonova NA, Sineoky SP. The piggy bac transposon as a tool in genetic engineering. Appl Biochem Microbiol. 2017;53:874–81.

Lau YL, Lee WC, Chen J, Zhong Z, Jian J, Amir A, et al. Draft genomes of Anopheles cracens and Anopheles maculatus: comparison of simian malaria and human malaria vectors in peninsular Malaysia. PLoS One. 2016;11:1–24.

Lerat E, Casacuberta J, Chaparro C, Vieira C. On the importance to acknowledge transposable elements in epigenomic analyses. Genes (Basel). 2019;10:258.

Logue K, Small ST, Chan ER, Reimer L, Siba PM, Zimmerman PA, et al. Whole-genome sequencing reveals absence of recent gene flow and separate demographic histories for *Anopheles punctulatus* mosquitoes in Papua New Guinea. Mol Ecol. 2015;24:1263–74.

Malicki M, Spaller T, Winckler T, Hammann C. DIRS retrotransposons amplify via linear, single-stranded cDNA intermediates. Nucleic Acids Res. 2020;48:4230–43.

Marinotti O, Cerqueira GC, De Almeida LGP, Ferro MIT, Da Silva Loreto EL, Zaha A, et al. The genome of *Anopheles darlingi*, the main neotropical malaria vector. Nucleic Acids Res. 2013;41:7387–400.

Mathiopoulos KD, Della Torre A, Predazzi V, Petrarca V, Coluzzi M. Cloning of inversion breakpoints in the *Anopheles gambiae* complex traces a transposable element at the inversion junction. Proc Natl Acad Sci U S A. 1998;95:12444–9.

Mccarthy EM, Mcdonald JF. LTR_STRUC: a novel search and identification program for LTR retrotransposons. Bioinformatics. 2003;19:362–7.

Miles A, Harding NJ, Bottà G, Clarkson CS, Antão T, Kozak K, et al. Genetic diversity of the African malaria vector *Anopheles gambiae*. Nature. 2017;552:96–100.

Mukabayire O, Besansky NJ. Distribution of T1, Q, Pegasus and mariner transposable elements on the polytene chromosomes of PEST, a standard strain of *Anopheles gambiae*. Chromosoma. 1996;104:585–95.

Neafsey DE, Waterhouse RM, Abai MR, Aganezov SS, Alekseyev MA, Allen JE, et al. Highly evolvable malaria vectors: the genomes of 16 Anopheles mosquitoes. Science. 2015;347 (6217):1258522.

Nelson MG, Linheiro RS, Bergman CM. McClintock: an integrated pipeline for detecting transposable element insertions in whole-genome shotgun sequencing data. G3. 2017;7:2763.

O'brochta DA, Subramanian RA, Orsetti J, Peckham E, Nolan N, Arensburger P, et al. hAT element population genetics in *Anopheles gambiae* s.l. in Mozambique. Genetica. 2006;127:185.

Ogola EO, Odero JO, Mwangangi JM, Masiga DK, Tchouassi DP. Population genetics of Anopheles funestus, the African malaria vector, Kenya. Parasit Vectors. 2019;12:15.

Ou S, Su W, Liao Y, Chougule K, Agda JRA, Hellinga AJ, et al. Benchmarking transposable element annotation methods for creation of a streamlined, comprehensive pipeline. Genome Biol. 2019;20:275.

Petersen M, Armisén D, Gibbs RA, Hering L, Khila A, Mayer G, et al. Diversity and evolution of the transposable element repertoire in arthropods with particular reference to insects. BMC Evol Biol. 2019;19:11.

Pettersson ME, Rochus CM, Han F, Chen J, Hill J, Wallerman O, et al. A chromosome-level assembly of the Atlantic herring genome-detection of a supergene and other signals of selection. Genome Res. 2019;29:1919–28.

Price AL, Jones NC, Pevzner PA. De novo identification of repeat families in large genomes. Bioinformatics. 2005;21:351–8.

Quesneville H, Nouaud D, Anxolabéhère D. Detection of new transposable element families in Drosophila melanogaster and *Anopheles gambiae* genomes. J Mol Evol. 2003;57:50–9.

Rebollo R, Romanish MT, Mager DL. Transposable elements: an abundant and natural source of regulatory sequences for host genes. Annu Rev Genet. 2012;46:21–42.

Robertson HM. The mariner transposable element is widespread in insects. Nature. 1993;362:241–5.

Romans P, Bhattacharyya RK, Colavita A. Ikirara, a novel transposon family from the malaria vector mosquito, *Anopheles gambiae*. Insect Mol Biol. 1998;7:1–10.

Salgueiro P, Moreno M, Simard F, O'brochta D, Pinto J. New insights into the population structure of *Anopheles gambiae* s.s. in the Gulf of Guinea Islands revealed by Herves transposable elements. PLoS One. 2013;8:e62964.

Schrader L, Schmitz J. The impact of transposable elements in adaptive evolution. Mol Ecol. 2019;28:1537–49.

Sessegolo C, Burlet N, Haudry A. Strong phylogenetic inertia on genome size and transposable element content among 26 species of flies. Biol Lett. 2016;12(8):20160407.

Shao H, Tu Z. Expanding the diversity of the IS630-Tc1-mariner superfamily: discovery of a unique DD37E transposon and reclassification of the DD37D and DD39D transposons. Genetics. 2001;159:1103–15.

Sharakhov IV, White BJ, Sharakhova MV, Kayondo J, Lobo NF, Santolamazza F, et al. Breakpoint structure reveals the unique origin of an interspecific chromosomal inversion (2La) in the *Anopheles gambiae* complex. Proc Natl Acad Sci U S A. 2006;103:6258–62.

Struchiner CJ, Massad E, Tu Z, Ribeiro JMC. The tempo and mode of evolution of transposable elements as revealed by molecular phylogenies reconstructed from mosquito genomes. Evolution. 2009;63:3136–46.

Subramanian RA, Arensburger P, Atkinson PW, O'brochta DA. Transposable element dynamics of the hAT element Herves in the human malaria vector *Anopheles gambiae* s.s. Genetics. 2007;176:2477–87.

Surendran SN, Sivabalakrishnan K, Sivasingham A, Jayadas TTP, Karvannan K, Santhirasegaram S, et al. Anthropogenic factors driving recent range expansion of the malaria vector. Front Public Health. 2019;7:53.

Tennessen JA, Ingham VA, Toé KH, Guelbéogo WM, Sagnon NF, Kuzma R, et al. A population genomic unveiling of a new cryptic mosquito taxon within the malaria-transmitting *Anopheles gambiae* complex. bioRxiv. 2020; https://doi.org/10.1101/2020.05.26.116988.

The Anopheles gambiae Genomes Consortium. Genome variation and population structure among 1142 mosquitoes of the African malaria vector species Anopheles gambiae and Anopheles coluzzii. Genome Res. 2020;30:1533–46.

Tu Z. Eight novel families of miniature inverted repeat transposable elements in the African malaria mosquito, *Anopheles gambiae*. Proc Natl Acad Sci U S A. 2001;98:1699–704.

Tu Z, Coates C. Mosquito transposable elements. Insect Biochem Mol Biol. 2004;34:631–44.

Tu Z, Li S. Mobile genetic elements of malaria vectors and other mosquitoes. In: Brindley PJ, editor. Mobile genetic elements in metazoan parasites. Austin: Landes Bioscience; 2009.

Tubío JMC, Naveira H, Costas J. Structural and evolutionary analyses of the Ty3/gypsy group of LTR retrotransposons in the genome of *Anopheles gambiae*. Mol Biol Evol. 2005;22:29–39.

Volohonsky G, Terenzi O, Soichot J, Naujoks DA, Nolan T, Windbichler N, et al. Tools for *Anopheles gambiae* transgenesis. G3 (Bethesda). 2015;5:1151–63.

Waterhouse RM, Aganezov S, Anselmetti Y, Lee J, Ruzzante L, Reijnders MJMF, et al. Evolutionary superscaffolding and chromosome anchoring to improve Anopheles genome assemblies. BMC Biol. 2020;18:1–20.

Wicker T, Sabot F, Hua-Van A, Bennetzen JL, Capy P, Chalhoub B, et al. A unified classification system for eukaryotic transposable elements. Nat Rev Genet. 2007;8:973–82.

World Health Organization. World malaria report. 2019.

World Health Organization. Lymphatic filariasis fact sheets. 2020.

Xu Z, Wang H. LTR-FINDER: an efficient tool for the prediction of full-length LTR retrotransposons. Nucleic Acids Res. 2007;35:265–8.

Zhou Y, Cahan SH. A novel family of terminal-repeat retrotransposon in miniature (TRIM) in the genome of the red harvester ant, *Pogonomyrmex barbatus*. PLoS One. 2012;7:e53401.

Zhou D, Zhang D, Ding G, Shi L, Hou Q, Ye Y, et al. Genome sequence of Anopheles sinensis provides insight into genetics basis of mosquito competence for malaria parasites. BMC Genomics. 2014;15:42.

Disentangling Types of Linked Selection Using Patterns of Nucleotide Variation in *Drosophila pseudoobscura*

Katharine L. Korunes, Kieran Samuk, and Mohamed A. F. Noor

Abstract A central goal in evolutionary genetics is to understand the forces that drive patterns of nucleotide variation within species. Recombination rate and nucleotide diversity are positively correlated across loci within many species. Selective sweeps and background selection both may contribute to this pattern by reducing nucleotide variation disproportionately in regions of low recombination. While there is unambiguous documentation of some forms of selective sweeps, it is less clear how much background selection and soft selective sweeps contribute. Here, we discuss the study of these different types of linked selection, paying particular attention to the large body of work devoted to linked selection in *Drosophila*. We discuss patterns of nucleotide variation in the classic model system *Drosophila pseudoobscura*, and we leverage this system to test for the effects of linked selection in the near absence of hard, complete selective sweeps. Our case study of variation in sets of loci shared between *Drosophila pseudoobscura* and *D. miranda* suggests that selection at linked sites may reduce nucleotide variation even in the absence of hard selective sweeps. We discuss how this work relates to recent approaches and challenges in the study of how selection at linked sites influences patterns of nucleotide variation.

Keywords Background selection · Linked selection · Recombination · Selective sweeps

1 Introduction

One of the most striking and widespread patterns in genomic data is the positive correlation between local recombination rates and nucleotide diversity (e.g., Nachman 2001; Tenaillon et al. 2001; Takahashi et al. 2004; Begun et al. 2007;

K. L. Korunes (✉) · K. Samuk · M. A. F. Noor
Biology Department, Duke University, Durham, NC, USA
e-mail: katharine.korunes@duke.edu

Julian R. Dupuis and Om P. Rajora (eds.), *Population Genomics: Insects*, Population Genomics [Om P. Rajora (Editor-in-Chief)],
https://doi.org/10.1007/13836_2021_87,

Cutter and Choi 2010; Burri et al. 2015). Since this pattern was observed in *Drosophila* 30 years ago (Aguade et al. 1989; Begun and Aquadro 1992; W. Stephan and Langley 1989), many studies have attempted to elucidate the evolutionary forces that drive this association. The parsimonious explanation of a neutral process, such as mutagenic effects of recombination, driving this association was undermined by the observation that sequence divergence between *Drosophila melanogaster* and its sister species, *D. simulans,* is not associated with recombination rates (Begun and Aquadro 1992; but see Halldorsson et al. 2019 regarding the mutagenic effects of recombination in humans). Instead, evidence supports the hypothesis that much of this pattern can be explained by natural selection eliminating variation in regions of low recombination (Begun and Aquadro 1992; Begun et al. 2007; McGaugh et al. 2012). Strong linkage in regions of low recombination allows positive and/or negative selection at a given locus to reduce nucleotide diversity at surrounding, linked loci. On a per-base pair level, this results in a greater reduction in diversity due to linked selection in regions of low vs. high recombination.

Hard selective sweeps result from positive selection when a single haplotype containing an advantageous mutation replaces multiple ancestral haplotypes, and alleles linked to the advantageous mutation become fixed along with the mutation as it spreads (Fig. 1a; Maynard Smith and Haigh 1974; Kaplan et al. 1989). However, selection at linked sites may also reduce nucleotide variation in the absence of hard selective sweeps. Linked selection can also occur via partial sweeps, soft sweeps, and background selection. Partial sweeps are selective sweeps that are incomplete, where a mutation has increased in frequency but has not reached fixation. Soft sweeps occur when multiple copies of an allele (on different backgrounds) contribute to a selective sweep, though these copies may or may not be identical by descent (Hermisson and Pennings 2005). Hermisson and Pennings (2005) introduced a model in which standing genetic variation is the basis of a soft sweep, such as when an allele segregating in the population becomes advantageous after a shift in selective pressures, such as a change in the environment (Fig. 1b). A soft sweep can also occur when an adaptive allele repeatedly enters the population (Pennings and Hermisson 2006). For example, migration or mutation might repeatedly introduce an advantageous allele into a population (Fig. 1c). Like hard sweeps, partial and soft sweeps may reduce diversity at nearby neutral sites as a result of a beneficial mutation increasing in frequency due to positive selection.

Negative selection can also reduce linked neutral variation through the action of background selection. Under background selection, multiple deleterious mutations arise over time, and selection eliminates the deleterious mutations along with closely linked neutral variation (Fig. 1d) (Charlesworth et al. 1993). Neutral mutations can only drift to high frequencies in a population if they exist on haplotypes that remain free of (or have the fewest/weakest) deleterious mutations.

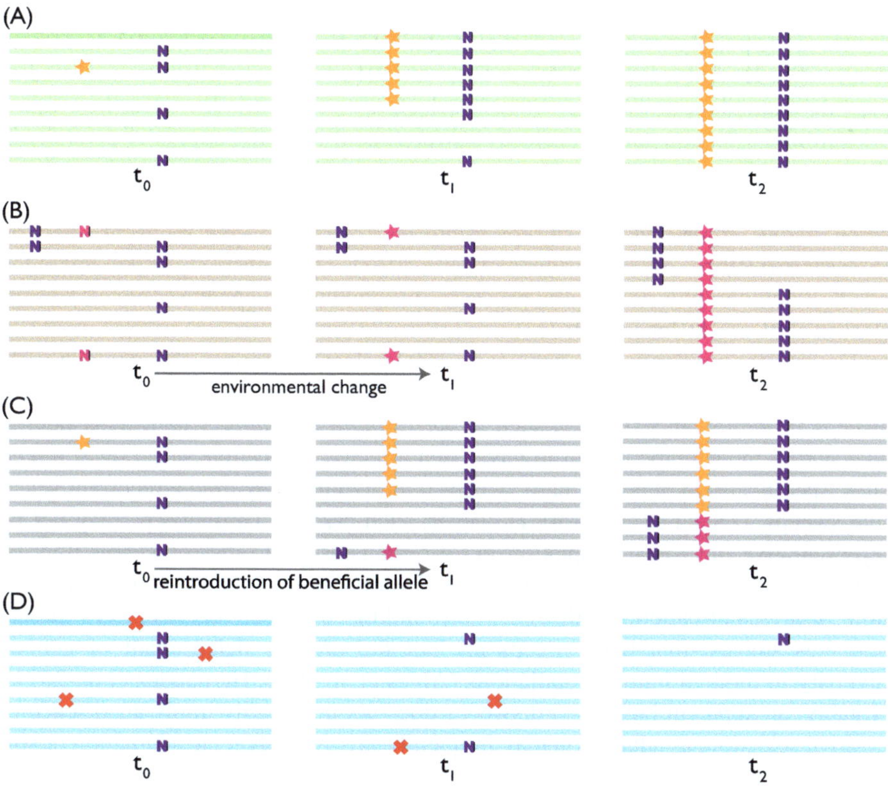

Fig. 1 Models of how selection influences linked variation. The illustrated population has a neutral polymorphism segregating at time t_0 at a site indicated by "N." Under the hard sweeps model (**a**), a novel advantageous mutation (gold star) arises at t_0, increases in frequency to yield a partial sweep at t_1, and has swept to fixation by t_2. (**b**, **c**) represent soft sweeps. In (**b**), a previously neutral mutation becomes advantageous (magenta star) due to an environmental change between t_0 and t_1, so the beneficial allele copies contributing to the sweep are identical by descent but segregating on different backgrounds due to recombination. In (**c**), the contributing allele enters the population recurrently via mutation or migration. Under background selection (**d**), haplotypes with deleterious mutations (red) are eliminated, which reduces linked neutral variation

1.1 Inferring the Relative Contributions of Types of Linked Selection

There is a long-standing debate over the relative contributions of different types of linked selection to shaping patterns of genetic diversity (e.g., Stephan 2010; Jensen 2014; Renzette et al. 2016). Because recombination reduces the strength of linkage, different modes of linked selection all reduce nucleotide variation disproportionately in regions of low recombination. However, the independent effects of different types of linked selection are difficult to disentangle. Many studies document regions of low nucleotide diversity centered on adaptive alleles that were previously rare or absent in the population (e.g., de Groot et al. 2002; Bersaglieri et al. 2004; Colosimo et al. 2005; Hoekstra et al. 2006; see Stephan 2016 for review). While such studies provide unambiguous evidence of hard selective sweeps, it is less clear how much background selection and soft selective sweeps contribute to patterns of diversity in the genome. Few studies seek empirical signatures of linked selection in the absence of hard sweeps, and these signatures are often more controversial. For example, some studies argue that soft sweeps might be a very prevalent mode of selection (Messer and Petrov 2013; Schrider and Kern 2017), while others argue that this is the result of demographic and neutral processes mimicking the signatures of soft sweeps (Harris et al. 2018; Jensen 2014).

Despite the evidence for many individual selective sweeps, the genome-wide prevalence of sweeps is still ambiguous. There is evidence for widespread selective sweeps in well-studied species such as *Drosophila simulans,* where most neutral sites across the genome appear to have been affected by positive selection at linked sites, specifically by repeated hard sweeps over evolutionary time ("recurrent selective sweeps") (Sella et al. 2009). However, such cases of hard sweeps may not be representative of typical adaptation in all organisms, and other processes may be more prevalent as drivers of reduced neutral variation in regions of low recombination. Analyses of patterns of diversity surrounding conserved regions in 179 human genomes suggest that hard sweeps are not the primary determinant of genetic diversity in modern human populations (Alves et al. 2012; Hernandez et al. 2011). Notably, an apparent lack of hard sweeps does not necessarily imply an absence of positive selection, as adaptation may frequently involve polygenic adaptation or soft sweeps (e.g., Pritchard et al. 2010). Similarly, recurrent selective sweeps appear to be rare in plants, and background selection may dominate as a driver of patterns of genetic diversity in many plant populations (reviewed in Slotte 2014).

1.2 The Dynamics of Background Selection

Other recent work focuses on the role of the strength of selection and underscores the importance of accounting for weakly deleterious mutations. In particular, background selection caused by weakly deleterious alleles has been highlighted as a key factor shaping patterns of genetic variation (Lohmueller et al. 2011). Though the effects of background selection are in part dictated by the strength of selection, this occurs very differently than it does under selective sweeps. Unlike selective sweeps, which are most pronounced in the case of strong selection and dominant mutations, the strongest reduction of linked diversity due to background selection occurs when selection is relatively weak and mutations are recessive (Charlesworth 2012a; Charlesworth et al. 1993; Cutter and Payseur 2013). Strongly deleterious alleles are kept at low frequencies in the population, while weakly deleterious alleles often persist longer in the population. This implies that if there is strong linkage between neutral sites and sites under selection, then weakly deleterious alleles can be recombined onto a greater number of haplotypes compared to strongly deleterious alleles before being eliminated (Nordborg et al. 1996). This observation means that we must use caution when inferring background selection based on constraint per se (i.e. as a proxy for the strength of negative selection). It is tempting to assume that genes that are more conserved among taxa, or under greater constraint, would experience greater background selection. However, as discussed, this is not necessarily true if highly constrained genes experience stronger selection.

In addition to the strength of selection, the genomic context of deleterious alleles, including other nearby sites under selection and the strength of linkage, affects background selection. Recent work to help distinguish types of linked selection includes simulations modeled after both humans and *Drosophila* showing that the spatial arrangement of constrained sites can dictate whether the valley of diversity caused by background selection mirrors signatures caused by positive selection (Schrider 2020). Similarly, any factors that increase linkage affect background selection. A striking example is the apparent role of background selection in shaping patterns of variation of the *Drosophila* X chromosome, as illustrated by a recent investigation of silent site diversity in *D. melanogaster* (Charlesworth 2012b), where mean silent site diversity in the ancestral population appears to be approximately the same for X-linked loci and autosomal loci (Andolfatto 2001; Hutter et al. 2007). The similarity of neutral diversity for X-linked and autosomal loci deviates from the null expectation based on the ratio of the effective population sizes (N_e) of the X chromosome and each autosome. This ratio of effective population sizes is expected to be ¾, given an equal sex ratio and a random distribution of the number of progeny of each sex (Wright 1931). According to the results from Charlesworth (2012a), background selection can explain the equality of N_e between the X chromosomes and the autosomes, if background selection is more effective on the autosomes. This explanation fits with the fact that there is no meiotic recombination in male *Drosophila,* so the X chromosome (of which 2/3 are present in females) has a greater effective recombination rate compared to the autosomes (of which half are in females). The observed X:autosome silent site diversity ratio was closer to the expectation in *D. pseudoobscura,* and recent data reveals a lower than expected ratio in *D. simulans* (Jackson et al. 2017). The evolutionary mechanisms underlying X:autosome differences remain poorly understood, but it is possible that the ratios in *D. pseudoobscura* and *D. simulans* are due to higher recombination rates in these species (Charlesworth 2012b; Jackson et al. 2017).

These observations of X:autosome neutral diversity illustrate the intricacies of how background selection operates under varying degrees of linkage. Relatedly, our ability to infer the effects of background selection is often complicated by demographic history (Torres et al. 2020), which often involves changes in N_e. The interplay between N_e and background selection can also be seen in asexual and selfing species, where linkage is often strong. Evidence suggests that background selection plays a prominent role in decreasing genetic diversity under uniparental inheritance (Agrawal and Hartfield 2016; Roze 2016). Similarly, background selection has been implicated as a driver of low observed variability on non-recombining chromosomes such as the dot chromosome of *D. melanogaster,* the neo-Y of

D. miranda (Kaiser and Charlesworth 2009), and recently, the Y chromosome in humans (Wilson Sayres et al. 2014). On non-recombining chromosomes or in low recombination genomic regions, interference among selected loci plays a major role in the action of background selection. Hill–Robertson interference (Felsenstein 1974) influences the interaction between recombination and selection, and interference among the sites that cause background selection can result in a larger role of drift in shaping allele frequencies (reviewed in Charlesworth and Campos 2014).

1.3 Recent Progress in Genome-Wide Scans for Selection

Genome-wide scans for positive selection often rely on the hard sweeps model, with neutrality as a null hypothesis (Vitti et al. 2013). However, as seen in a recent implementation of the SweepFinder2 framework, some emerging methods for genome-wide scans for sweeps account for background selection (DeGiorgio et al. 2016; Huber et al. 2016). The controversy over the role of selective sweeps necessitates a baseline prediction of genetic diversity patterns under background selection alone. The standard methodology for distinguishing sweeps and background selection typically relies on detecting shifts in site frequency spectra, analyzed through statistics such as Tajima's D (Tajima 1989), Fu and Li's D (Fu and Li 1993), or Fay and Wu's H (Fay and Wu 2000). This approach often yields inconclusive results, partially because these statistics typically assume a standard neutral model and may neglect the effects of demography or population structure.

In recent years, the rapid accumulation of genomic data has inspired new possible approaches to this problem. McVicker et al. (2009) inferred a map of background selection effects in hominid genome evolution using polymorphism data from five primate species, including humans. McVicker et al. (2009) estimated the effects of purifying selection on neutral diversity as a function of selection coefficients, recombination rates, and deleterious mutation rates. At the megabase scale, their map of background selection effects explains a large proportion of observed diversity patterns across the genome, though the associated estimate of the deleterious mutation rate from the data is substantially higher than other estimates from the literature (Kondrashov 2003; Nachman and Crowell 2000).

Similarly, predictions from background selection models alone may explain a large proportion of observed variation in *Drosophila melanogaster*. According to one estimate, background selection may explain as much as 70% of the observed variance in nucleotide diversity across individual *D. melanogaster* (Comeron 2014). Evidence continues to indicate that genome-wide patterns of variation are broadly consistent with expectations of background selection in many species (Charlesworth 2012b; Charlesworth and Campos 2014; Comeron 2014; Hernandez et al. 2011; Lohmueller et al. 2011; McVicker et al. 2009; Pouyet et al. 2018; Renzette et al. 2016). Recent work even suggests that background selection may affect neutral divergence between distantly related species (Phung et al. 2016).

The recent attempts to infer genome-wide effects of background selection bolster the argument that background selection *could* be a major predictor of genome-wide diversity. While studies such as McVicker et al. (2009) and Comeron (2014) underscore that a background selection model can explain a large proportion of observed diversity patterns across polymorphism data (in primates and *D. melanogaster,* respectively), a common limitation of this type of analysis is the lack of a relative assessment of the explanatory power of a sweeps-only model or a joint model (background selection and sweeps together). Recent work aims to simultaneously estimate the effects of both positive and negative selection in genomic data (Elyashiv et al. 2016). These emerging modeling approaches have advanced our understanding of how different selective forces contribute to lower diversity in regions of low recombination. However, there is essentially no direct empirical evidence of the operation of background selection, and no standard methodology for detecting it. Notably, Elyashiv et al. (2016) and McVicker et al. (2009) both estimated a higher deleterious mutation rate than direct empirical studies have reported. Discrepancies such as these underscore the need to continue improving models of linked selection and to inform them with empirical observations.

2 A Case Study from *D. pseudoobscura*: A Test for the Effects of Linked Selection in the Near Absence of Recent Hard Selective Sweeps

Differentiating the effects of different types of linked selection is critical to understanding the extent to which linked selection influences genome-wide patterns of variation. Further, identifying the evolutionary forces acting locally on specific regions of the genome is central to delineating which portions may be evolving neutrally or close to neutrally. There is growing interest in identifying such neutrally evolving sequences to remove biases in demographic inference (Pouyet et al. 2018; Schrider et al. 2016). To leverage empirical evidence towards differentiating the effects of types of linked selection, we present a new potential approach for assessing the effects of selection at linked sites not attributable to recent hard, complete sweeps on within-species patterns of genetic diversity. We test this method using the model system *Drosophila pseudoobscura* and its close relative, *D. miranda.* Previous work shows that *D. pseudoobscura* exhibits a positive association between recombination rate and nucleotide diversity (Kulathinal et al. 2008; McGaugh et al. 2012). Background selection, selective sweeps, or a combination of these forces may explain this association. We attempt to distinguish these forces by comparing the *D. pseudoobscura* genome to closely related genomes. *D. miranda* serves as an appropriate point of comparison, because it is a close relative but (unlike *D. persimilis*) does not hybridize with *D. pseudoobscura*. An additional species, *D. lowei*, provides an outgroup for polarizing differences as ancestral or derived when examining substitutions between *D. pseudoobscura* and *D. miranda*.

We examine the effects of linked selection in the absence of hard sweeps by looking at variation in genes lacking fixed differences between *Drosophila pseudoobscura* and *D. miranda*. This approach greatly reduces the probability that a hard sweep has occurred recently in the subset of genes examined. Here, we present our assessments of nucleotide variation in this set of loci, and we use the findings and caveats (elaborated in *Case Study: Results* and *Discussion*) of this case study to stimulate discussion on potential future directions in differentiating the effects of different types of linked selection. This examination of variation within *Drosophila pseudoobscura* suggests that removing the effect of recent, hard sweeps does not dramatically alter the strength of the relationship between neutral variation and recombination rates, supporting the idea that selection at linked sites may reduce nucleotide variation even in the absence of hard selective sweeps.

2.1 *Case Study: Methods*

2.1.1 Ortholog Identification

We examined variation between *D. pseudoobscura* and *D. miranda* using whole-genome sequence data from 29 *D. pseudoobscura* and 11 *D. miranda* strains (McGaugh et al. 2012; Samuk et al. 2020; data accessions provided in Table 1). *D. miranda* is not independently annotated, but it diverged from *D. pseudoobscura* within the past ~2 million years (Wang and Hey 1996), facilitating alignment of the *D. miranda* genomes to the *D. pseudoobscura* genome assembly. One *D. lowei* genome was included for use as an outgroup. We aligned the sequences from these 41 genomes to the most recent *D. pseudoobscura* reference assembly (Dpse_3.04: GCA_000001765.2) using BWA-0.7.5a (Li and Durbin 2009). We used GATK v3.8 to call SNPs (McKenna et al. 2010; Van der Auwera et al. 2013). Prior to downstream analyses, we filtered based on GATK's hard filtering recommendations to exclude sites with QualByDepth (QD) < 2.0, FisherStrand Bias (FS) > 60, and StrandOddsRatio (SOR) > 3.0, MQ < 40, MQRankSum < -12.5, ReadPosRankSum < -8. For each of *D. pseudoobscura's* 16,959 annotated genes published by Flybase (http://flybase.org, Full Annotation Release 3.04), we used the annotated gene span to identify the variant data at that locus for each of the 41 genomes. Scripts used to analyze the data are available at https://github.com/kkorunes/Dpseudoobscura_LinkedSelection).

2.1.2 Recombination Map

Recombination rates were obtained from two independent studies of crossover rates within *D. pseudoobscura* (McGaugh et al. 2012; Samuk et al. 2020). LD-based recombination maps are also available for these species (Smukowski Heil et al.

Table 1 Genomic data

Strain	NCBI accession
Drosophila lowei	
Lab3Lowei	SRX091467: SRR330416, SRR330418
Drosophila miranda	
MA28	SRX950183: SRR1873751
MAO101-4	SRX950187: SRR1873752
MAO3-3	SRX950188: SRR1873753
MAO3-4	SRX950189: SRR1873754
MAO3-5	SRX950190: SRR1873755
MAO3-6	SRX950211: SRR2042916
ML14	SRX965452: SRR1925723
ML16	SRX965455: SRR1925728
ML6f	SRX965460: SRR1925734
SP138	SRX965461: SRR1925735
SP235	SRX965462: SRR1925736
Drosophila pseudoobscura	
AFC12	SRX091462: SRR330321, SRR330322, SRR330323, SRR330324
FS18	SRX091310: SRR330103, SRR330104, SRR330105, SRR330106
MAT32	SRX091461: SRR330317, SRR330318, SRR330319, SRR330320
MATTL	SRX091324: SRR330129, SRR330130, SRR330131, SRR330197
MSH9	SRX091465: SRR330329, SRR330330, SRR330331, SRR330333
MSH24	SRX091463: SRR330325, SRR330326, SRR330327, SRR330328
PP1134	SRX091323: SRR330125, SRR330126, SRR330127, SRR330128
PP1137	SRX091311: SRR330107
A24	SRX7842600: SRR11230564
M15	SRX7842599: SRR11230565
M20	SRX7842598: SRR11230566
A30	SRX7842597: SRR11230567
A57	SRX7842596: SRR11230568
A49	SRX7842595: SRR11230569
A48	SRX7842594: SRR11230570
A12	SRX7842593: SRR11230571
M27	SRX7842591: SRR11230573
Flag14	SRX7842590: SRR11230574
A19	SRX7842589: SRR11230575
M17	SRX7842588: SRR11230576
A60	SRX7842587: SRR11230577
M14	SRX7842586: SRR11230578
MV2-25	SRX7842585: SRR11230579
A6	SRX7842584: SRR11230580
M6	SRX7842583: SRR11230581
M13	SRX7842582: SRR11230582
A56	SRX7842581: SRR11230583
A14	SRX7842580: SRR11230584
A47	SRX7842579: SRR11230585

2015) and are highly correlated with the empirical recombination maps generated by McGaugh et al. (2012). Our use of the empirical recombination maps avoids issues of circularity from measuring sequence diversity using LD-based recombination maps. The recombination dataset from McGaugh et al. 2012 consists of two fine-scale recombination maps for two separate pairs of inbred lines of *D. pseudoobscura* (McGaugh et al. 2012). We averaged these two nearly identical replicate maps from McGaugh et al. (2012), which together cover much of chromosome 2 and the right arm of the X chromosome (chromosome XR). The recombination dataset generated by Samuk et al. (2020) covers much of chromosome 2, chromosome 4, and both arms of the X chromosome (XL and XR). For each chromosome, we used the recombination map with the greatest marker density, which led to the selection of chromosome 2 from McGaugh et al. (2012) and chromosomes 4, XL, and XR from Samuk et al. (2020). Together, the recombination datasets used amount to approximately 65% (~99 mb) of the ~153 mb *D. pseudoobscura* genome assembly. For each of the 231 genomic intervals denoted by flanking markers (averaging 431 kb apart), the maps provide a recombination rate reported in Kosambi centiMorgans per megabase (cM/mb). We used the coordinates of the markers to determine which interval contains each of the genes under examination.

2.1.3 Analysis of Fixed Differences and Neutral Variation

For each gene, we compared the predicted gene span (including untranslated regions (UTRs), coding sequences, and introns) across all 41 individuals to identify fixed differences between *D. pseudoobscura* and *D. miranda*. To reduce possible effects of selective sweeps in or near the gene, we included the introns and UTRs since evidence suggests that selection regularly acts upon these regions (e.g., Kohn et al. 2004; Andolfatto 2005). To assess the effect of extending the screen to include nearby regions beyond the UTRs, we repeated the analyses including 50 bp upstream and 50 bp downstream of each gene span in our screen for fixed differences. Evidence suggests that LD decays rapidly in *Drosophila* (Langley et al. 2000), and screening the gene span itself would exclude many sweeps of advantageous mutations near the coding region. However, some (e.g., strongly selected) sweeps that occurred outside of the UTR boundaries could still affect nucleotide diversity within the gene. We also tested larger region sizes to assess whether extending our filter further upstream and downstream would affect the results. We found that extending the upstream/downstream regions does not change the directions of the association between recombination rate and nucleotide diversity, but the power to detect a relationship is reduced (Table 2). Thus, we primarily focus on results from screening the genes themselves.

To increase the number of genes we can examine while maintaining the criterion of "absence of recent selective sweeps," we compared the loci to the outgroup species, *D. lowei,* and we used a simple parsimony argument to determine which species likely possesses the ancestral allele for each fixed difference. The lineage with the ancestral allele at a given locus was considered to be the lineage less likely

Table 2 The effect of upstream/downstream region size on measures of neutral diversity in association with recombination rate

Loci[a]	Correlation (Pearson's R) between fourfold degenerate site π and local recombination rate
All genes	$n = 6{,}640$, $R = 0.19$, $p < 2.2 \times 10^{-16}$
Autosomal	$n = 3{,}723$, $R = 0.21$, $p < 2.2 \times 10^{-16}$
X chromosome	$n = 2{,}917$, $R = 0.08$, $p = 4.32 \times 10^{-6}$
No sweeps (within the gene)	$n = 151$, $R = 0.25$, $p = 0.0017$
Autosomal	$n = 107$, $R = 0.15$, $p = 0.1114$
X chromosome	$n = 44$, $R = 0.41$, $p = 0.0059$
No sweeps (gene + 50 bp buffer)	$n = 121$, $R = 0.27$, $p = 0.0028$
Autosomal	$n = 85$, $R = 0.17$, $p = 0.1127$
X chromosome	$n = 36$, $R = 0.42$, $p = 0.0099$
No sweeps (gene + 100 bp buffer)	$n = 77$, $R = 0.21$, $p = 0.0602$
Autosomal	$n = 49$, $R = 0.19$, $p = 0.2029$
X chromosome	$n = 28$, $R = 0.24$, $p = 0.2090$

[a]Limited to genes within the recombination map. "No sweeps" indicates that these loci had no derived fixed differences in *D. pseudoobscura*

to have been affected by recent sweeps, whereas the lineage with the derived allele may have experienced a sweep at that locus. For example, if a gene has several differences between *D. pseudoobscura* and *D. miranda,* but all differences are derived in *D. miranda*, then that *D. pseudoobscura* locus is presumably free of the effects of recent hard sweeps. Following this argument, only when *all* fixed differences were derived in *D. miranda* did we analyze neutral variation at that locus within *D. pseudoobscura.* Using this strategy to identify hard sweeps (Fig. 2), we have obtained a set of *D. pseudoobscura* loci where we can examine the effects of linked selection in the relative absence of hard selective sweeps.

To assess neutral variation, we computed fourfold degenerate site π (custom scripts provided in accompanying Github repository linked above), since synonymous sites represent a category of sites that evolve under relatively minimal constraint (Andolfatto 2005). For genes that have multiple predicted isoforms, we analyzed the longest isoform entry for each gene. R version 3.6.1 was used to perform statistical tests, including linear regressions and analysis of covariance (ANCOVA) to assess differences in the slope and y-intercepts of the regression lines of local recombination rate and neutral diversity between the sets of loci (R Core Team, 2019, "R: A Language and Environment for Statistical Computing." R Foundation for Statistical Computing, Vienna, Austria. https://www.R-project.org/).

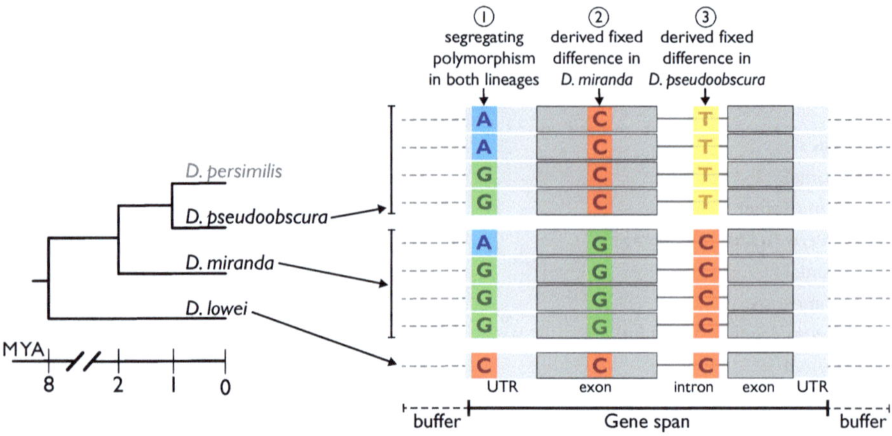

Fig. 2 Schematic of the logic underlying sweep exclusion. A cladogram (left) shows the phylogenetic context of the analyzed species. For the examined orthologs shared between *D. pseudoobscura* and *D. miranda*, we compared each locus across the 41 genomes described above. As an example (right), we depict 4 *D. pseudoobscura,* 4 *D. miranda*, and the outgroup *D. lowei* sequence to illustrate our strategy for excluding hard sweeps in predicted gene spans. This example gene has a within-species polymorphism (site 1) and two fixed differences between *D. pseudoobscura* and *D. miranda* (sites 2 and 3). By comparing to the outgroup *D. lowei,* we find that *D. pseudoobscura's* fixed "C" at site 2 matches the likely ancestral state. In contrast, site 3 has a fixed, derived "T" in *D. pseudoobscura.* This fixed difference at site 3 represents a possible recent sweep in *D. pseudoobscura* and would disqualify this whole gene from the set of *D. pseudoobscura* loci unlikely to have been affected by recent sweeps.

2.2 *Case Study: Results*

Of the examined orthologs shared between *D. pseudoobscura* and *D. miranda,* 6,640 loci are contained within the recombination map. These 6,640 loci exhibit a positive association between recombination rate and neutral diversity (fourfold degenerate site π) within *D. pseudoobscura* (Table 2; Fig. 3a), in agreement with previous studies within *D. pseudoobscura* (Kulathinal et al. 2008; McGaugh et al. 2012). This significant positive relationship exists for both autosomal and X chromosome loci (Fig. 3b, c).

Next, we examined how screening out fixed differences affects the positive relationship between recombination rate and neutral diversity. The effects of hard selective sweeps on within-species nucleotide variation can be minimized by screening *D. pseudoobscura* loci for derived fixed differences in their coding regions, introns, and untranslated regions (Fig. 2). We identified 151 *D. pseudoobscura* loci with no derived fixed differences within the gene's coding sequence, introns, or untranslated regions. This set of loci is unlikely to have experienced recent hard sweeps within the gene span, since we have reduced the possibility that a novel mutation arose and spread to fixation within *D. pseudoobscura* after the divergence of *D. pseudoobscura* and *D. miranda* from their common ancestor. These 151 loci

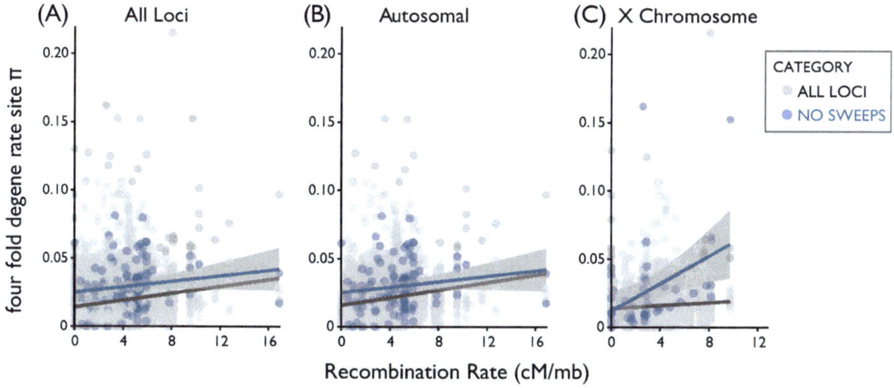

Fig. 3 Neutral diversity (fourfold degenerate site π) regressed onto local recombination rate. This relationship is shown for all loci (gray) and for the subset of loci with no fixed differences (*D. pseudoobscura* vs. *D. miranda*) derived in *D. pseudoobscura* (dark blue), shown for (**a**) the full set of loci, (**b**) autosomal loci, and (**c**) loci on the X chromosome

exhibit a positive association between recombination rate and neutral diversity (fourfold degenerate site π) within *D. pseudoobscura* (Table 2; Pearson's $R = 0.25$, $p = 0.0017$). The set of 151 genes is a relatively small proportion of the total number of loci, but this proportion should not be interpreted as a percentage of loci not affected by sweeps. This approach does not yield a binary between loci affected by sweeps vs. not affected by sweeps. Instead, the set of 151 loci represents a subset with reduced possible effects of selective sweeps in or near the gene. Some loci may have acquired fixed differences by processes other than hard sweeps, and screening for fixed differences may not capture all hard sweeps, as discussed further below.

We next discuss how the positive association between recombination rate and neutral diversity observed in the set of genes without derived fixed differences might differ from that observed in the full set of loci. Because hard sweeps reduce neutral variation, a selective sweeps model predicts less neutral variation compared to predicted nucleotide variation in the absence of hard sweeps. Removing effects of hard sweeps may also result in a weaker correlation between neutral variation and recombination. If background selection and/or soft sweeps have detectable effects on neutral variation, we expect that the correlation between neutral variation and recombination rate in the absence of hard sweeps will be positive but weaker than the correlation in the presence of hard sweeps.

When we examine the correlation between neutral diversity within *D. pseudoobscura* and recombination rate at loci less likely to have experienced hard sweeps (Fig. 3, "no sweeps"), the association does not appear especially different from the full set of loci. For autosomal loci, the slope of the regression line of fourfold degenerate site π against local recombination rate did not differ significantly between the full set of loci and the subset of loci without hard sweeps

(Fig. 3b; ANCOVA, $F = 0.514$, $\text{df} = 1, p = 0.47$ for the interaction of recombination rate by locus category). However, the y-intercept was higher for the set of loci without hard sweeps compared to the set of all autosomal loci (ANCOVA, $F = 27.05$, $\text{df} = 1$, $p = 2.08 \times 10^{-7}$, for locus category effect). The higher y-intercept indicates an overall higher level of neutral variation in the absence of potential recent hard sweeps. In contrast, the subset of X chromosome loci exhibited significantly different slopes (Fig. 3c; ANCOVA, $F = 31.43$, $\text{df} = 1, p = 2.25 \times 10^{-8}$ for the interaction of recombination rate x locus category) between all X chromosome loci and the subset of loci without hard sweeps). Given the limited sample size of X chromosome loci, we approach X vs. autosome comparisons with caution in this case study, but we emphasize that this is an important area for future work.

Finally, we considered whether these differences observed between the loci less likely to have experienced hard sweeps and the full set of loci might occur by chance. We compared the observed outcomes for each chromosome to the range of possible outcomes for any random subset of loci. To compare the results to the distribution of possible outcomes, we took random subsamples of the full set of genes (10,000 sets of loci randomly sampled with replacement), where the number of sampled genes was equal to the number of loci in the original data set without hard sweeps (107 autosomal and 44 X chromosome loci). By randomly resampling from the full set of genes from each chromosome, we find that the differences observed between the full set and the set without the effects of sweeps could result from sampling. Bootstrapping and examining the difference between correlation coefficients tell us about the probability that the observed *difference* in correlation coefficients might be observed by chance. Relative to an approach of comparing observed correlations directly, this approach of sampling the difference in correlation coefficients is more robust to wide confidence intervals around correlation coefficients.

Using the correlation between fourfold degenerate site π and recombination rate in the set of loci with no hard sweeps, we compared this value to the correlation observed in each random sample. Figure 4 shows the distribution of differences between observed Pearson's R in each random sample and observed Pearson's R in the set of loci with no hard sweeps. The means of both the autosomal and X chromosome distributions in Fig. 4 differ from zero (autosomal t-test: $t = 49.007$, $\text{df} = 9{,}999$, $p < 2.2 \times 10^{-16}$; X chromosome t-test: $t = -134.4$, $\text{df} = 9{,}999$, $p < 2.2 \times 10^{-16}$), suggesting that the difference in observed correlation in the set of loci without hard sweeps is unlikely to have occurred by chance. However, the distributions in Fig. 4 include zero, suggesting that the association between neutral variation and recombination in the set of loci without sweeps falls within the distribution of randomly selected sets of loci (Fig. 4), Together, these observations suggest that removing the effects of hard sweeps does not drastically change the relationship between neutral diversity and recombination rate, though it may alter the strength of this relationship.

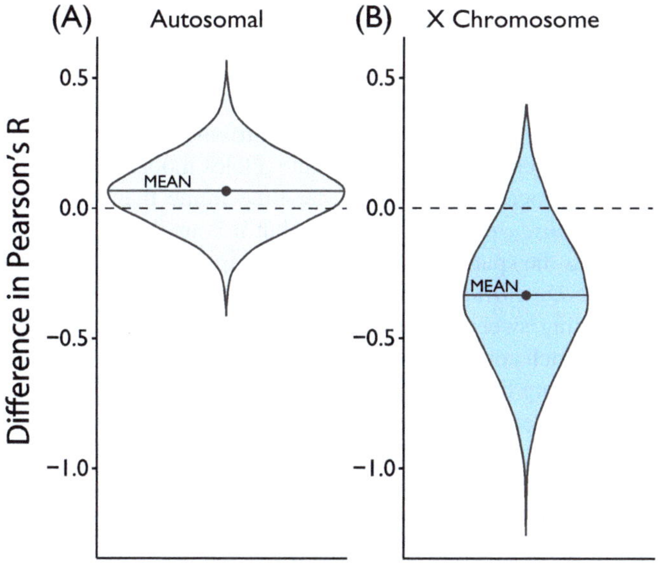

Fig. 4 Comparison of the set of loci with no hard sweeps to the sampling distribution of correlation coefficients between fourfold degenerate site π and recombination rate. Using the correlation between fourfold degenerate site π and recombination rate in the set of loci with no hard sweeps ("no sweeps R"), we compared this value to the correlation observed in random samples. Here, we show the distribution of difference between observed Pearson's *R* in each random sample and the "no sweeps R" for autosomal (**a**) and X chromosome (**b**) loci. Each distribution above includes 10,000 sets of loci that were randomly sampled with replacement from the full set of autosomal or X chromosome loci. The randomly sampled sets contained the same number of genes as the set of loci with no hard sweeps for that genomic region (107 autosomal loci and 44 X chromosome loci)

3 Discussion

The case study presented here aims to empirically examine the effects of linked selection via background selection and soft sweeps on within-species patterns of genetic diversity using a strategy that reduces the effects of recent hard sweeps. We demonstrate how this strategy can be applied to sister taxa to identify which loci have potentially experienced recent hard sweeps. This method can be applied to any pair of diverging taxa where sequence data is available, and the taxa are sufficiently similar at the nucleotide level to allow for direct comparison of coding sequences. That said, we next explore the important limitations of this approach and use these limitations to underscore areas where future work is needed.

First, the approach used in our case study is perhaps conservative in that it excludes all genes with fixed differences, even though many of these loci may be affected by background selection or soft sweeps, and several may in fact have never experienced a hard sweep. In fact, because background selection decreases the effective population size (N_e) at a locus and increases the relative effect of drift, fixed differences are likely to arise under the influence of background selection. Lineage sorting—or the sorting of ancestral polymorphism by chance—may also yield fixed differences that did not involve sweeps. Additionally, we cannot exclude that ancient hard sweeps, predating the split of the two species, may have long-lasting effects on neutral diversity that still contribute to the association between recombination and diversity. Under an infinite alleles model, recovery of heterozygosity after a sweep occurs in a far shorter time frame than the divergence time between *D. pseudoobscura* and *D. miranda*, but it is unclear how this translates to recovery of π across the span of a gene. Another important note about this particular case study is that *D. miranda* does not hybridize with *D. pseudoobscura*. The strategy for excluding sweeps would need to be re-evaluated in species pairs with recent gene flow, which could allow new mutations in either lineage to be exchanged with the other, reducing the possibility that a sweep would be detectable based on a fixed difference. Further, we cannot definitively exclude effects from hard sweeps that occurred beyond the untranslated regions of the genes studied. Though LD decays rapidly in *Drosophila*, even in regions of low recombination (Langley et al. 2000), it is difficult to exclude the possibility that LD within a gene may have been affected by a strong or recent sweep in a very nearby region.

Obtaining empirical evidence of linked selection beyond hard sweeps remains challenging. As reviewed above (see Sect. 1.3), genome-wide patterns of variation in many taxa are broadly consistent with expectations of background selection being a major driver of genome-wide patterns of diversity (e.g., McVicker et al. 2009; Hernandez et al. 2011; Lohmueller et al. 2011; Charlesworth 2012b; Charlesworth and Campos 2014; Comeron 2014; Renzette et al. 2016; Pouyet et al. 2018). However, model-based approaches have often lacked comparative assessments of the explanatory power of sweeps-only models vs. joint models of background selection and sweeps together (though see Elyashiv et al. 2016). Modeling approaches also depend on expectations built from empirical observations, but we have little empirical data on linked selection outside of hard selective sweeps. To our knowledge, there are currently no standard approaches for gathering direct empirical evidence of the operation of background selection. Despite the limitations of the approach applied in the case study above, it provides an empirical strategy for examining how selection reduces linked neutral variation after minimizing possible effects of recent hard sweeps. Much work remains to resolve the long-standing debate over the relative contributions of different types of linked selection to patterns of genetic diversity (e.g., Stephan 2010; Jensen 2014; Renzette et al. 2016). To fully understand the impacts and interactions of types of linked selection, we need to thoughtfully combine theoretical and empirical approaches.

Methods such as the one discussed in the case study above are necessary to gather important empirical observations to inform and test theoretical work toward disentangling the multiple types of linked selection that affect patterns of variation throughout *Drosophila* genomes and in insect genomes more broadly. Simulations allow us to vary one parameter at a time, but empirical studies allow us to develop our intuition of the range of various parameters in nature and create simulations that model after empirically observed patterns. For instance, recent simulations show that understanding how the spatial arrangement of genes changes background selection may help in learning how its signatures differ from those of sweeps (Schrider 2020). Similarly, we know that the landscape of recombination shapes the action of linked selection, so we need empirical assessments of recombination rate. Flies of the *pseudoobscura* group examined in the case study above are established model systems for the evolutionary genetics of recombination, and multiple empirical studies have analyzed and discussed the *D. pseudoobscura* recombination landscape and its evolution (Korunes and Noor 2019; Kulathinal et al. 2008; Samuk et al. 2020; Smukowski and Noor 2011; Smukowski Heil et al. 2015; Stevison et al. 2011; Stevison and Noor 2010). For understanding linked selection in diverse insect systems, we need broader sampling of recombination maps and their variation within insect species. For example, several species of social insects have been discovered to have some of the highest recombination rates among plants and animals, and the effects of recombination and linked selection in social insects may be an exciting area of future work (Jones et al. 2019; Liu et al. 2017; Meznar et al. 2010; Sirviö et al. 2011; Wallberg et al. 2015; Wilfert et al. 2007). As recombination rates vary greatly across species and even within species (Stapley et al. 2017), empirical studies of recombination and recombination rate variation are likely to play a large role in the future of linked selection studies.

4 Future Perspectives

In the field of population genomics, both within insects and more broadly, the challenges highlighted here exemplify recent advances and important areas for future work in the study of linked selection and its roles in driving patterns of nucleotide variation. Recent methodological and empirical work has improved our understanding of linked selection, but we are just beginning to appreciate the complex interactions of genetic variants with each other, with the recombination landscape, and with environments. While there has been major progress in understanding and documenting some forms of selective sweeps, understanding background selection and soft selective sweeps remains an area with substantial room for growth. Here, we have explored the study of these different types of linked selection, with particular emphasis on the body of work devoted to linked selection in *Drosophila*. As demonstrated with our case study of *Drosophila pseudoobscura*, many challenges remain in distinguishing types of linked selection. Moving forward, the field of insect population genetics must emphasize developing new empirical methods for interpreting signals of linked selection, jointly considering models of background selection and sweeps together, and understanding recombination rate variation. New empirical insights will continue to emerge with increased genomic sampling of diverse taxa, including the sampling of species with diverse recombination landscapes. As discussed above, the role of background selection in non-recombining regions and the diversity of sex determination systems in insects suggest that an area of exciting future development may be linked selection on non-recombining or dosage-dependent sex chromosomes. Future empirical work in these areas will help inform methodological and theoretical advances as we aim to understand the impacts and interactions of types of linked selection.

References

Agrawal AF, Hartfield M. Coalescence with background and balancing selection in systems with bi- and uniparental reproduction: contrasting partial asexuality and selfing. Genetics. 2016;202 (1):313–26. https://doi.org/10.1534/genetics.115.181024.

Aguade M, Miyashita N, Langley CH. Reduced variation in the yellow-achaete-scute region in natural populations of *Drosophila melanogaster*. Genetics. 1989;122(3):607–15. http://www.pubmedcentral.nih.gov/articlerender.fcgi?artid=1203734&tool=pmcentrez&rendertype=abstract

Alves I, Šrámková Hanulová A, Foll M, Excoffier L. Genomic data reveal a complex making of humans. PLoS Genet. 2012;8(7):e1002837. https://doi.org/10.1371/journal.pgen.1002837.

Andolfatto P. Contrasting patterns of X-linked and autosomal nucleotide variation in Drosophila melanogaster and *Drosophila simulans*. Mol Biol Evol. 2001;18(3):279–90. http://www.ncbi.nlm.nih.gov/pubmed/11230529

Andolfatto P. Adaptive evolution of non-coding DNA in Drosophila. Nature. 2005;437 (7062):1149–52. https://doi.org/10.1038/nature04107.

Begun DJ, Aquadro CF. Levels of naturally occurring DNA polymorphism correlate with recombination rates in *D. melanogaster*. Nature. 1992;356(6369):519–20. https://doi.org/10.1038/356519a0.

Begun DJ, Holloway AK, Stevens K, Hillier LW, Poh Y-P, Hahn MW, et al. Population genomics: whole-genome analysis of polymorphism and divergence in *Drosophila simulans*. PLoS Biol. 2007;5(11):e310. https://doi.org/10.1371/journal.pbio.0050310.

Bersaglieri T, Sabeti PC, Patterson N, Vanderploeg T, Schaffner SF, Drake JA, et al. Genetic signatures of strong recent positive selection at the lactase gene. Am J Hum Genet. 2004;74 (6):1111–20. https://doi.org/10.1086/421051.

Burri R, Nater A, Kawakami T, Mugal CF, Olason PI, Smeds L, et al. Linked selection and recombination rate variation drive the evolution of the genomic landscape of differentiation across the speciation continuum of Ficedula flycatchers. Genome Res. 2015;25(11):1656–65. https://doi.org/10.1101/gr.196485.115.

Charlesworth B. The effects of deleterious mutations on evolution at linked sites. Genetics. 2012a;190(1):5–22. https://doi.org/10.1534/genetics.111.134288.

Charlesworth B. The role of background selection in shaping patterns of molecular evolution and variation: evidence from variability on the Drosophila X chromosome. Genetics. 2012b;191 (1):233–46. https://doi.org/10.1534/genetics.111.138073.

Charlesworth B, Campos JL. The relations between recombination rate and patterns of molecular variation and evolution in Drosophila. Annu Rev Genet. 2014;48(1):383–403. https://doi.org/10.1146/annurev-genet-120213-092525.

Charlesworth B, Morgan MT, Charlesworth D. The effect of deleterious mutations on neutral molecular variation. Genetics. 1993;134(4):1289–303. http://www.genetics.org/content/134/4/1289.short

Colosimo PF, Hosemann KE, Balabhadra S, Villarreal G, Dickson M, Grimwood J, et al. Widespread parallel evolution in sticklebacks by repeated fixation of ectodysplasin alleles. Science. 2005;307(5717):1928–33. https://doi.org/10.1126/science.1107239.

Comeron JM. Background selection as baseline for nucleotide variation across the Drosophila genome. PLoS Genet. 2014;10(6):e1004434. https://doi.org/10.1371/journal.pgen.1004434.

Cutter AD, Choi JY. Natural selection shapes nucleotide polymorphism across the genome of the nematode *Caenorhabditis briggsae*. Genome Res. 2010;20(8):1103–11. https://doi.org/10.1101/gr.104331.109.

Cutter AD, Payseur BA. Genomic signatures of selection at linked sites: unifying the disparity among species. Nat Rev Genet. 2013;14(4):262–74. https://doi.org/10.1038/nrg3425.

de Groot NG, Otting N, Doxiadis GGM, Balla-Jhagjhoorsingh SS, Heeney JL, van Rood JJ, et al. Evidence for an ancient selective sweep in the MHC class I gene repertoire of chimpanzees. Proc Natl Acad Sci. 2002;99(18):11748–53. https://doi.org/10.1073/pnas.182420799.

DeGiorgio M, Huber CD, Hubisz MJ, Hellmann I, Nielsen R. SweepFinder2: increased sensitivity, robustness and flexibility. Bioinformatics. 2016;32(12):1895–7. https://doi.org/10.1093/bioinformatics/btw051.

Elyashiv E, Sattath S, Hu TT, Strutsovsky A, McVicker G, Andolfatto P, et al. A genomic map of the effects of linked selection in Drosophila. PLoS Genet. 2016;12(8):e1006130. https://doi.org/10.1371/journal.pgen.1006130.

Fay JC, Wu C-I. Hitchhiking under positive Darwinian selection. Genetics. 2000;155(3):1405–13.

Felsenstein J. The evolutionary advantage of recombination. Genetics. 1974;78(2):737–56.

Fu YX, Li WH. Statistical tests of neutrality of mutations. Genetics. 1993;133(3):693–709.

Halldorsson BV, Palsson G, Stefansson OA, Jonsson H, Hardarson MT, Eggertsson HP, et al. Characterizing mutagenic effects of recombination through a sequence-level genetic map. Science. 2019;363(6425):eaau1043. https://doi.org/10.1126/science.aau1043.

Harris RB, Sackman A, Jensen JD. On the unfounded enthusiasm for soft selective sweeps II: examining recent evidence from humans, flies, and viruses. PLoS Genet. 2018;14(12):e1007859. https://doi.org/10.1371/journal.pgen.1007859.

Hermisson J, Pennings PS. Soft sweeps. Genetics. 2005;169(4):2335–52. https://doi.org/10.1534/genetics.104.036947.

Hernandez RD, Kelley JL, Elyashiv E, Melton SC, Auton A, McVean G, et al. Classic selective sweeps were rare in recent human evolution. Science. 2011;331(6019):920–4. https://doi.org/10.1126/science.1198878.

Hoekstra HE, Hirschmann RJ, Bundey RA, Insel PA, Crossland JP. A single amino acid mutation contributes to adaptive beach mouse color pattern. Science. 2006;313(5783):101–4. https://doi.org/10.1126/science.1126121.

Huber CD, DeGiorgio M, Hellmann I, Nielsen R. Detecting recent selective sweeps while controlling for mutation rate and background selection. Mol Ecol. 2016;25(1):142–56. https://doi.org/10.1111/mec.13351.

Hutter S, Li H, Beisswanger S, De Lorenzo D, Stephan W. Distinctly different sex ratios in African and European populations of *Drosophila melanogaster* inferred from chromosomewide single nucleotide polymorphism data. Genetics. 2007;177(1):469–80. https://doi.org/10.1534/genetics.107.074922.

Jackson BC, Campos JL, Haddrill PR, Charlesworth B, Zeng K. Variation in the intensity of selection on codon bias over time causes contrasting patterns of base composition evolution in Drosophila. Genome Biol Evol. 2017;9(1):102–23. https://doi.org/10.1093/gbe/evw291.

Jensen JD. On the unfounded enthusiasm for soft selective sweeps. Nat Commun. 2014;5:5281. https://doi.org/10.1038/ncomms6281.

Jones JC, Wallberg A, Christmas MJ, Kapheim KM, Webster MT. Extreme differences in recombination rate between the genomes of a solitary and a social bee. Mol Biol Evol. 2019;36(10):2277–91. https://doi.org/10.1093/molbev/msz130.

Kaiser VB, Charlesworth B. The effects of deleterious mutations on evolution in non-recombining genomes. Trends Genet. 2009;25(1):9–12. https://doi.org/10.1016/j.tig.2008.10.009.

Kaplan NL, Hudson RR, Langley CH. The "hitchhiking effect" revisited. Genetics. 1989;123(4):887–99. http://www.pubmedcentral.nih.gov/articlerender.fcgi?artid=1203897&tool=pmcentrez&rendertype=abstract

Kohn MH, Fang S, Wu C-I. Inference of positive and negative selection on the 5′ regulatory regions of Drosophila genes. Mol Biol Evol. 2004;21(2):374–83. https://doi.org/10.1093/molbev/msh026.

Kondrashov AS. Direct estimates of human per nucleotide mutation rates at 20 loci causing Mendelian diseases. Hum Mutat. 2003;21(1):12–27. https://doi.org/10.1002/humu.10147.

Korunes KL, Noor MAF. Pervasive gene conversion in chromosomal inversion heterozygotes. Mol Ecol. 2019;28(6):1302–15. https://doi.org/10.1111/mec.14921.

Kulathinal RJ, Bennett SM, Fitzpatrick CL, Noor MAF. Fine-scale mapping of recombination rate in Drosophila refines its correlation to diversity and divergence. Proc Natl Acad Sci. 2008;105(29):10051–6. https://doi.org/10.1073/pnas.0801848105.

Langley CH, Lazzaro BP, Phillips W, Heikkinen E, Braverman JM. Linkage disequilibria and the site frequency spectra in the su(s) and su(wa) regions of the *Drosophila melanogaster* X chromosome. Genetics. 2000;156(4):1837–52. http://www.genetics.org/content/156/4/1837.short

Li H, Durbin R. Fast and accurate short read alignment with burrows-wheeler transform. Bioinformatics. 2009;25(14):1754–60. https://doi.org/10.1093/bioinformatics/btp324.

Liu H, Jia Y, Sun X, Tian D, Hurst LD, Yang S. Direct determination of the mutation rate in the bumblebee reveals evidence for weak recombination-associated mutation and an approximate rate constancy in insects. Mol Biol Evol. 2017;34(1):119–30. https://doi.org/10.1093/molbev/msw226.

Lohmueller KE, Albrechtsen A, Li Y, Kim SY, Korneliussen T, Vinckenbosch N, et al. Natural selection affects multiple aspects of genetic variation at putatively neutral sites across the human genome. PLoS Genet. 2011;7(10):e1002326. https://doi.org/10.1371/journal.pgen.1002326.

Maynard Smith J, Haigh J. The hitch-hiking effect of a favourable gene. Genet Res. 1974;23(1):23–35. http://www.ncbi.nlm.nih.gov/pubmed/4407212

McGaugh SE, Heil CSS, Manzano-Winkler B, Loewe L, Goldstein S, Himmel TL, et al. Recombination modulates how selection affects linked sites in Drosophila. PLoS Biol. 2012;10(11):e1001422. https://doi.org/10.1371/journal.pbio.1001422.

McKenna A, Hanna M, Banks E, Sivachenko A, Cibulskis K, Kernytsky A, et al. The genome analysis toolkit: a MapReduce framework for analyzing next-generation DNA sequencing data. Genome Res. 2010;20(9):1297–303. https://doi.org/10.1101/gr.107524.110.

McVicker G, Gordon D, Davis C, Green P. Widespread genomic signatures of natural selection in hominid evolution. PLoS Genet. 2009;5(5):e1000471. https://doi.org/10.1371/journal.pgen.1000471.

Messer PW, Petrov DA. Population genomics of rapid adaptation by soft selective sweeps. Trends Ecol Evol. 2013;28(11):659–69. https://doi.org/10.1016/j.tree.2013.08.003.

Meznar ER, Gadau J, Koeniger N, Rueppell O. Comparative linkage mapping suggests a high recombination rate in all honeybees. J Hered. 2010;101(Suppl 1):S118–26. https://doi.org/10.1093/jhered/esq002.

Nachman MW. Single nucleotide polymorphisms and recombination rate in humans. Trends Genet. 2001;17(9):481–5. https://doi.org/10.1016/S0168-9525(01)02409-X.

Nachman MW, Crowell SL. Estimate of the mutation rate per nucleotide in humans. Genetics. 2000;156(1):297–304.

Nordborg M, Charlesworth B, Charlesworth D. The effect of recombination on background selection. Genet Res. 1996;67(2):159–74. http://www.ncbi.nlm.nih.gov/pubmed/8801188

Pennings PS, Hermisson J. Soft sweeps II – molecular population genetics of adaptation from recurrent mutation or migration. Mol Biol Evol. 2006;23(5):1076–84. https://doi.org/10.1093/molbev/msj117.

Phung TN, Huber CD, Lohmueller KE. Determining the effect of natural selection on linked neutral divergence across species. PLoS Genet. 2016;12(8):e1006199. https://doi.org/10.1371/journal.pgen.1006199.

Pouyet F, Aeschbacher S, Thiéry A, Excoffier L. Background selection and biased gene conversion affect more than 95% of the human genome and bias demographic inferences. Elife. 2018;7: e36317. https://doi.org/10.7554/eLife.36317.

Pritchard JK, Pickrell JK, Coop G. The genetics of human adaptation: hard sweeps, soft sweeps, and polygenic adaptation. Curr Biol. 2010;20(4):R208–15. https://doi.org/10.1016/j.cub.2009.11.055.

Renzette N, Kowalik TF, Jensen JD. On the relative roles of background selection and genetic hitchhiking in shaping human cytomegalovirus genetic diversity. Mol Ecol. 2016;25(1):403–13. https://doi.org/10.1111/mec.13331.

Roze D. Background selection in partially selfing populations. Genetics. 2016;203(2):937–57. https://doi.org/10.1534/genetics.116.187955.

Samuk K, Manzano-Winkler B, Ritz KR, Noor MAF. Natural selection shapes variation in genome-wide recombination rate in Drosophila pseudoobscura. Curr Biol. 2020;30(8):1517–1528.E6. https://doi.org/10.1016/j.cub.2020.03.053.

Schrider DR. Background selection does not mimic the patterns of genetic diversity produced by selective sweeps. BioRxiv. 2020; https://doi.org/10.1101/2019.12.13.876136.

Schrider DR, Kern AD. Soft sweeps are the dominant mode of adaptation in the human genome. Mol Biol Evol. 2017;34(8):1863–77. https://doi.org/10.1093/molbev/msx154.

Schrider DR, Shanku AG, Kern AD. Effects of linked selective sweeps on demographic inference and model selection. Genetics. 2016;204(3):1207–23. https://doi.org/10.1534/genetics.116.190223.

Sella G, Petrov DA, Przeworski M, Andolfatto P. Pervasive natural selection in the Drosophila genome? PLoS Genet. 2009;5(6):e1000495. https://doi.org/10.1371/journal.pgen.1000495.

Sirviö A, Johnston JS, Wenseleers T, Pamilo P. A high recombination rate in eusocial Hymenoptera: evidence from the common wasp Vespula vulgaris. BMC Genet. 2011;12:95. https://doi.org/10.1186/1471-2156-12-95.

Slotte T. The impact of linked selection on plant genomic variation. Brief Funct Genomics. 2014;13 (4):268–75. https://doi.org/10.1093/bfgp/elu009.

Smukowski Heil CS, Ellison C, Dubin M, Noor MAF. Recombining without hotspots: a comprehensive evolutionary portrait of recombination in two closely related species of Drosophila. Genome Biol Evol. 2015;7(10):2829–42. https://doi.org/10.1093/gbe/evv182.

Smukowski CS, Noor MAF. Recombination rate variation in closely related species. Heredity. 2011;107(6):496–508. https://doi.org/10.1038/hdy.2011.44.

Stapley J, Feulner PGD, Johnston SE, Santure AW, Smadja CM. Variation in recombination frequency and distribution across eukaryotes: patterns and processes. Philos Trans R Soc B. 2017;372(1736) https://doi.org/10.1098/rstb.2016.0455.

Stephan W. Genetic hitchhiking versus background selection: the controversy and its implications. Philos Trans R Soc Lond B Biol Sci. 2010;365(1544):1245–53. https://doi.org/10.1098/rstb.2009.0278.

Stephan W. Signatures of positive selection: from selective sweeps at individual loci to subtle allele frequency changes in polygenic adaptation. Mol Ecol. 2016;25(1):79–88. https://doi.org/10.1111/mec.13288.

Stephan W, Langley CH. Molecular genetic variation in the centromeric region of the X chromosome in three *Drosophila ananassae* populations. I. Contrasts between the vermilion and forked loci. Genetics. 1989;121(1):89–99. http://www.genetics.org/content/121/1/89.abstract

Stevison LS, Noor MAF. Genetic and evolutionary correlates of fine-scale recombination rate variation in *Drosophila persimilis*. J Mol Evol. 2010;71(5–6):332–45. https://doi.org/10.1007/s00239-010-9388-1.

Stevison LS, Hoehn KB, Noor MAF. Effects of inversions on within- and between-species recombination and divergence. Genome Biol Evol. 2011;3:830–41. https://doi.org/10.1093/gbe/evr081.

Tajima F. Statistical method for testing the neutral mutation hypothesis by DNA polymorphism. Genetics. 1989;123(3):585–95. http://www.pubmedcentral.nih.gov/articlerender.fcgi?artid=1203831&tool=pmcentrez&rendertype=abstract

Takahashi A, Liu Y-H, Saitou N. Genetic variation versus recombination rate in a structured population of mice. Mol Biol Evol. 2004;21(2):404–9. https://doi.org/10.1093/molbev/msh030.

Tenaillon MI, Sawkins MC, Long AD, Gaut RL, Doebley JF, Gaut BS. Patterns of DNA sequence polymorphism along chromosome 1 of maize (*Zea mays* ssp. *mays* L.). Proc Natl Acad Sci. 2001;98(16):9161–6. https://doi.org/10.1073/pnas.151244298.

Torres R, Stetter MG, Hernandez RD, Ross-Ibarra J. The temporal dynamics of background selection in non-equilibrium populations. Genetics. 2020;214(4):1019–30. https://doi.org/10.1534/genetics.119.302892.

Van der Auwera GA, Carneiro MO, Hartl C, Poplin R, del Angel G, Levy-Moonshine A, et al. From FastQ data to high-confidence variant calls: the genome analysis toolkit best practices pipeline. Curr Protoc Bioinformatics. 2013;43(1):11.10.1–11.10.33. https://doi.org/10.1002/0471250953.bi1110s43.

Vitti JJ, Grossman SR, Sabeti PC. Detecting natural selection in genomic data. Annu Rev Genet. 2013;47(1):97–120. https://doi.org/10.1146/annurev-genet-111212-133526.

Wallberg A, Glémin S, Webster MT. Extreme recombination frequencies shape genome variation and evolution in the honeybee, *Apis mellifera*. PLoS Genet. 2015;11(4):e1005189. https://doi.org/10.1371/journal.pgen.1005189.

Wang RL, Hey J. The speciation history of Drosophila pseudoobscura and close relatives: inferences from DNA sequence variation at the period locus. Genetics. 1996;144(3):1113–26. http://www.genetics.org/content/144/3/1113.abstract

Wilfert L, Gadau J, Schmid-Hempel P. Variation in genomic recombination rates among animal taxa and the case of social insects. Heredity. 2007;98(4):189–97. https://doi.org/10.1038/sj.hdy.6800950.

Wilson Sayres MA, Lohmueller KE, Nielsen R. Natural selection reduced diversity on human Y chromosomes. PLoS Genet. 2014;10(1):e1004064. https://doi.org/10.1371/journal.pgen.1004064.

Wright S. Evolution in mendelian populations. Genetics. 1931;16(2):97–159.

Part II
Insect Pest Management and Pesticide Resistance

Incorporating Population Genomics into Integrated Insect Pest Management: A Reduced-Representation Sequencing Approach

Erin O. Campbell, Kanishka Senevirathna, and Boyd A. Mori

Abstract Insects have fundamental roles in the world, helping to maintain ecosystem functioning and stability. However, a minority of insects are pests and significantly threaten biodiversity, natural resources, and food security. World trade has increased the spread of invasive insects, and combined with increased abiotic stresses due to climate change, global agricultural, forestry, and horticultural production is under immense pressure. To counteract the threat of these pests and support the development of integrated pest management strategies, biosurveillance strategies are being employed. Biosurveillance strategies aim to collect and understand information about pest ecology and evolution and support integrated pest management by answering fundamental questions on pest origin, behavior, and movement potential. Genomic biosurveillance, leveraging high-throughput sequencing and reduced-representation sequencing (RRS) techniques, can be used to identify patterns and processes involved in movement and adaptation of insects, which is not possible to capture with traditionally used techniques (e.g., trapping). Reduced-representation sequencing methods have gained popularity due to their technical flexibility, relatively low cost, and ability to produce genome-wide data without prior knowledge of specific genomes. This chapter provides an overview of RRS methodology and its applications in insect population genetics, focusing on its potential for insect pest management in forestry, horticulture, and agroecosystems.

Keywords Adaptation · Biosurveillance · High-throughput sequencing · Insect pests · Integrated pest management · Invasive species · Phylogeography · Population genomics · Reduced representation sequencing · Risk intelligence

---

E. O. Campbell (✉)
Canadian Food Inspection Agency, Ottawa, ON, Canada

Department of Biological Sciences, University of Alberta, Edmonton, AB, Canada
e-mail: erin.campbell@inspection.gc.ca

K. Senevirathna · B. A. Mori
Department of Agricultural, Food, and Nutrition Sciences, University of Alberta, Edmonton, AB, Canada

Julian R. Dupuis and Om P. Rajora (eds.), *Population Genomics: Insects*, Population Genomics [Om P. Rajora (Editor-in-Chief)], https://doi.org/10.1007/13836_2024_110,

1 General Introduction

Global trade has significantly increased the number of invasive alien species worldwide, threatening biodiversity, natural resources, food security, and negatively impacting national economies owing to costs associated with reduced production, quarantines, and pest management (Hulme 2009, 2021; Matheson and McGaughran 2022; Paini et al. 2016; Tay and Gordon 2019; Venette and Hutchison 2021). Due to their intimate interactions with plants, insects are a serious threat to global forest, agricultural, and horticultural ecosystems that produce essential food, commodities, and services. In addition, certain resource cultivation and management practices, including the planting of monocultures and use of pesticides, can make managed ecosystems vulnerable to both native and invasive insect pests by providing large areas of uninterrupted habitat, contribute to the development of insecticide resistance, and reduce overall biodiversity (Altieri 1999; Denholm et al. 2002; Gagic et al. 2021; Root 1973). The effects of climate change may further compromise already-stressed ecosystems or open previously unsuitable habitat for insect pests, allowing them to expand or shift their ecological niches and increase their potential for destruction (Cuddington et al. 2018; Jactel et al. 2019; Lehmann et al. 2020; Skendžić et al. 2021, but see Bates et al. 2020).

To mitigate the dual crises of climate change and increased global threat of insect pests, **biosurveillance** (see Box 1 for bolded glossary terms) strategies may be employed to collect and understand information about pest ecology and evolution (Hulme 2014; Roe et al. 2018). Pest biosurveillance typically addresses, in broad terms, three primary questions: *where did it come from? what is it doing?* and *where is it going?* These questions are essential for the early detection and proactive control of pest species and feed into broader **integrated pest management (IPM)** frameworks that seek to reduce the overall biological, ecological, cultural, and economic impacts of pests (Kogan 1998; Stern et al. 1959; Stenberg 2017). Determining the geographic origin of invasive pests (*where did it come from?*) can provide vital information about likely invasion routes, contribute to strategic monitoring and control initiatives, or be used as justification for imposing regulatory restrictions to reduce range expansion. Once established however, species will interact with their new environment in myriad ways (*what is it doing?*), which may influence their trajectory and impact as pests; for instance, contact with close relatives may lead to hybridization, and novel selective pressures such as access to new potential hosts and climatic zones may lead to adaptation. These interactions can contribute to changes in population density, range extent, or ecological niche shifts and influence whether and at what scale a pest is likely to become a significant threat (*where is it going?*) (Hall 2016; McGaughran et al. 2024).

Most typically, insect biosurveillance involves visual or trapping surveys in areas vulnerable to new introductions or significant damage by pests, such as along political borders, in areas with high densities of host plants (i.e., agricultural fields, orchards, forests, etc.), or near ports of entry. Insect traps primarily target adult life stages, while visual surveys may be conducted by inspecting plant material for all life stages or for characteristic damage caused by the target species. Newer methods for biosurveillance include the use of portable loop-mediated isothermal amplification (LAMP), recombinase polymerase amplification (RPA), or the sampling of environmental DNA (Agarwal et al. 2020; Shashank et al. 2023; Valentin et al. 2020; Young et al. 2021). While such biosurveillance approaches are useful for detecting movement into new areas and in some instances monitoring changes in population density, they often do not provide sufficient information about which populations may be acting as sources for spread, nor do they address how populations and species are interacting on the landscape or adapting to changing conditions. Luckily, these patterns and processes often have genetic signatures associated with them (Cristescu 2015; Ryan et al. 2019), and **genomic biosurveillance** can take advantage of this, in part by profiling population- and species-specific differences in allele frequencies to provide a molecular dimension useful for addressing our three primary biosurveillance questions (Roe et al. 2018).

Massively parallel innovations for DNA and RNA sequencing, collectively called **high-throughput sequencing (HTS)**, have enabled the production of molecular data on an unprecedented scale. These advances have not only made it increasingly simple to sequence entire genomes and transcriptomes, but have also led to the development of a variety of genome-sampling techniques that may be collectively referred to as **reduced representation sequencing (RRS)** (summarized in Campbell et al. 2018). These techniques rapidly gained popularity after 2007 with the publication of the now widely known restriction site-associated DNA sequencing method (RADseq, Miller et al. 2007; Baird et al. 2008) and have since given rise to a large number of modified methods, demonstrating the significant potential of RRS for investigating an array of ecological and evolutionary questions (Andrews et al. 2016; Davey and Blaxter 2011; Hohenlohe et al. 2010a, b; Rivera-Colón et al. 2021).

The popularity of RRS can be partly attributed to its technical flexibility, relatively low cost, and ability to produce hundreds to thousands of markers per sample, far exceeding what has conventionally been possible for studies using microsatellites or gene sequences. However, its greatest appeal is arguably that most RRS methods do not require prior knowledge of the specific genome(s) being investigated to produce genome-wide data. Instead, these techniques take advantage of restriction enzymes, which target short DNA sequences that occur in great numbers throughout the genome (Altshuler et al. 2000). This not only bypasses the time-consuming challenge of developing large numbers of species-specific probes or primers but enables genetically and geographically fine-scale studies of non-model organisms which lack sequenced reference genomes. This is particularly useful for insects, which are taxonomically hyper-diverse and often poorly understood, despite their disproportionate ecological and economic significance (Aukema et al. 2011; Jordan et al. 2021; Losey and Vaughn 2006; Paini et al. 2016). Management efforts have begun to leverage RRS as a tool for genomic biosurveillance to track and profile insect pest populations (Bilodeau et al. 2019; Estoup and Guillemaud 2010; Picq et al. 2018), but its application to IPM has typically lagged behind that of other fields like conservation genetics (McCartney et al. 2019; Roe et al. 2018).

The purpose of this chapter is thus to provide a synopsis of the general methodology and use of RRS for insect population genomics, with particular attention to applications for insect pest management in forestry, horticulture, and agroecosystems. We provide a brief technical summary of common approaches for the production of RRS data, outline current best practices and limitations of these approaches, and discuss several case studies that emphasize ongoing opportunities for the incorporation of these data into insect-focused IPM initiatives.

2 The Fundamentals of RRS for Population Genetics

The primary basis for most RRS techniques lies in the use of Type II **restriction enzymes** for digesting (also called "cutting") DNA (Altshuler et al. 2000). These enzymes target specific **recognition sequences** – short strings of otherwise anonymous DNA sequence – that are ubiquitous across biological life (Andrews et al. 2016; Di Felice et al. 2019; Williams 2003). After digesting genomic DNA with restriction enzymes, most RRS methods ligate **adapters** to the ends of each fragment. These adapters typically contain one or more unique **barcodes** and **sequencing primers** on either end of the target fragment, which enable individual samples to be **multiplexed** on the **flowcell** and then bioinformatically separated (or **demultiplexed**) after sequencing is complete. After ligation, restriction fragments may be additionally randomly sheared (as in RADseq (Baird et al. 2008)), and a **size selection** step may be performed to generate the reduced representation library; sometimes neither (see GBS-based approaches (Elshire et al. 2011; Poland et al. 2012)) or only one of these steps are performed (see ddRAD (Peterson et al. 2012)). The adapter ligation schema is one of the most variable components of RRS and is commonly modified to facilitate multiplexing needs (Jiang et al. 2016; Torkamaneh et al. 2019). Samples are pooled together, and amplification via PCR then produces many copies of each fragment before sequencing. Collectively, these procedures are referred to as library preparation (see Davey et al. (2011) for a general overview), though different RRS strategies may change the order, add, or remove steps in this process (Campbell et al. 2018). Once library preparation is complete, sequencing is performed on a high-throughput, short-read sequencer to produce reads that are typically 100–150 bp long.

Box 1 Glossary Terms

Biosurveillance is the process of gathering essential knowledge on biological threats, including insects, and the risk they pose to the environment, human health, and economy. It focuses on developing effective surveillance, prevention, and management strategies and aids decision-makers in pest management (Roe et al. 2018).

(continued)

Box 1 (continued)

Integrated Pest Management (IPM) is a set of strategies which revolve around a set of key principles which encompass: (1) prevention and suppression, (2) monitoring, (3) decision-making, (4) intervention, and (5) evaluation (Barzman et al. 2015). Implementation of these principles requires integration of cultural, behavioral, biological, and chemical tactics, which rely on knowledge of the biology and ecology of insect pests and their natural enemies (e.g., predators and parasitoids) (Stern et al. 1959; Kogan 1998; Dara 2019). The aim of IPM is to reduce pest populations to tolerable levels while minimizing the impact on non-target organisms and the environment.

Genomic biosurveillance is the process of gathering essential genomic data, generally through high-throughput sequencing technologies, which helps to identify potential detrimental organisms, including insects pests. From the data, genomic signatures can be used to identify pests, their origins, routes of invasion, and adaptations resulting from changes in distribution or response to management strategies, which can help inform current and future management strategies (Hamelin and Roe 2020).

High-throughput sequencing (HTS) (also called next-generation sequencing) is a method of DNA sequencing that enables the simultaneous analysis of millions of DNA fragments in a single run. There are several HTS platforms including those from Illumina Inc., PacBio (Pacific Biosciences of California), and Oxford NanoPore Technologies. All HTS relies on template preparation, sequencing, and bioinformatic processing (i.e., alignment and assembly). It is a rapid and cost-effective approach to sequencing large amounts of genetic material, allowing for the analysis of complex biological samples and the detection of genetic variation (Metzker 2010).

Reduced representation sequencing (RRS) is a DNA sequencing method that uses restriction enzymes to digest genomes into fragments, prior to sequencing. This technique reduces genome complexity and generates genome-wide data by surveying numerous loci across populations or species (Altshuler et al. 2000; Campbell et al. 2018).

Restriction enzymes are enzymes (originally isolated from bacteria) that are used in molecular biology to cut DNA molecules at specific recognition sequences (sites), which produce DNA fragments with known sequence at each end that can be analyzed or manipulated in various ways.

Recognition sequences (or recognition sites) are specific DNA sequences, usually 4–8 bp long, that are recognized and cleaved by restriction enzymes. The specific sequence determines the cutting site of the restriction enzyme.

(continued)

Box 1 (continued)

Adapters, in the context of HTS, are short synthetic DNA molecules that include platform-specific sequences for recognition by the sequencing instrument and are ligated to primers and to the ends of DNA fragments to be sequenced. Adapters can also contain unique **barcodes** (or indexes) which are unique sequences used to distinguish different samples in a multiplexed sequencing run (Mir et al. 2013; Hu et al. 2021).

Sequencing primers are short, single-stranded DNA molecules that anneal to a known region of DNA and provide a starting point for DNA polymerase to synthesize a complementary strand of DNA during sequencing. In the context of HTS, sequencing primers are typically proprietary and made to match pieces of synthetic DNA ligated to DNA fragments during library preparation.

Multiplexing, in genomics, refers to pooling multiple DNA samples together for simultaneous sequencing. Each individual DNA sample is ligated to a unique barcode (or index) which allows each sample to be identified and separated from each other during downstream data analysis (Hu et al. 2021).

A **flowcell**, on Illumina sequencing platforms, is a glass slide with lanes that contain oligonucleotides that complement adaptors ligated to DNA fragments. DNA fragments are amplified through several rounds of sequencing reactions. The flowcell is designed to allow the DNA to be analyzed in parallel, which makes the sequencing process faster and more efficient (Slatko et al. 2018).

Demultiplexing refers to the process of separating sequence data from a sequencing run that was generated from multiplexed samples. This process relies on the unique barcodes (or indexes) that were attached to each individual DNA sample prior to sequencing to separate the multiplexed samples into individual samples (Mir et al. 2013).

Size selection refers to the process of isolating DNA fragments of a particular size range using various techniques, including gel electrophoresis, magnetic bead-based separation, or column-based purification (Podnar et al. 2014).

Tags refer to the short DNA fragments that are created by restriction enzyme digestion of genomic DNA.

Sequencing coverage, in the context of HTS, refers to the percentage of the total genome that is being sequenced (Sims et al. 2014).

Sequencing depth refers to the number of times a particular nucleotide or fragment of a DNA molecule has been read during sequencing (Sims et al. 2014).

(continued)

Box 1 (continued)

Ascertainment bias refers to the systematic error or distortion that occurs when data is collected in a way that is not empirically representative (Albrechtsen et al. 2010). In RRS, this may result from improper study design and can cause locus dropout, biases in allele frequencies, estimated genetic diversity, or other deviations in the data that may vary from expected theoretical distributions.

2.1 A Balancing Act: The Importance of Sequencing Depth and Coverage

RRS library preparation typically begins with determining how many loci, or **tags**, adjacent to recognition sequences should be cut (**sequencing coverage**) and how many times each tag should be sequenced (**sequencing depth**, sometimes also referred to as depth of coverage or, confusingly, just coverage). These variables are perhaps the most essential technical component of an RRS-based study. While high-throughput sequencers have relatively high raw error rates, sequencing each tag multiple times reduces the impact of individual, random sequencing errors and produces more accurate estimates of heterozygosity, improving genotyping accuracy overall (Fox et al. 2014; Nazareno and Knowles 2021; Pfeiffer et al. 2018; Stoler and Nekrutenko 2021; Wang et al. 2012). Sequencing coverage, on the other hand, directly controls how much of the genome is sampled and therefore influences dataset power. Because individual biallelic SNPs contain relatively little genetic information compared to multi-allelic microsatellites or gene sequences, greater numbers of RRS loci are required to resolve both fine-scale genomic and geographic patterns (e.g., Sunde et al. 2020). As with sequencing depth, expected or known genetic diversity of the study system will influence decisions about sequencing coverage, as lower rates of standing polymorphism may necessitate sampling a greater number of genomic loci to capture enough signal for downstream analyses (e.g., Davey et al. 2011).

Thus, a properly designed RRS study needs to balance sequencing depth and coverage to minimize sequencing error, maximize genotyping accuracy and dataset power, and ensure it remains cost-efficient (Beissinger et al. 2013). These parameters can be modified to fit the needs of a particular study in several ways. Most directly, the use of common- versus rare-cutting restriction enzymes in single- and multi-digest RRS protocols can be adjusted to obtain desired sequence depth and coverage (Baird et al. 2008; Graham et al. 2015; Peterson et al. 2012; Poland et al. 2012). In addition, selective primers can be used after adapter ligation to preferentially amplify a subset of restriction fragments (Elshire et al. 2011; Poland et al. 2012; Sonah et al. 2013), and fragment analyzers can be used to manually perform size selection by isolating a range of fragment lengths prior to sequencing; each of these techniques (or combinations thereof) can be used to manipulate the coverage-to-depth ratio.

Software can now also perform in silico digestions of reference genomes, when available, using different combinations of restriction enzymes so that users may empirically estimate sequencing depth and coverage prior to sequencing (Herrera et al. 2015; Lepais and Weir 2014; Mora-Márquez et al. 2016; Torkamaneh et al. 2019; Yan et al. 2021). Unfortunately, for organisms that lack sufficiently closely related reference genomes, investigators must rely on theoretical estimates of depth and coverage (Box 2) and then essentially "sequence and see" to determine whether realized (i.e., post-sequencing) coverage and depth is sufficient. These realized values can vary from estimates due to amplification and/or sequencing biases, genome characteristics including high copy number, repetitive elements, DNA methylation, or other factors that may contribute to the unequal distribution or cutting of restriction sites across the genome (Beissinger et al. 2013; Davey et al. 2013; Harismendy et al. 2009). In the case of insects, when a reference genome is not available, this exploratory approach may be not only expensive and time consuming, but can also waste potentially precious specimen material for species that are particularly small. Currently, relatively few insect pests have highly contiguous reference genomes available, though this is rapidly changing. Genome sequencing technology is improving and dropping in cost, and global sequencing efforts such as the Wellcome Sanger Institute Tree of Life project and the Ag100 Pest Initiative have increased the number of high quality, freely available genomes for insects – including for some of the world's most significant insect pests (Childers et al. 2021).

Box 2 Maximizing Your Dataset: Estimating Sequencing Depth and Genome Coverage

In order to plan for adequate depth and coverage, one needs to know the approximate genome size of the study taxon, how often a restriction enzyme is likely to cut in the genome, the typical output capacity in basepairs (bp) of the sequencer being used, and how many specimens will be multiplexed together.

Estimating the frequency of particular restriction enzyme recognition sites throughout the genome is relatively straightforward. The frequency of recognition sites can be estimated as 4^n, where "4" represents the four nucleotides that make up DNA (A, C, G, and T) and "n" represents the length of the recognition site of interest. If the approximate genome size is known, you can then determine how many unique sequence tags are likely to be in the genome. Flow cytometry and *k*-mer analysis can be used to estimate genome size (Hanrahan and Johnston 2011; He et al. 2016), but in the absence of this data, the genome size of other closely related taxa can be used as a proxy for RRS.

For example, using a genome size of 500 megabases (Mb), a fairly typical size for an insect, we can compare estimates of the number of tags that can be cut using either the *PstI* restriction enzyme, which has a 6 bp recognition sequence, or the *MspI* restriction enzyme, which has a 4 bp recognition sequence:

(continued)

Box 2 (continued)

$$\textit{PstI}: 4^6 = \text{a tag every 4,096 bp, on average;}$$
$$500{,}000{,}000\ \text{bp} \div 4{,}096\ \text{bp} = \text{approx.}122{,}070\ \text{tags in the genome}$$

$$\textit{MspI}: 4^4 = \text{a tag every 256 bp, on average;}$$
$$500{,}000{,}000\ \text{bp} \div 256\ \text{bp} = \text{approx. } 1{,}953{,}125\ \text{tags in the genome}$$

It should be noted that this calculation assumes equal proportions and even distributions of each nucleotide throughout the genome. Of course, we know that this is generally *not* the case, so these numbers represent simplified estimates. But this approach is still highly valuable for initial RRS project planning, which relies more on the general magnitude of these numbers rather than exact calculations. Using these two restriction enzymes as an example, it should also be apparent that shorter recognition sequences can be orders of magnitude more common in the genome than longer sequences. Accordingly, the use of restriction enzymes that target short recognition sequences will produce greater genome coverage than restriction enzymes that target longer, less frequent recognition sites.

Continuing with the example above, if each tag is sequenced to a length of 100 bp on a sequencing platform that can output 25 Gb of data, then the theoretical sequencing coverage and depth for a single, 500 Mb genome can be estimated:

$$\textit{PstI}: 122{,}070\ \text{tags} \times 100\ \text{bp} = 12{,}207{,}000\ \text{bp}$$
$$12{,}207{,}000\ \text{bp} \div 500{,}000{,}000\ \text{bp} = 0.024, \text{ or } 2.4\%\ \text{genome coverage}$$
$$25{,}000{,}000{,}000\ \text{bp} \div 12{,}207{,}000\ \text{bp} = \text{average sequencing depth of } 2{,}048$$
$$\textit{MspI}: 1{,}953{,}125\ \text{tags} \times 100\ \text{bp} = 195{,}312{,}500\ \text{bp}$$
$$195{,}312{,}500\ \text{bp} \div 500{,}000{,}000\ \text{bp} = 0.39, \text{ or } 39\%\ \text{genome coverage}$$
$$25{,}000{,}000{,}000\ \text{bp} \div 195{,}312{,}500 = \text{average sequencing depth of } 128$$

Note that in both examples, the total number of sequenced basepairs is the same – 25 Gb – but that the genome digested with *PstI* has lower coverage than the genome digested with *MspI* and much greater sequencing depth, illustrating the relationship between these two variables when sequencing output is finite. Additionally, if users wish to multiplex many samples into a single sequencing run, sequencing depth will necessarily decrease to account for this:

(continued)

Box 2 (continued)

$$\textit{PstI}: \text{starting depth of } 2{,}048 \div 100 \text{ samples} = \text{final average depth of } 20.48 \text{ per sample}$$

$$\textit{MspI}: \text{starting depth of } 128 \div 100 \text{ samples} = \text{final average depth of } 1.28 \text{ per sample}$$

Note the difference in final average depth for both examples. The final sequencing depth of the dataset digested with *MspI* is insufficient for accurate genotyping, and so it will be necessary to make some adjustments that will mitigate this issue. For example, swapping restriction enzymes or performing a size selection will reduce coverage and increase depth while maintaining multiplexing capacity, while reducing the number of multiplexed samples will free up more sequencing output and increase depth without sacrificing much coverage. Or sequencing this data on a platform with a higher sequencing capacity, if feasible, will preserve genome coverage, sequencing depth, *and* specimen multiplexing.

2.2 Optimizing RRS Experimental Design

As a complexity-reduction strategy, RRS methods are most useful when genome sizes are too large to feasibly sequence entire genomes or when whole genomes are not necessary or desirable for a particular study (Davey et al. 2011; Eaton and Overcast 2020). The cost of RRS library preparation is also still significantly less expensive than producing whole genome data, though this will likely change as the cost of whole genome sequencing continues to decrease. From an experimental design perspective, because these methods rely on shared restriction enzyme recognition sites between populations or species, RRS is best applied to studies focused on populations or closely related species, as the number of these shared sites is inversely proportional to evolutionary divergences between taxa (Davey et al. 2013). If taxa are too distantly related and have exceptionally high genetic diversity, or the sampling design does not adequately capture natural levels of population genetic variation, this can lead to **ascertainment bias**. For RRS methods, ascertainment bias can contribute to non-random patterns of missing data via locus dropout and may tend to retain conserved markers or those with common polymorphisms over markers with rarer minor alleles; these patterns may ultimately skew assessments of genetic diversity, local adaptation, and estimates of introgression between taxa (Arnold et al. 2013; Cordeiro et al. 2020; Gautier et al. 2013; Lee et al. 2018). This is also true for aligning sequence reads to reference genomes during SNP calling; recent work demonstrates that alignment to genomes of increasing phylogenetic distance can reduce alignment quality and the number of retained SNPs and inhibit observations of geographically fine-scale population structure (Bohling 2020; Shafer et al. 2017). Therefore, even when sequencing depth and coverage have been properly considered, sampling strategy and the approach to SNP calling can have large impacts on final assessments of genetic diversity between populations.

RRS approaches also typically require significantly more starting DNA than what is necessary for gene, microsatellite, or even whole genome re-sequencing (Andrews et al. 2016). For particularly small insects, or if non-invasive sampling of voucher material is necessary (e.g. sampling of a single leg from museum specimens), it may be advisable to perform whole genome amplification prior to RRS library preparation to increase the starting concentration of DNA (Cruaud et al. 2018). Another consideration is DNA quality; degraded samples (typical of museum specimens and also samples collected as part of forestry and agriculture biomonitoring programs) are not as easily sequenced with RRS because highly sheared DNA leaves fewer restriction enzyme recognition sites intact or may produce low-quality raw sequences (Graham et al. 2015). However, some RRS methods have been developed to work with degraded or ancient samples (Schmid et al. 2017; Suchan et al. 2016).

The potential for data consistency and reproducibility also warrants careful consideration. RRS targets high numbers of anonymous portions of the genome, an approach that differs from other sequencing techniques that either target known loci or re-sequence the entire genome. Not all of these targeted regions will be consistently cut and/or sequenced, and there is some randomness governing which genomic loci will ultimately be present in any particular RRS sequencing run. Because of this, there are some key technical features that need to remain consistent to maximize data reproducibility between studies or for long-term projects. Restriction enzyme choice is the most primary consideration for RRS, as the use of different enzymes will fundamentally alter the sequence data. As discussed previously, the use of selective primers, mechanical shearing, size selection, or single versus double digest approaches can also influence which and how many genomic loci end up being sequenced.

Additionally, differences in RRS data output and/or composition can occur between sequencing runs even when the library preparation method does not change (called run or lane effects) due to either human or machine-induced variation (e.g., variable efficiency of digestion reactions, PCR amplification biases, differences in fragment cluster density, or reagent washing across the flowcell used during sequencing). Therefore, understanding how these techniques may impact the resulting sequence data, as well as any future compatibility requirements of the data, is necessary to ensure sequencing efforts are as reproducible as possible. A few studies have shown that careful bioinformatic filtering can compensate for some run-specific or even method-specific artefacts, with the caveat that dataset compatibility should be carefully assessed and planned for when combining data, rather than be taken for granted (Campbell et al. 2017; DaCosta and Sorenson 2014; Graham et al. 2020).

2.3 *Considerations for RRS Data Processing*

Historically, molecular ecology has been limited by how much data could feasibly be produced, both in terms of cost and time. However, HTS has effectively removed this constraint, and now the computational demands and expertise necessary for handling data on this scale are more likely to be limiting. Sequencing advances have necessitated the rapid development of new computing solutions for the HTS era, and a working knowledge of various programming languages and informatics tools for data handling, processing, and analysis is virtually essential for those wishing to implement HTS techniques such as RRS into their research. This can be daunting, and the bioinformatics component of working with RRS data is often not only the largest barrier to new users, but can also be a major source of error if RRS locus assembly and SNP calling is done improperly (Díaz-Arce and Rodríguez-Ezpeleta 2019; Shafer et al. 2017). Fortunately, there are now many different software pipelines for processing RRS datasets, as well as several guides describing best practices for experimental design and data processing (Catchen et al. 2011; Cerca et al. 2021; Eaton and Overcast 2020; McCartney-Melstad et al. 2019; Nazareno and Knowles 2021; Paris et al. 2017; Qu et al. 2019; Rochette et al. 2019; Torkamaneh et al. 2016).

3 Applying RRS to Insect Pest Biosurveillance and Management

Most fundamentally, population genetics identifies genetic patterns that may result from historical and contemporary habitat configurations as well as interactions within and between species. These patterns may arise via neutral forces (e.g., genetic drift) or as a response to natural selection and may lead to population-specific differences in nucleotide diversity and allele frequency distributions, or even larger structural variants such as genomic inversions and copy number variants (CNVs) (Dorant et al. 2020; Hohenlohe et al. 2010a, b; Mérot et al. 2020).

Accurately profiling changes across the genome can provide insights into the relative impact of genetic bottlenecks, admixture, or adaptation on pest populations and relate these patterns to risk intelligence to determine management priorities. From a pest management perspective, failure to accurately recognize and describe these patterns can have serious consequences, resulting in wastage of management resources, unnecessary quarantines, or the misuse of insecticides (Cordeiro et al. 2020; Doellman et al. 2020). This can create delays in implementing effective intervention or eradication strategies and exacerbate the impact of insect pests on food supplies and natural resources (Raszick et al. 2021; Roe et al. 2018). Reduced representation sequencing has become increasingly used in this context to investigate genetic structure, phylogeography, and genomic adaptation of insect pests at both small and large geographic scales (Blackburn et al. 2017; Campbell et al. 2020;

Erlandson et al. 2019; Lumley et al. 2020; McCartney et al. 2019; Picq et al. 2023; Taylor et al. 2021), establishing an essential foundation of practical knowledge for genomic biosurveillance efforts and the development of early intervention strategies in agricultural and forest ecosystems.

3.1 Characterization of Insect Pest Population Structure

Perhaps the most fundamental application of RRS data for population genetics is its use for assessing genetic variation both within and between closely related taxa to identify genetic populations. In pest management frameworks, this can have significant applied value as it is often the first step in understanding the molecular ecology of both native and invasive pests. For instance, the *Choristoneura* spruce budworm species complex in North America has been the focus of several studies that used RRS to assess population structure, admixture, and adaptation in *Choristoneura fumiferana* and two ecologically divergent subspecies of the closely related *C. occidentalis* (Blackburn et al. 2017; Brunet et al. 2017; Larroque et al. 2019; Lumley et al. 2020; Nelson et al. 2021; Picq et al. 2023). Insights from this work include observations of genomic differentiation in *C. fumiferana* that closely track changes in climate zone and host plant distribution across its range, as well as the identification of both autosomal and sex-linked "islands of differentiation" between populations containing genes associated with dispersal, diapause, and overwintering success (Lumley et al. 2020; Nelson et al. 2021), traits that are of fundamental interest for management.

Genomic investigation of the spongy moth, *Lymantria dispar*, has recently clarified the global population structure and phylogeography of this species complex and emphasized potential gaps in current molecular biosurveillance strategies (Picq et al. 2023). Broadly speaking, females of the European subspecies *L. d. dispar*, an established invasive in North America, are flightless, while morphologically indistinguishable subspecies in Asia (*L. d. asiatica* and *L. d. japonica*) have flighted females; however, there is considerable variation in flight capacity in parts of Eurasia where *L. d. dispar* and *L. d. asiatica* meet. Picq et al. (2023) used RRS data to identify a genetic cline between these subspecies that coincides with variation in flight capacity (Fig. 1) and additionally found that many specimens from this region had mitochondrial DNA consistent with *L. d. dispar*, even when nuclear DNA was consistent with *L. d. asiatica*. This has potentially significant consequences for biosecurity in North America, as current spongy moth molecular diagnostic tools use mitochondrial DNA to differentiate subspecies (Stewart et al. 2016, 2019). The work of Picq et al. (2023) suggests that mitochondrial DNA on its own may not be a sufficient proxy for flight capacity, depending upon where an intercepted specimen originated from, and highlights the need for continued work to identify the genes associated with female flight.

Associating population genomic structure with both historical and contemporary landscape composition can also elucidate the role of landscape in facilitating or

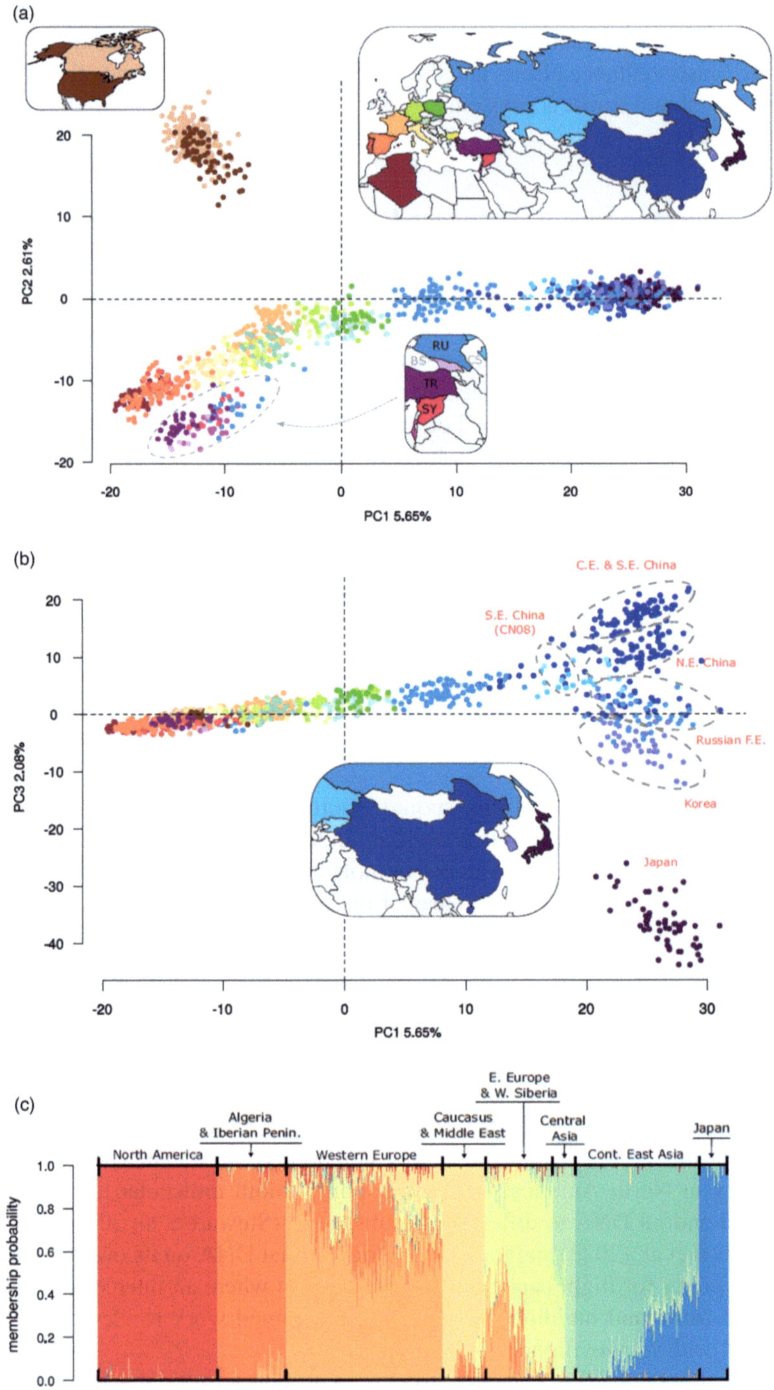

Fig. 1 Population structure analysis. Principal component analysis (PCA) based on 2,125 neutral SNPs derived from 1,288 *Lymantria dispar* individuals. (**a**) PC1 versus PC2 plot. Individuals (dots) are colored according to their country of origin (25 different countries). (**b**) PC1 versus PC3 plot.

impeding gene flow and local adaptation and may provide an important temporal context for pest management given the rapid change of landscape composition in managed ecosystems over the past few hundred years (Anderson et al. 2010). Crossley et al. (2019) used the Colorado potato beetle, *Leptinotarsa decemlineata*, to explore the relative importance of both historical and contemporary landscape variables on present-day genetic structuring of *L. decemlineata* populations in the United States. Using a landscape genomics approach with GBS data, this study determined that genetic diversity was more influenced by contemporary, rather than historical, landscape composition, which appeared to restrict movement between populations (Fig. 2). This not only suggests that expected temporal lags in changes to genetic diversity resulting from rapid landscape modification may not always be significant, but also that land use itself may be leveraged to control the spread of pests in some cases, opening up new considerations for how agricultural practices may support IPM strategies.

This type of work may also have important implications for other studies on lesser-known pest species. For instance, an RRS-based population genetic study of the canola flower midge, *Contarinia brassicola*, in the Canadian prairies recovered genetic structuring in some populations that were quite different from other, nearby sampling locations, and which did not seem to be driven by isolation by distance or other habitat variables considered in the study (Campbell et al. 2020). This pest was only formally described in 2019 and is likely native to Canada (Mori et al. 2019), but little is known about its distribution outside of canola (*Brassica napus* and *B. rapa*)-growing regions and, because this species had only been observed developing on canola, this study did not consider alternative hosts. The authors suggested that canola-centric sampling may have impacted the recovered genetic patterns and that the divergent populations may in fact represent genetic spillover from alternative Brassicaceae hosts. Given the rapid increase in canola production in this region over the past hundred years, if *C. brassicola* is indeed endemic to Canada, it would have likely utilized other hosts prior to large-scale canola production. If *C. brassicola* is found on other host plants, adopting a temporal perspective amidst rapid landscape changes may further clarify the extent to which landscape composition may influence the genetic diversity of this species.

Fig. 1 (continued) The dotted ellipses indicate populations from (i) Korea, (ii) the Russian Far East and extreme North-Eastern China (CN04), (iii) North-Eastern China, (iv) Central-Eastern and South-Eastern China, and (v) the southernmost Chinese population (CN08). (**c**) Barplot showing probability of membership of each sample to each of eight geographic clusters, as computed by ADMIXTURE. Samples along the *x* axis are ordered from west to east according to their longitudinal coordinates, with possible exceptions for some of the more southerly samples from the Caucasus/middle-east, which were placed between those of Western and Eastern Europe. *Lymantria dispar dispar* broadly ranges across Europe, west of the Ural Mountains, in North Africa, and North America, while *L. dispar asiatica* ranges across continental Asia, the Ural Mountains, and the Russian Far East, and *L. dispar japonica* is restricted to parts of Japan. (Figure and caption reproduced from Picq et al. (2023) (CC BY 4.0, https://creativecommons.org/licenses/by/4.0/). The caption has been modified for clarity)

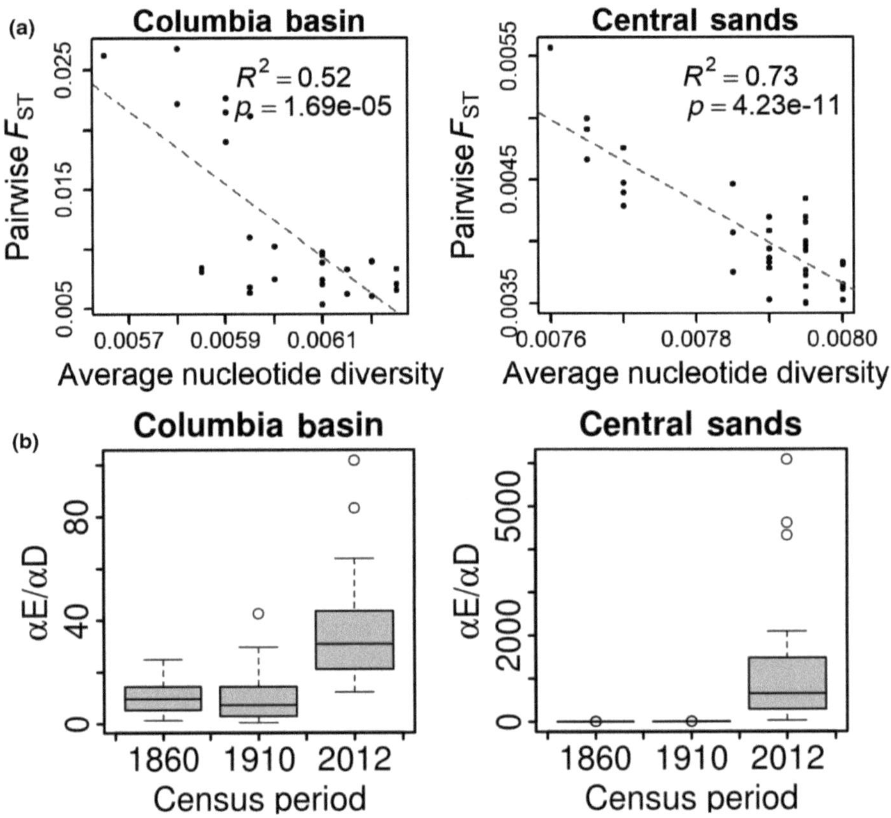

Fig. 2 Patterns of genetic differentiation in Colorado potato beetle, *Leptinotarsa decemlineata*, correlate with contemporary, not historic, potato land cover (**a**) Relationship between genetic differentiation (pairwise F_{ST}) and diversity (average nucleotide diversity) of *L. decemlineata* populations in Oregon, Washington, and Wisconsin. (**b**) BEDASSLE estimates of the effect size of landscape resistance (LRT) relative to geographic distance (αE/αD) on allele frequency differences in the Columbia Basin of Oregon and Washington (left) and the Central Sands of Wisconsin (right) in census periods of 1860, 1910, and 2012. Boxplots represent the distribution of final parameter estimates across 30 independent Markov chains run for four million steps each. (Figure and caption reproduced from Crossley et al. (2019) (CC BY 4.0, https://creativecommons.org/licenses/by/4.0/). The caption has been modified for clarity)

3.2 *Tracing Invasion Routes, Colonization History, and Identifying the Geographic Origin of Insect Pests*

The assessment of population structure using RRS data can also be scaled up to reconstruct global invasion routes and determine the origin of invasive populations by assessing genetic similarity between populations, identifying genetic signatures of population bottlenecks and range expansions resulting from invasion events, and may also be used to characterize complex scenarios of spread including secondary introduction via **bridgehead** events (Blumenfeld et al. 2021; Garnas et al. 2016). Ryan et al. (2019)

demonstrate this nicely with their expansive, citizen science-assisted study of the cabbage white butterfly, *Pieris rapae*, a cosmopolitan pest of cruciferous crops. Using RRS data, Ryan et al. (2019) reconstructed the invasion history of *P. rapae* and in doing so detected an increase in effective population size in ancestral European populations that occurred near the end of the last glacial maximum; this timing broadly coincides with the development of agriculture in this region and suggests that population expansions occurred within the ancestral range of *P. rapae* prior to the colonization of invasive ranges. This study goes on to date subsequent diversification between native European and invasive Asian subspecies to approximately 1,200 years ago following the increased cultivation of brassicaceous crops and identifies human-mediated transport along the Silk Road and other heavy trade routes as likely sources for multiple introductions of *P. rapae* over the last thousand years. Investigation of more recently established invasive *P. rapae* populations (within the last 200 years) additionally supports plausible bridgehead events both within the United States and from the United States to New Zealand.

Reconstruction of the invasion history of the termite pest *Coptotermes formosanus* using RRS techniques found evidence of multiple introductions into the United States (Blumenfeld et al. 2021), including a bridgehead event where an invasive population in Hawaii (itself the product of admixture between two distinct Asian source populations) acted as the likely origin of an invasive population in the southeastern United States (Fig. 3). A second, distinct incursion into Florida from China appeared admixed with the southeastern invasion, indicating a complex

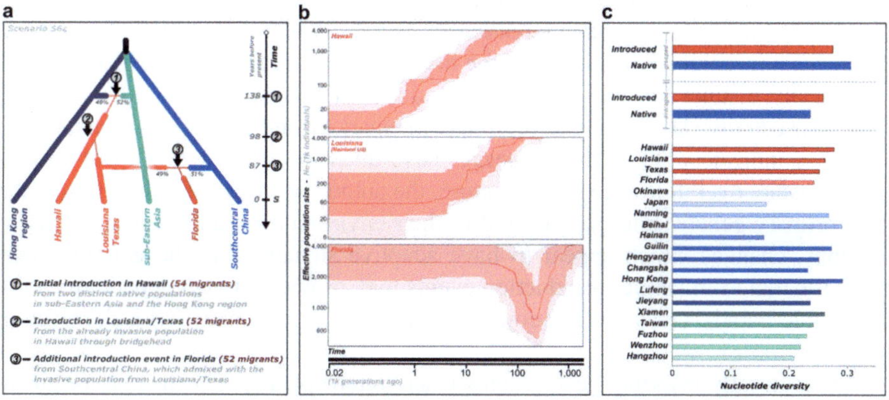

Fig. 3 The invasion history of *Coptotermes formosanus*. (**a**) Graphical representation of the most likely invasion history scenario for *C. formosanus* out of Asia tested through approximate Bayesian computation random forest (ABC RF) machine learning. Thin dotted lines represent bottleneck events. Estimated duration of each bottleneck event and all of the posterior parameter estimates are provided in Supplementary Table 11 of Blumenfeld et al. (2021). Time is not to scale, with S indicating sampling time. (**b**) Estimation of the variation of effective population size through time for three invasive localities using Stairway Plot 2. The solid red line is the estimate of the median effective population size, and the light and dark red shaded areas represent the 95% and 75% confidence intervals, respectively. (**c**) Nucleotide diversity in the introduced and native range. All native and introduced localities were first analyzed while grouped together, then independently. (Figure and caption reproduced from Blumenfeld et al. (2021) (CC BY 4.0, https://creativecommons.org/licenses/by/4.0/). The caption has been modified for clarity)

invasion history that had not been sufficiently characterized in prior work using mitochondrial data or microsatellites. Moreover, Blumenfeld et al. (2021) also found that invasive populations in the United States did not appear to have significantly reduced within-population genetic diversity compared to native Asian populations, despite predicting that the bridgehead event would cause a greater reduction in genetic diversity due to the two sequential bottlenecks that invasive populations would have likely experienced. The authors suggest these findings point to secondary invasions as a means of increasing genetic diversity in recently established invasive populations.

If this is true, then repeated introductions in a single region may have a potentially protective effect on newly established populations and may influence population stability and persistence in novel habitat. These insights contribute to a greater understanding of the so-called genetic paradox of invasions – that genetic bottlenecks often accompany introduction events, which should pose several challenges to the successful establishment of invasives given reductions in genetic diversity and increased inbreeding, reducing the capacity for adaptation (Allendorf and Lundquist 2003; Schrieber and Lachmuth 2016). Despite this, many invasive species have become serious pests, though the existence of repeated incursions and their influence upon population establishment and adaptation of invasive pests are often not known.

In cases where pest invasion history is better documented, it is clear that secondary invasions, while perhaps a boon to recently established invasive pests, are not always prerequisites for invasion success. The fall webworm, *Hyphantria cunea*, was introduced to China approximately 40 years ago through a single invasion event and experienced a severe genetic bottleneck as a result (Cao et al. 2016). Since then, *H. cunea* has rapidly expanded its range in China and has further diversified into two genetically distinct lineages that occur in different geographic regions. Dai et al. (2024) used ddRAD to generate genome-wide markers and search for candidate loci that may indicate whether these lineages have diverged as a result of local adaptation. Despite overall low genetic structure, the authors were able to identify that the two primary genetic lineages differed in their environmental and geographic associations, changes which appear to have adaptive signatures in the genome. This suggests that even single-introduction events may provide enough genetic diversity to enable rapid, varied adaptation in species as a response to local selection pressures.

In recurrently invading pest systems where intercepted specimens may originate from multiple potential source locations, RRS data can be used to not only identify population structure and reconstruct invasions, but also to develop molecular diagnostic tools to determine the origin of intercepted specimens. The Mexican fruit fly, *Anastrepha ludens*, is serious pest of fruit native to Central America and Mexico. Commercialization of fruit production in the southern United States has resulted in repeated invasions of *A. ludens* into this region. Using a ddRAD approach, Dupuis et al. (2019) investigated the population genomic structure of *A. ludens* across its native and invasive range. Using thousands of SNPs, they recovered four genetically distinct populations across the species range. The authors then identified a smaller set of SNPs that were able to re-create the same geographic patterns as the larger dataset, and used it to produce and validate a diagnostic SNP panel that could genotype intercepted *A. ludens* and provide information about their likely origin. From a regulatory science perspective, this type of information is essential for identifying the most common routes of invasion for pests associated with trade and can help identify risk, refine quarantine areas, or otherwise inform IPM strategies to reduce the impact of insect pests worldwide.

Given the broad, often global, geographic scope that these types of studies require, RRS methods can offer an affordable option for sequencing many populations and uncovering complex patterns of invasion and spread, improving our understanding of how invasive species may establish themselves in new regions.

3.3 Identification of Cryptic Variation and Genomic Adaptation

Perhaps one of the biggest challenges in IPM lies in detecting the mechanisms that contribute to pest adaptation and response to new selective pressures. While physiological responses to novel pressures – for instance, exposure to insecticides or to novel climates – may be tested experimentally, molecular profiling is necessary to actually identify the mechanisms that underlie these responses and develop marker-based biosurveillance methods to profile field populations. From a risk intelligence perspective, this is valuable information that can help predict spread and ecological or economic impact, and RRS provides viable opportunities to leverage genome-wide information for risk assessment, pest management, and regulation.

For instance, Dupuis et al. (2018) investigated genomic signatures of selection in the melon fly, *Zeugodacus cucurbitae*, to determine whether invasive populations may benefit from different control strategies. This polyphagous fruit pest is native to the Indian subcontinent but has become established in several other regions of the world, including Africa, Hawaii, and other parts of Asia. Using ddRAD-derived SNPs, this study produced a range-wide assessment of population structure and also identified outlier loci in invasive populations of *Z. cucurbitae* in Africa and Hawaii to determine whether these populations exhibit different patterns of natural selection. This study identified six primary genomic clusters of *Z. cucurbitae* across its range and also frequently recovered admixture between populations. They also found that in comparisons of populations from Africa or Hawaii to those from Asia, highly differentiated outlier SNPs were mostly the same (Fig. 4). In fact, these populations only contained a single unique outlier locus that was not shared with other populations. These results suggest that selective pressures in African and Hawaiian populations are broadly similar, and therefore, those similar management strategies are likely to be equally effective in both localities.

Genomic assessments of adaptation can also elucidate patterns of natural selection at smaller geographic scales. Andrews et al. (2020) used RADseq-derived SNPs to identify and determine the geographic range of cryptic species complexes in

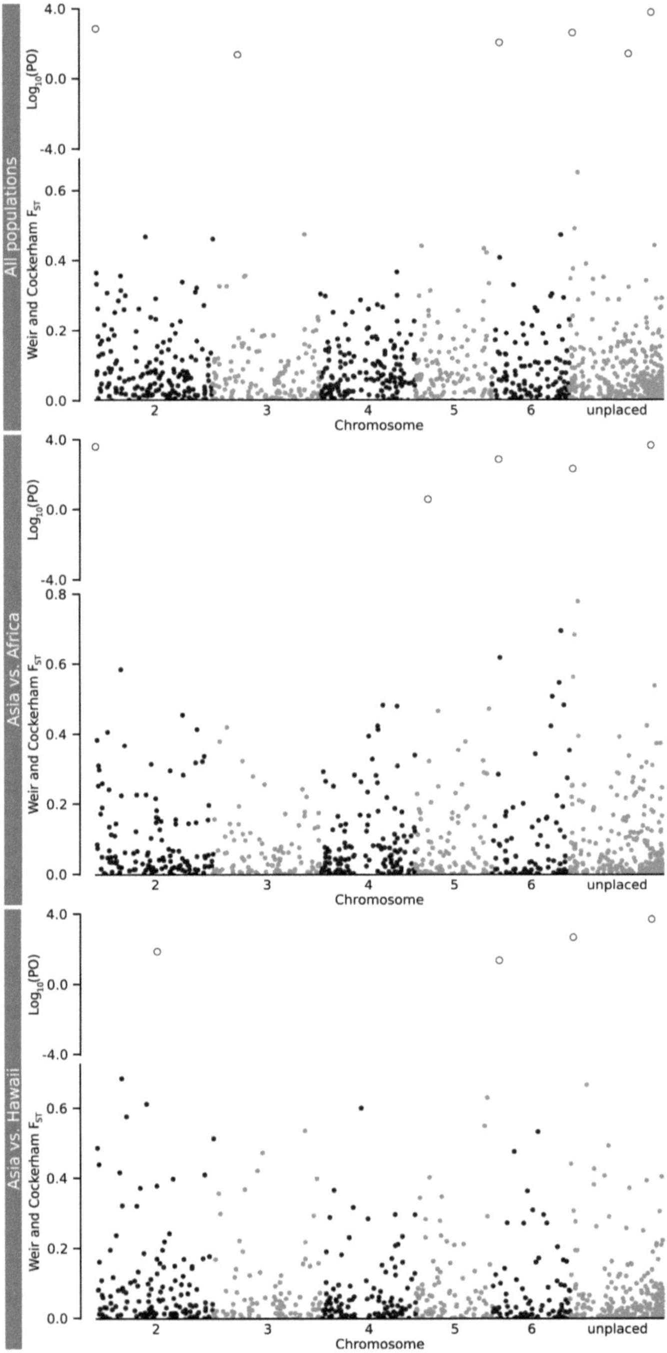

Fig. 4 Genome-wide plots of Weir and Cockerham's F_{ST} (lower plots) and Log_{10} posterior odds (PO) of BayeScan outlier loci (upper plots) for *Zeugodacus cucurbitae*. Outliers predicted by BayeScan are indicated by larger, white filled circles and are broadly similar in comparisons between Africa and Asia and Hawaii and Asia. (Figure and caption reproduced from Dupuis et al. (2018), Springer Nature, 2018, reproduced with permission from SNCSC)

Limonius californicus and *L. infuscatus*, two species of wireworm (Coleoptera: Elateridae) that are significant agricultural pests in temperate and sub-tropical regions. Within parts of Canada and the United States, this study found that sympatric lineages within each species appear not to interbreed, suggesting that the number of species in these complexes based on morphological data may be underestimated; in an IPM context, this discovery may directly lead to changes in management if these divergent lineages are found to represent distinct species. Additionally, genome scan analyses indicated that wireworms sampled from plots of land that had been continuously treated with insecticide were more genetically divergent than samples taken from plots that had not been recently treated with insecticide, suggesting a genomic response to insecticide treatment. The identification of differentiated genomic regions provides an essential starting point for understanding the genetic mechanisms involved in local adaptation. If there are genetic polymorphisms that correlate to phenotypes of interest, those markers can be used to develop assays that profile natural populations.

Chen et al. (2019) used a similar approach for the cotton bollworm, *Helicoverpa armigera*, to profile indoxacarb insecticide resistance between strains. After conducting multiple crossing events between specimens with different indoxacarb resistance profiles, a GBS approach was used to sequence crossed individuals to produce a linkage map and conduct a genome-wide association study (GWAS) that would correlate resistance profiles to genotype. The authors found several loci located primarily within a single autosomal region that were associated with the indoxacarb resistant phenotype, effectively identifying the genomic location and the genes likely involved in the development of insecticide resistance in this species. Chen et al. (2019) next used the identified markers to develop an assay for detecting insecticide resistance in field populations, demonstrating how this type of work can directly contribute to enhanced biosurveillance through the profiling of traits relevant for pest management and control programs.

Such patterns of local adaptation may lead to population structuring that reflects variable organismal response to selective pressures; however, for species with large geographic ranges and/or broad ecological niches, high gene flow may act as a homogenizing force that can make it a challenge to identify markers or genomic regions that may be involved in adaptive responses (Yeaman and Whitlock 2011; Tigano and Friesen 2016). In these cases, correlating genetic markers to other sources of data, such as morphological or ecological variables, may improve the recovery of candidate markers associated with local adaptation. Yadav et al. (2019) used a landscape genomic approach to identify patterns of adaptation in the grasshopper species *Phaulacridium vittatum*, a widespread agricultural pest in Australia that experiences high rates of gene flow due to its large range and capacity for long-distance dispersal. This species has known morphological differences that associate with local habitat and climatic differences across its range, but these traits had not been previously tied to genomic patterns that suggested they arose via local adaptation. Using ddRAD-derived SNPs, the authors parsed out neutral markers to assess genome-wide landscape connectivity, enabling them to identify putatively non-neutral markers correlated with differences in environmental tolerances and morphology across the species range.

This spatial approach increased the power to detect genomic patterns in *P. vittatum* that reflected weak restrictions in gene flow based on environmental factors that may otherwise be missed using non-spatial analyses. A similar landscape genetic approach using GBS-derived SNPs was used to assess genomic adaptation in the Colorado potato beetle, *L. decemlineata* (Crossley et al. 2017). This study confirmed that insecticide resistance varied across its range in the United States based on insecticide usage and that resistance in *L. decemlineata* is a polygenic response likely facilitated by a combination of standing genetic variability and variability arising from strong selective sweeps due to insecticide exposure.

4 Challenges and Opportunities for the Use of RRS to Support Insect Pest Biosurveillance Strategies

The application of RRS and other genomic technologies in IPM is gaining momentum and providing new opportunities for pest biosurveillance. The adoption of such techniques greatly enhances risk intelligence through the acquisition of new knowledge about insect pest ecology and evolution, including faster and more robust identification of pest species, determination of origin and invasion dynamics, and the assessment of local adaptation (Bilodeau et al. 2019; Roe et al. 2018). Despite these opportunities, these technologies also present several challenges for the integration of such data into management or regulatory frameworks.

4.1 Identifying the "Realized Niche" of RRS in Population Genomics

While RRS-based methods are currently among the most cost-effective methods to produce large quantities of genome-wide data for organisms with moderate to large-sized genomes, it is likely that other methods for genotyping SNPs will continue to drop in cost and become more widely used in population genetics over time. Despite this, we suggest that RRS methods serve a specific niche that will likely ensure their continued use in population genetics, even as other sequencing technologies improve and become more affordable.

For instance, whole genome sequencing (WGS) is the most likely successor to RRS-based population genomics, but it is not yet cost-effective enough to sequence samples at sufficient depths for robust genotyping. To compensate, WGS studies must either sample many specimens per population or rely on pooling data together (e.g., using pool-seq methods) and limit analyses to assessing population variation rather than individual variation (Kurland et al. 2019; Li et al. 2011; Lou et al. 2020). For studies that are interested in identifying potential dispersal and colonization events, hybridization, or inferring the distribution of particular phenotypes in natural populations, individual genotypes are a necessity.

Whole genome sequencing also poses analytical challenges, as low-coverage sequencing often requires users to work with genotype likelihoods – probabilistic estimates of genotypes that reflect uncertainties in the data – rather than genotypes themselves (Nielsen et al. 2011). In contrast, because RRS is a complexity-reduction strategy, it allows for deeper sequencing and more robust genotyping using fewer specimens. Thus, even though RRS methods sequence only a fraction of the genome, they often produce enough data to work with and identify genotypes that can be integrated into diagnostic tools to support genomic biosurveillance.

Finally, as we have already discussed, the greatest use of RRS data in population genomics thus far has focused on profiling variation in the form of single SNPs. However, it is also possible to use RRS to profile larger genomic features such multi-SNP haplotypes, indels, and copy number variants (CNVs) in populations. This is an approach not yet applied to insect population genomics or biosurveillance but has been recently used in human and other plant and animal systems when SNPs alone are insufficient for identifying genomic patterns associated with particular phenotypes (Dorant et al. 2020; McKinney et al. 2017a, b). These types of structural variants may more directly influence gene expression or other functional processes than individual SNPs (Gamazon and Stranger 2015; Rowan et al. 2019) and may provide new insights into how polymorphisms relate to and influence phenotypes that are of interest from a pest management or regulatory perspective. This is an area which will benefit from additional study using natural systems, and the economic and ecological importance of insects makes them viable candidates for such work.

4.2 *Closing the Gaps Between Research, Policy, and Stakeholders: Integrating Population Genomics and Pest Management*

The development and deployment of IPM tools for plant health is multi-dimensional, often involving academic institutions, industry, community, all levels of government, and plant protection organizations that work to establish national and international lines of communication for phytosanitary standards, regulation, and imports/exports (Bilodeau et al. 2019). Because of this complexity, there are many challenges associated with the uptake of genomics-informed strategies to inform biosurveillance and risk intelligence. Bilodeau et al. (2019) have nicely summarized many of these issues and also proposed general workflows that may improve the integration of genomic data into biosurveillance and, by extension, other aspects of IPM (Fig. 5). Bilodeau et al. (2019) also emphasize that this integration needs to be cost-effective and efficient, have a clear policy framework, and is greatly benefitted by ongoing engagement with end users to ensure their needs are met and they have the knowledge necessary to use these tools effectively. To this end, both collaboration and communication at multiple stages are essential for ensuring the uptake of genomics-informed IPM.

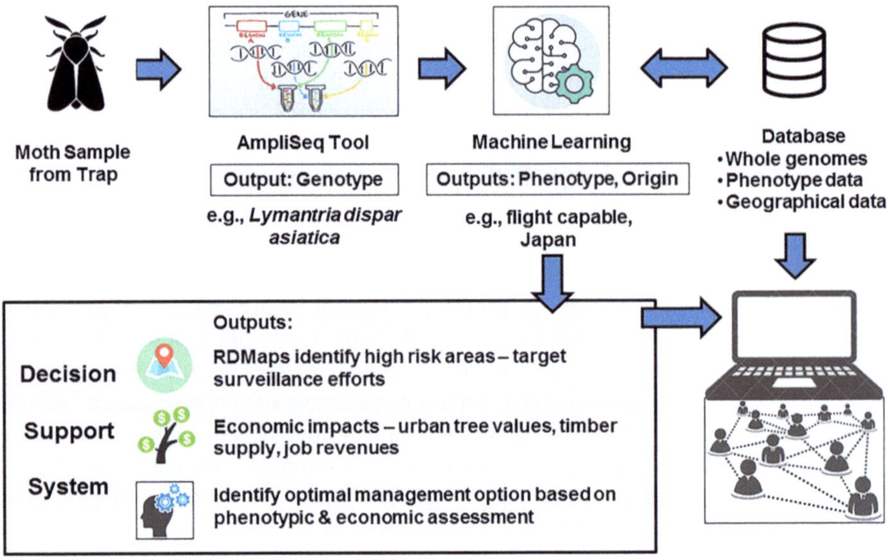

Fig. 5 Flowchart describing the movement of sample information from acquisition to the end user community for one of four key forest invasive alien species (FIAS) species under development in the Biosurveillance of Alien Forest Enemies (biosafe) research project. (Figure and caption reproduced from Bilodeau et al. (2019), Springer Nature, 2019, reproduced with permission from SNCSC. The caption has been modified for clarity)

Stakeholders may also vary in terms of how or what type of data is gathered as part of IPM strategies and how this data is used for decision making, and so there is an imperative to understand the social and cultural lenses that influence IPM locally, regionally, and globally (Deguine et al. 2021; Parsa et al. 2014). This may be a significant pinchpoint for genomics-informed IPM, which not only requires equipment, financial resources, and expertise that may be difficult to acquire, but may seem abstract or even incompatible with traditional land use and management practices. In conservation and environmental science, there has been some progress on the integration of place-based learning and Traditional Ecological Knowledge (TEK) as an essential context for genomics (Bowles et al. 2022; Rayne et al. 2022), and there is likely potential to extend this context into IPM strategies as well.

5 Conclusion

Biosurveillance, supported by reduced-representation sequencing (RSS), can help to answer fundamental questions for integrated pest management of insects in forestry, horticulture, and agroecosystems. RSS can not only be used to simply characterize insect pest population structure, but also trace invasion routes, colonization history, and identify geographic origin of insect pests and cryptic variation and genomic adaptation, which can fundamentally influence integrated insect pest management. Although whole genome resequencing is becoming considerably less expensive, RRS can still fill a niche where individual genotypes are needed, and can be leveraged to support the development of diagnostic tools, as well as the potential to profile larger genomic regions most cost effectively. Ultimately, to incorporate genomics into biosurveillance, clear communication between relevant stakeholders and how to use genomic data for decision making is imperative to take full advantage of this technology.

References

Agarwal A, Cunningham JP, Valenzuela I, Blacket MJ. A diagnostic LAMP assay for the destructive grapevine insect pest, phylloxera (*Daktulosphaira vitifoliae*). Sci Rep. 2020;10: 21229.

Albrechtsen A, Nielsen FC, Nielsen R. Ascertainment biases in SNP chips affect measures of population divergence. Mol Biol Evol. 2010;27:2534–47.

Allendorf FW, Lundquist LL. Introduction: population biology, evolution, and control of invasive species. Conserv Biol. 2003;17:24–30.

Altieri MA. The ecological role of biodiversity in agroecosystems. Agric Ecosyst Environ. 1999;74: 19–31.

Altshuler D, Pollara VJ, Cowles CR, Van Etten WJ, Baldwin J, Linton L, et al. An SNP map of the human genome generated by reduced representation shotgun sequencing. Nature. 2000;407: 513–6.

Anderson CD, Epperson BK, Fortin M-J, Holderegger R, James PMA, Rosenberg MS, et al. Considering spatial and temporal scale in landscape-genetic studies of gene flow. Mol Ecol. 2010;19:3565–75.

Andrews KR, Good JM, Miller MR, Luikart G, Hohenlohe PA. Harnessing the power of RADseq for ecological and evolutionary genomics. Nat Rev Genet. 2016;17:81–92.

Andrews KR, Gerritsen A, Rashed A, Crowder DW, Rondon SI, van Herk WG, et al. Wireworm (Coleoptera: Elateridae) genomic analysis reveals putative cryptic species, population structure, and adaptation to pest control. Commun Biol. 2020;3:489.

Arnold B, Corbett-Detig RB, Hartl D, Bomblie K. RADseq underestimates diversity and introduces genealogical biases due to non-random haplotype sampling. Mol Ecol. 2013;22:3179–90.

Aukema JE, Leung B, Kovacs K, Chivers C, Britton KO, Englin J, et al. Economic impacts of non-native forest insects in the continental United States. PLoS One. 2011;6:e24587.

Baird NA, Etter PD, Atwood TS, Currey MC, Shiver AL, Lewis ZA, et al. Rapid SNP discovery and genetic mapping using sequenced RAD markers. PLoS One. 2008;3:e3376.

Barzman M, Bàrberi P, Birch ANE, Boonekamp P, Dachbrodt-Saaydeh S, Graf B, et al. Eight principles of integrated pest management. Agron Sustain Dev. 2015;35:1199–215.

Bates OK, Ollier S, Bertelsmeier C. Smaller climatic niche shifts in invasive than non-invasive alien ant species. Nat Commun. 2020;11:5213.

Beissinger TM, Hirsch CN, Sekhon RS, Foerster JM, Johnson JM, Muttoni G, et al. Marker density and read depth for genotyping populations using genotyping-by-sequencing. Genetics. 2013;193:1073–81.

Bilodeau P, Roe AD, Bilodeau G, Blackburn GS, Cui M, Cusson M, et al. Biosurveillance of forest insects: part II – adoption of genomic tools by end user communities and barriers to integration. J Pest Sci. 2019;92:71–82.

Blackburn GS, Brunet BMT, Muirhead K, Cusson M, Béliveau C, Levesque RC, et al. Distinct sources of gene flow produce contrasting population genetic dynamics at different range boundaries of a *Choristoneura* budworm. Mol Ecol. 2017;26:6666–84.

Blumenfeld AJ, Eyer P-A, Husseneder C, Mo J, Johnson LNL, Wang C, et al. Bridgehead effect and multiple introductions shape the global invasion history of a termite. Commun Biol. 2021;4:196.

Bohling J. Evaluting the effect of reference genome divergence on the analysis of empirical RADseq datasets. Ecol Evol. 2020;10:7585–601.

Bowles E, Jeon H-B, Marin K, MacLeod P, Farser DJ. Freshwater fisheries monitoring in northern ecosystems using Indigenous ecological knowledge, genomics, and life history: insights for community decision-making. FACETS. 2022;7:1214–43.

Brunet BMT, Blackburn GS, Muirhead K, Lumley LM, Boyle B, Lévesque RC, et al. Two's company, three's a crowd: new insights on spruce budworm species boundaries using genotyping-by-sequencing in an integrative species assessment (Lepidoptera: Tortricidae). Syst Entomol. 2017;42:317–28.

Campbell EO, Davis CS, Dupuis JR, Muirhead K, Sperling FAH. Cross-platform compatibility of *de novo*-aligned SNPs in a nonmodel butterfly genus. Mol Ecol Resour. 2017;17:e84–93.

Campbell EO, Brunet BMT, Dupuis JR, Sperling FAH. Would an RRS by any other name sound as RAD? Methods Ecol Evol. 2018;9:1920–7.

Campbell EO, Dupuis JR, Holowachuk J, Hladun S, Vankosky MA, Mori BA. Disjunction between canola distribution and the genetic structure of its recently described pest, the canola flower midge (*Contarinia brassicola*). Ecol Evol. 2020;10:13284–96.

Cao LJ, Wei SJ, Hoffmann AA, Wen JB, Chen M. Rapid genetic structuring of populations of the invasive fall webworm in relation to spatial expansion and control campaigns. Divers Distrib. 2016;22:1276–87.

Catchen JM, Amores A, Hohenlohe P, Cresko W, Postlethwait J. Stacks: building and genotyping loci de novo from short-read sequences. G3 (Bethesda). 2011;1:171–82.

Cerca J, Maurstad MF, Rochette NC, Rivera-Colón AG, Rayamajhi N, Catchen JM, et al. Removing the bad apples: a simple bioinformatics method to improve loci-recovery in de nova RADseq data for non-model organisms. Methods Ecol Evol. 2021;12:805–17.

Chen Y, Bird L, Woolley L, Walsh T, Gordon K, Herron G. Linkage mapping in indoxacarb resistance locus in *Helicoverpa armigera* (Lepidoptera: Noctuidae) by genotype-by-sequencing. Pest Manag Sci. 2019;76:617–27.

Childers AK, Geib SM, Sim SB, Poelchau MF, Boates BS, Simmonds TJ, et al. The USDA-ARS Ag100Pest Initiative: high-quality genome assemblies for agricultural pest arthropod research. Insects. 2021;12:626.

Cordeiro EMG, Pantoja-Gomez LM, de Paiva JB, Nascimento ARB, Omoto C, Michel AP, et al. Hybridization and introgression between *Helicoverpa armigera* and *H. zea*: an adaptational bridge. BMC Evol Biol. 2020;20:61.

Cristescu ME. Genetic reconstructions of invasion history. Mol Ecol. 2015;24:2212–25.

Crossley MS, Chen YH, Groves RL, Schoville SD. Landscape genomics of Colorado potato beetle provides evidence of polygenic adaptation to insecticides. Mol Ecol. 2017;26:6284–300.

Crossley MS, Rondon SI, Schoville SD. Patterns of genetic differentiation in Colorado potato beetle correlate with contemporary, not historic, potato land cover. Evol Appl. 2019;12:804–14.

Cruaud A, Groussier G, Genson G, Sauné L, Polaszek A, Rasplus J-Y. Pushing the limits of whole genome amplification: successful sequencing of RADseq library from a single microhymenopteran (Chalcidoidea, *Trichogramma*). PeerJ. 2018;6:e5640.

Cuddington K, Sobek-Swant S, Crosthwaite JC, Lyons DB, Sinclair BJ. Probability of emerald ash borer impact for Canadian cities and North America: a mechanistic model. Biol Invasions. 2018;20:2661–77.

DaCosta JM, Sorenson MD. Amplification biases and consistent recovery of loci in a double-digest RAD-seq protocol. PLoS One. 2014;9:e106713.

Dai J-X, Cao L-J, Chen J-C, Yang F, Shen X-J, Ma L-J, Hoffmann AA, Chen M, Wei S-J. Testing for adaptive changes linked to range expansion following a single introduction of the fall webworm. Mol Ecol. 2024;33:e17038.

Dara SK. The new integrated pest management paradigm for the modern age. J Integr Pest Manag. 2019;10:12.

Davey JW, Blaxter ML. RADseq: next-generation population genetics. Brief Funct Genomics. 2011;9:416–23.

Davey JW, Hohenlohe PA, Etter PD, Boone JQ, Catchen JM, Blaxter ML. Genome-wide genetic marker discovery and genotyping using next-generation sequencing. Nat Rev Genet. 2011;12: 499–510.

Davey JW, Cezard T, Fuentes-Utrilla P, Eland C, Gharbi K, Blaxter ML. Special features of RAD sequencing data: implications for genotyping. Mol Ecol. 2013;22:3151–64.

Deguine J-P, Aubertot J-N, Flore RJ, Lescourret F, Wyckhuys KAG, Ratnadass A. Integrated pest management: good intentions, hard realities. A review. Agron Sustain Dev. 2021;41:38.

Denholm I, Devine GJ, Williamson MS. Insecticide resistance on the move. Science. 2002;297: 2222–3.

Díaz-Arce N, Rodríguez-Ezpeleta N. Selecting RAD-seq data analysis parameters for population genetics: the more the better? Front Genet. 2019;10:533.

Di Felice F, Micheli G, Camilloni G. Restriction enzymes and their use in molecular biology: an overview. J Biosci. 2019;44:38.

Doellman MM, Hood GR, Gersfeld J, Driscoe A, Xu CCY, Sheehy RN, et al. Identifying diagnostic genetic markers for a cryptic invasive agricultural pest: a test case using the apple maggot fly (Diptera: Tephritidae). Ann Entomol Soc Am. 2020;113:246–56.

Dorant Y, Cayuela H, Wellband K, Laporte M, Rougemont Q, Mérot C, et al. Copy number variants outperform SNPs to reveal genotype-temperature association in a marine species. Mol Ecol. 2020;29:4765–82.

Dupuis JR, Sim SB, San Jose M, Leblanc L, Hoassain MA, Rubinoff D, et al. Population genomics and comparisons of selective signatures in two invasions of melon fly, *Bactrocera cucurbitae* (Diptera: Tephritidae). Biol Invasions. 2018;20:1211–28.

Dupuis JR, Ruiz-Arce R, Barr NB, Thomas DB, Geib SM. Range-wide population genomics of the Mexican fruit fly: toward development of pathway analysis tools. Evol Appl. 2019;12:1641–60.

Eaton DAR, Overcast I. ipyrad: interactive assembly and analysis of RADseq datasets. Bioinformatics. 2020;36:2592–4.

Elshire RJ, Glaubitz JC, Poland JA, Kawamoto K, Buckler ES, Mitchell SE. A robust, simple genotyping-by-sequencing (GBS) approach for high diversity species. PLoS One. 2011;6: e19379.

Erlandson MA, Mori BA, Coutu C, Holowachuk J, Olfert OO, Gariepy TD, et al. Examining population structure of a bertha armyworm, *Mamestra configurata*, (Lepidoptera: Noctuidae), outbreak in western North America: implications for gene flow and dispersal. PLoS One. 2019;14:e0218993.

Estoup A, Guillemaud T. Reconstructing routes of invasion using genetic data: why, how and so what? Mol Ecol. 2010;19:4113–30.

Fox EJ, Reid-Bayliss KS, Emond MJ, Loeb LA. Accuracy of next generation sequencing platforms. Next Gener Seq Appl. 2014;1:1000106. https://doi.org/10.4172/jngsa.1000106.

Gagic V, Holding M, Venables WN, Hulthen AD, Schellhorn NA. Better outcomes for pest pressure, insecticide use, and yield in less intensive agricultural landscapes. PNAS. 2021;118: e2018100118.

Gamazon ER, Stranger BE. The impact of copy number variation on gene expression. Brief Funct Genomics. 2015;14:352–7.

Garnas JR, Auger-Rozenberg M-A, Roques A, Bertelsmeier C, Wingfield MJ, Saccaggi DL, et al. Complex patterns of global spread in invasive insects: eco-evolutionary and management consequences. Biol Invasions. 2016;18:935–52.

Gautier M, Gharbi K, Cezard T, Foucaud J, Kerdelhué C, Pudlo P, et al. The effect of RAD allele dropout on the estimation of genetic variation within and between populations. Mol Ecol. 2013;22:3165–78.

Graham CF, Glenn TC, McArthur AG, Boreham DR, Kieran T, Lance S, et al. Impacts of degraded DNA on restriction enzyme associated DNA sequencing (RADseq). Mol Ecol Resour. 2015;15: 1304–15.

Graham CF, Boreham DR, Manzon RG, Stott W, Wilson JY, Somers CM. How "simple" methodological decisions affect interpretation of population structure based on reduced representation library DNA sequencing: a case study using lake whitefish. PLoS One. 2020;15:e0226608.

Hall RJ. Hybridization helps colonizers become conquerors. Proc Natl Acad Sci USA. 2016;113: 9963–4.

Hamelin RC, Roe AD. Genomic biosurveillance of forest invasive alien enemies: a story written in code. Evol Appl. 2020;13:95–115.

Hanrahan SJ, Johnston JS. New genome size estimates of 134 species of arthropods. Chromosom Res. 2011;19:809.

Harismendy O, Ng PC, Strausberg RL, Wang X, Stockwell TB, Beeson KY, et al. Evaluation of next generation sequencing platforms for population targeted sequencing studies. Genome Biol. 2009;10:R32.

He K, Lin K, Wang G, Li F. Genome sizes of nine insect species determined by flow cytometry and *k*-mer analysis. Front Physiol. 2016;7:569.

Herrera S, Reyes-Herrera PH, Shank TM. Predicting RAD-seq marker numbers across the eukaryotic tree of life. Genome Biol Evol. 2015;7:3207–25.

Hohenlohe PA, Bassham S, Etter PD, Stiffler N, Johnson EA, Cresko WA. Population genomics of parallel adaptation in threespine stickleback using sequenced RAD tags. PLoS Genet. 2010a;6: e1000862.

Hohenlohe PA, Phillips PC, Cresko WA. Using population genomics to detect selection in natural populations: key concepts and methodological considerations. Int J Plant Sci. 2010b;71:1059–71.

Hu T, Chitnis N, Monos D, Dinh A. Next-generation sequencing technologies: an overview. Hum Immunol. 2021;82:801–11.

Hulme PE. Trade, transport and trouble: managing invasive species pathways in an era of globalization. J Appl Ecol. 2009;46:10–8.

Hulme PE. An introduction to plant biosecurity: past, present and future. In: Gordh G, McKirdy S, editors. The handbook of plant biosecurity. Dordrecht: Springer International Publishing AG, part of Springer Nature; 2014. p. 1–25.

Hulme PE. Unwelcome exchange: international trade as a direct and indirect driver of biological invasions worldwide. One Earth. 2021;4:666–79.

Jactel H, Koricheva J, Castagneyrol B. Responses of forest pests to climate change: not to simple. Curr Opin Insect Sci. 2019;35:103–8.

Jiang Z, Wang H, Michal JJ, Zhou X, Liu B, Solberg Woods LC, et al. Genome wise sampling sequencing for SNP genotyping: methods, challenges and future developments. Int J Biol Sci. 2016;12:100–8.

Jordan A, Patch HM, Grozinger CM, Khanna V. Economic dependence and vulnerability of United States agricultural sector on insect-mediate pollination services. Environ Sci Technol. 2021;55: 2243–53.

Kogan M. Integrated pest management: historical perspectives and contemporary developments. Annu Rev Entomol. 1998;43:243–70.

Kurland S, Wheat CW, de la Paz Celorio Mancera M, Kutschera VE, Hill J, Andersson A, et al. Exploring a pool-seq only approach for gaining population genomic insights in nonmodel species. Ecol Evol. 2019;9:11448–63.

Larroque J, Legault S, Johns R, Lumley L, Cusson M, Renaut S, et al. Temporal variation in spatial genetic structure during population outbreaks: distinguishing among different potential drivers of spatial synchrony. Evol Appl. 2019;12:1931–45.

Lee KM, Kivelä SM, Ivanov V, Hausmann A, Kaila L, Wahlberg N, et al. Information dropout patterns in restriction site associated DNA phylogenomics and comparison with multilocus Sanger data in a species-rich moth genus. Syst Biol. 2018;67:925–39.

Lehmann P, Ammunét T, Barton M, Battisti A, Eigenbrode SD, Jepsen JU, et al. Complex responses of global insect pests to climate warming. Front Ecol Environ. 2020;18:141–50.

Lepais O, Weir JT. SimRAD: an R package for simulation-based prediction of the number of loci expected in RADseq and similar genotyping by sequencing approaches. Mol Ecol Resour. 2014;14:1314–21.

Li Y, Sidore C, Kang HM, Boehnke M, Abecasis GR. Low-coverage sequencing: implications for design of complex trait association studies. Genome Res. 2011;21(6):940–51.

Losey JE, Vaughn M. The economic value of ecological services provided by insects. BioScience. 2006;56:311–23.

Lou R, Jacobs A, Wilder AP, Therkildesn NO. A beginner's guide to low-coverage whole genome sequencing for population genomics. Mol Ecol. 2020;30:5966–93.

Lumley LM, Pouliot E, Laroche J, Boyle B, Brunet BMT, Levesque RC, et al. Continent-wide population genomic structure and phylogeography of North America's most destructive conifer defoliator, the spruce budworm (*Choristoneura fumiferana*). Ecol Evol. 2020;10:914–27.

Matheson P, McGaughran A. Genomic data is missing for many highly invasive species, restricting our preparedness for escalating incursion rates. Sci Rep. 2022;12:13987.

McCartney MA, Mallez S, Gohl DM. Genome projects in invasion biology. Conserv Genet. 2019;20:1201–22.

McCartney-Melstad E, Gidiş M, Shaffer HB. An empirical pipeline for choosing the optimal clustering threshold in RADseq studies. Mol Ecol Resour. 2019;19:1195–204.

McGaughran A, Dhami MK, Parvizi E, Vaughan AL, Gleeson DM, Hodgins KA, et al. Genomic tools in biological invasions: current state and future frontiers. Mol Biol Evol. 2024;16:evad230.

McKinney GJ, Seeb JE, Seeb LW. Managing mixed-stock fisheries: genotyping multi-SNP haplotypes increases power for genetic stock identification. Can J Fish Aquat Sci. 2017a;74:429–34.

McKinney GJ, Waples RK, Seeb LW, Seeb JE. Paralogs are revealed by proportion of heterozygotes and deviations in read ratios in genotyping-by-sequencing data from natural populations. Mol Ecol. 2017b;17:656–69.

Mérot C, Oomen RA, Tigano A, Wellenreuther M. A roadmap for understanding the evolutionary significance of structural genomic variation. Trends Ecol Evol. 2020;35:561–72.

Metzker ML. Sequencing technologies – the next generation. Nat Rev Genet. 2010;11:31–46.

Miller MR, Dunham JP, Amores A, Cresko WA, Johnston EA. Rapid and cost-effective polymorphism identification and genotyping using restriction site associated DNA (RAD) markers. Genome Res. 2007;17:240–8.

Mir K, Neuhaus K, Bossert M, Schober S. Short barcodes for next generation sequencing. PLoS One. 2013;8:e82933.

Mora-Márquez F, García-Olivares V, Emerson BC, López de Heredia U. DDRADSEQTOOLS: a software package for in silico simulation and testing of double-digest RADseq experiments. Mol Ecol Resour. 2016;17:230–46.

Mori BA, Andreassen L, Heal JD, Dupuis JR, Soroka JJ, Sinclair BJ. A new species of *Contarinia* Rondani (Diptera: Cecidomyiidae) that induces flower galls on canola (*Brassicaceae*) in the Canadian prairies. Can Entomol. 2019;151:131–48.

Nazareno AG, Knowles LL. There is no "rule of thumb": genomic filter settings for a small plant population to obtain unbiased gene flow estimate. Front Plant Sci. 2021;12:677009.

Nelson TD, MacDonald ZG, Sperling FAH. Moths passing in the night: phenological and genomic divergences within a forest pest complex. Evol Appl. 2021;15:166–80.

Nielsen R, Paul JS, Albrechtsen A, Song YS. Genotype and SNP calling from next-generation sequencing data. Nat Rev Genet. 2011;12:443–51.

Paini DR, Sheppard AW, Cook DC, De Barro PJ, Worner SP, Thomas MB. Global threat to agriculture from invasive species. PNAS. 2016;113:7575–9.

Paris JR, Stevens JR, Catchen JM. Lost in parameter space: a road map for stacks. Methods Ecol Evol. 2017;8:1360–73.

Parsa S, Morse S, Bonifacio A, Chancellor TCB, Condori B, Crespo-Pérez V, et al. Obstacles to integrated pest management adoption in developing countries. Proc Natl Acad Sci. 2014;111: 3889–94.

Peterson BK, Weber JN, Kay EH, Fisher HS, Hoekstra HE. Double digest RADseq: an inexpensive method for de novo SNP discovery and genotyping in mode and non-model species. PLoS One. 2012;7:e37135.

Pfeiffer F, Gröber C, Blannk M, Händler K, Beyer M, Schultze JL, et al. Systematic evaluation of error rates and causes in short samples in next-generation sequencing. Sci Rep. 2018;8:10950.

Picq S, Keena M, Havill N, Stewart D, Pouliot E, Boyle B, et al. Assessing the potential of genotyping-by-sequencing-derived single nucleotide polymorphisms to identify the geographic origins of intercepted gypsy moth (*Lymantria dispar*) specimens: a proof-of-concept study. Evol Appl. 2018;11:325–39.

Picq S, Wu Y, Martemyanov VV, Pouliot E, Pfister SE, Hamelin R, et al. Range-wide population genomics of the spongy moth, *Lymantria dispar* (Erebidae): implications for biosurveillance, subspecies classification and phylogeography of a destructive moth. Evol Appl. 2023;16:638–56.

Podnar J, Deiderick H, Huerta G, Hunicke-Smith S. Next-generation sequencing RNA-seq library construction. Curr Protoc Mol Biol. 2014;106:4–21.

Poland GA, Brown PJ, Sorrells ME, Jannink J. Development of high-density genetic maps for barley and wheat using a novel two-enzyme genotyping-by-sequencing approach. PLoS One. 2012;7:e32253.

Qu W-M, Liang N, Wu Z-K, Zhao Y-G, Chu D. Minimum sample sizes for invasion genomics: empirical investigation in an invasive whitefly. Ecol Evol. 2019;10:38–49.

Raszick TJ, Dickens CM, Perkin LC, Tessnow AE, Suh CP-H, Ruiz-Arce R, et al. Population genomics and phylogeography of the boll weevil, *Anthonomus grandis* Boheman (Coleoptera: Curculionidae), in the United States, northern Mexico, and Argentina. Evol Appl. 2021;14: 1778–93.

Rayne A, Blair S, Dale M, Flack B, Jollows J, Moraga R, et al. Weaving place-based knowledge for culturally significant species in the age of genomics: looking to the past to navigate the future. Evol Appl. 2022;15:751–72.

Rivera-Colón AG, Rochette NC, Catchen JM. Simulation with RADinitio improves RADseq experimental design and sheds light on sources of missing data. Mol Ecol Resour. 2021;21: 363–78.

Rochette N, Rivera-Colón A, Catchen J. Stacks 2: analytical methods for paired-end sequencing improve RADseq-based population genomics. Mol Ecol. 2019;28:4737–54.

Roe AD, Torson AS, Bilodeau G, Bilodeau P, Blackburn GS, Cui MM, et al. Biosurveillance of forest insects: part I – integration and application of genomic tools to the surveillance of non-native forest pests. J Pest Sci. 2018;92:51–70.

Root RB. Organization of a plant-arthropod association in simple and diverse habitats: the fauna of collards (*Brassica oleracea*). Ecol Monogr. 1973;43:95–124.

Rowan BA, Heavens D, Feuerborn TR, Tock AJ, Henderson IR, Weigel D. An ultra high-density Arabidopsis thaliana cross-over map that refines the influences of structural variation and epigenetic features. Genetics. 2019;213:771–87.

Ryan SF, Lombaert E, Espeset A, Vila R, Talavera G, Dincă V, et al. Global invasion history of the agricultural pest butterfly *Pieris rapae* revealed with genomics and citizen science. Proc Natl Acad Sci. 2019;116:20015–24.

Schmid S, Genevest R, Gobet E, Suchan T, Sperisen C, Tinner W, et al. HyRAD-X, a versatile method combining exome capture and RAD sequencing to extract genomic information from ancient DNA. Methods Ecol Evol. 2017;8:1374–88.

Schrieber K, Lachmuth S. The genetic paradox of invasions revisited: the potential role of inbreeding x environment interactions in invasion success. Biol Rev. 2016;92:939–52.

Shafer AB, Peart CR, Tusso S, Maayan I, Brelsford A, Wheat CW, et al. Bioinformatic processing of RAD-seq data dramatically impacts downstream population genetic inference. Methods Ecol Evol. 2017;8:907–17.

Shashank PR, Parker BM, Rananaware SR, Plotkin D, Couch C, Yang LG, et al. CRISPR-based diagnostics detects invasive insect pests. Mol Ecol Resour. 2023;24:e13881.

Sims D, Sudbery I, Ilott NE, Heger A, Ponting CP. Sequencing depth and coverage: key considerations in genomic analyses. Nat Rev Genet. 2014;15:121–32.

Skendžić S, Zovko M, Živković IP, Lešić V, Lemić D. Effect of climate change of introduced and native agricultural invasive insect pests in Europe. Insects. 2021;12:985.

Slatko BE, Gardner AF, Ausubel FM. Overview of next-generation sequencing technologies. Curr Protoc Mol Biol. 2018;122:e59.

Sonah H, Bastien M, Iquira E, Tardivel A, Légaré B, Boyle B, et al. An improved genotyping-by-sequencing (GBS) approach offering increased versatility and efficiency of SNP discovery and genotyping. PLoS One. 2013;8:e54603.

Stenberg J. A conceptual framework for integrated pest management. Trends Plant Sci. 2017;22: 759–69.

Stern VM, Smith RF, van den Bosch R, Hagen KS. The integration of chemical and biological control of the posted alfalfa aphid: the integrated control concept. Hilgardia. 1959;29:81–101.

Stewart D, Zahiri R, Djoumad A, Freschi L, Lamarche J, Holden D, et al. A multi-species TaqMan PCR assay for the identification of Asian gypsy moths (*Lymantria* spp.) and other invasive lymantriines of biosecurity concern to North America. PLoS One. 2016;11:e0160878.

Stewart D, Nisole A, Djoumad A, Zahiri R, Lamarche J, Levesque RC, et al. A needle in a haystack: a multigene TaqMan assay for the detection of Asian gypsy moths in bulk pheromone trap samples. Biol Invasions. 2019;21:1843–56.

Stoler N, Nekrutenko A. Sequencing error profiles of Illumina sequencing instruments. NAR Genom Bioinform. 2021;3:lqab019. https://doi.org/10.1093/nargab/lqab019.

Suchan T, Pitteloud C, Gerasimova NS, Kostikova A, Schmid S, Arrigo N, et al. Hybridization capture using RAD probes (HyRAD), a new tool for performing genomic analysis on collection specimens. PLoS One. 2016;11:e0151651.

Sunde J, Yıldırım Y, Tibbline P, Forsman A. Comparing the performance of microsatellites and RADseq in population genetic studies: analysis of data for pike (*Esox lucius*) and a synthesis of previous studies. Front Genet. 2020;11:218.

Tay WT, Gordon KHJ. Going global – genomic insights into insect invasions. Curr Opin Insect Sci. 2019;31:123–30.

Taylor KL, Hamby KA, DeYonke AM, Fritz ML. Genome evolution in an agricultural pest following adoption of transgenic crops. Proc Natl Acad Sci. 2021;118:e2020853118.

Tigano A, Friesen VL. Genomics of local adaptation with gene flow. Mol Ecol. 2016;25:2144–64.

Torkamaneh D, Laroche J, Belzile F. Genome-wide SNP calling from genotyping by sequencing (GBS) data: a comparison of seven pipelines and two sequencing technologies. PLoS One. 2016;11:e0161333.

Torkamaneh D, Laroche J, Boyle B, Belzile F. DepthFinder: a tool to determine the optimal read depth for reduced-representation sequencing. Bioinformatics. 2019;36:26–32.

Valentin RE, Fonseca DM, Gable S, Kyle KE, Hamilton GC, Nielsen AL, et al. Moving eDNA surveys onto land: strategies for active eDNA aggregation to detect invasive forest insects. Mol Ecol Resour. 2020;20:746–55.

Venette RC, Hutchison WD. Invasive insect species: global challenges, strategies, and opportunities. Front Insect Sci. 2021;1:650520.

Wang VX, Blades N, Ding J, Sultana R, Parmigiani G. Estimation of sequencing error rates in short reads. BMC Bioinformatics. 2012;13:185.

Williams RJ. Restriction endonucleases: classification, properties, and applications. Mol Biotechnol. 2003;23:225–43.

Yadav S, Stow AJ, Dudaniec RY. Detection of environmental and morphological adaptation despite high landscape genetic connectivity in a pest grasshopper (*Phaulacridium vittatum*). Mol Ecol. 2019;28:3395–412.

Yan J, Vétek G, Pal C, Zhang J, Gmati R, Fan Q-H, et al. ddRAD sequencing: an emerging technology added to the biosecurity toolbox for tracing the origin of brown marmorated stink bug, *Halyomorpha halys* (Hemiptera: Pentatomidae). BMC Genomics. 2021;22:355.

Yeaman S, Whitlock MC. The genetic architecture of adaptation under migration-selection balance. Evolution. 2011;65:1897–911.

Young RG, Milián-Garcia Y, Yu J, Bullas-Appleton E, Hanner RH. Biosurveillance for invasive insect pest species using an environmental DNA metabarcoding approach and high salt trap collection fluid. Ecol Evol. 2021;11:1558–69.

Detecting Natural Selection Against Resistance of a Flea Beetle to Host Plant Defences Using a Population Genomics Approach

Kim M. C. A. Vermeer, Patrick Verbaarschot, Marcel Dicke, and Peter W. De Jong

Abstract Background: The flea beetle *Phyllotreta nemorum* is polymorphic for resistance to the defence of the atypical host plant *Barbarea vulgaris* G-type. We investigated the factors influencing the geographical distribution of this resistance in a structured population in Denmark, where the proportion of flea beetles resistant to the defences of *B. vulgaris* is remarkably low for individuals collected on other host plants than *B. vulgaris*. We tested the hypothesis that the resistance trait is selected against when beetles live on non-defended host plants that are geographically separated from a patch of defended host plants. New primers were developed for 11 polymorphic microsatellite loci of *Phyllotreta nemorum* which, in addition to 10 previously developed microsatellite markers, were used to assess the genetic population structure nearby Kværkeby.

Results: Outlier detection methods were applied to different measures of pairwise genetic differentiation between samples at neutral behaving microsatellite loci as well as the resistance trait and yielded the resistance trait as a candidate subject to selection. This also applied to the comparison between flea beetle subpopulations on non-defended host plants at different geographical distances from a patch of *B. vulgaris* G-type plants.

Conclusions: These results suggest that the different frequencies of flea beetle resistance on patches of non-defended host plants cannot be explained by mere dispersal limitations, but that a locus-specific effect, such as directional selection, must also play a role. The possible mechanism of such selection is discussed, along with the comparison of different genetic differentiation measures and mutation models in population genomic studies.

K. M. C. A. Vermeer · P. Verbaarschot · M. Dicke · P. W. De Jong (✉)
Laboratory of Entomology, Wageningen University, Wageningen, The Netherlands
e-mail: Kim.Vermeer@wur.nl; Patrick.verbaarschot@wur.nl; Marcel.dicke@wur.nl; Peter.dejong@wur.nl

Julian R. Dupuis and Om P. Rajora (eds.), *Population Genomics: Insects*, Population Genomics [Om P. Rajora (Editor-in-Chief)],
https://doi.org/10.1007/13836_2025_111,

Keywords *Brassicaceae* · *Chrysomelidae* · F-statistics · Local adaptation · *Phyllotreta nemorum* · Population genetics

1 Background

Coevolution is one of the key processes influencing and generating biodiversity (Thompson 1999, 2005, 2009; Thompson and Cunningham 2002; Wheat et al. 2007; Laine 2009; Forde et al. 2004; Hu et al. 2008; Sagan 1966; Cornell and Hawkins 2003). It involves the reciprocal evolutionary interaction between species. Coevolution seems more complex than formerly thought; reciprocal interactions are not ubiquitous throughout populations as previously depicted, but populations are hierarchically structured (Thompson 2005; Wright 1938, 1951, 1968; Agrawal et al. 2001; Avise 2004; Fox 1981). Coevolution sensu stricto only occurs in some subpopulations within overlapping distributions of the interacting species. Moreover, the mode, direction and strength of selection is now thought to vary across geographical regions. Selection on traits that influence reciprocal species interactions may, for example, depend on the presence or absence of a third species (Benkman et al. 2001). In 2005, Thompson published his conceptual framework of the geographic mosaic of coevolution (Thompson 2005). This framework integrates multidisciplinary approaches to enable the study of the dynamics and structure of coevolving interactions. Local adaptation is a crucial part of the raw material for coevolution as a geographical mosaic (Thompson 2005); the basic template for the geographic mosaic of coevolution is created by local (co)evolutionary adaptation. Excellent examples of local adaptation have been found in the study of natural resistance of insect herbivores to plant defences (Cornell and Hawkins 2003; Ehrlich and Raven 1964; Strong et al. 1984; Pellmyr 2003; Mopper and Strauss 1998).

One of the challenges emerging from Thompson's conceptual framework is to explain the observed geographical distribution of alleles at adaptive loci in terms of (local) selection versus processes that affect the whole genome such as migration and genetic drift (Vermeer et al. 2011). In Thompson's geographic mosaic theory of coevolution (Thompson 2005), the latter (genome-wide processes) alter the spatial distribution of traits, leading to one of the major attributes of the geographic mosaic of coevolution: 'trait remixing'. Trait remixing is predicted to influence the occurrence of traits amongst populations, and only in some populations will traits of interacting species be well matched. As a result, reciprocal selection only occurs in so-called 'coevolutionary hotspots' within the spatial distribution of an interaction, which are embedded in a broader network of 'coldspots' where selection can occur but is not reciprocal. In his theoretical framework, Thompson presents specific evolutionary hypotheses and ecological predictions which make the theory amenable to empirical testing. Although suggestions for such tests have been proposed (Gomulkiewicz et al. 2007), empirical tests of the predictions of the geographic mosaic theory are still extremely rare. We recently proposed to employ a population genomics approach to attribute the distribution of alleles at adaptive loci to local selection or trait remixing, thus enabling the detection of coevolutionary hotspots (Vermeer et al. 2011, see Box 1). In the present chapter, we apply this approach to an empirical study of a model system involving (co)evolutionary interactions between a herbivorous insect, *Phyllotreta nemorum* L. (Coleoptera: Chrysomelidae), and one of its host plants, *Barbarea vulgaris* R. Br. (Brassicales: Brassicaceae) (Nielsen 1997a; De Jong et al. 2000; Nielsen and de Jong 2005).

Box 1 The use of a population genomics approach to answer ecological questions

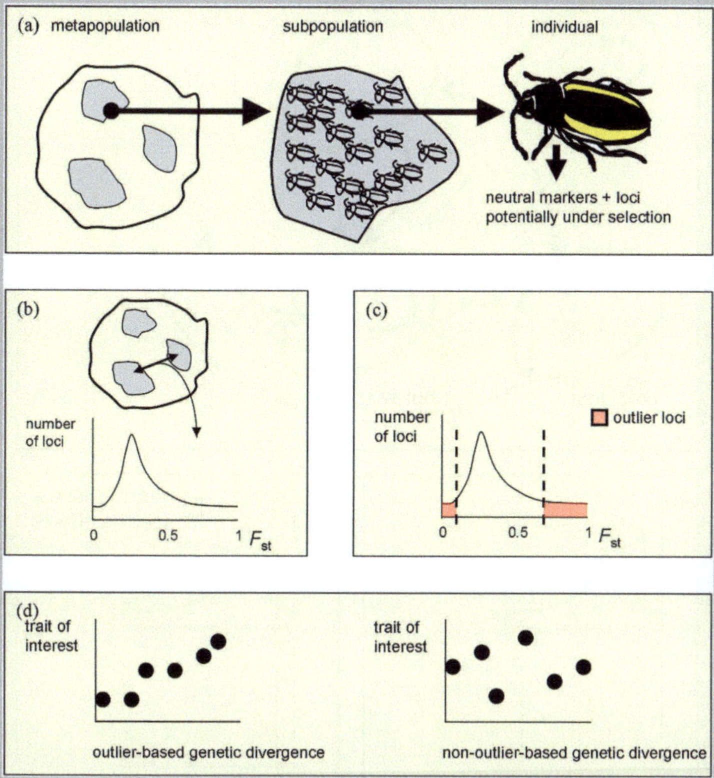

A. The population genomics approach consists of four steps. (a) Sampling numerous (neutral) molecular markers, such as microsatellites, as well as testing loci that are suspected to be under selection in individuals of different tentative subpopulations. (b) A frequency distribution is drawn

(continued)

Box 1 (continued)

of a measure of the variation between different samples, in this example the Fst value for each locus. (c) Screen the frequency distribution for statistical outliers. (d) Validation step to verify whether the outlier is caused by selection (or by hitch-hiking with/on a genome region that is under selection). In this example, non-outlier- and outlier-based genetic distances are correlated with differences in the ecological trait of interest. Reproduced with permission from Vermeer et al. (2011).

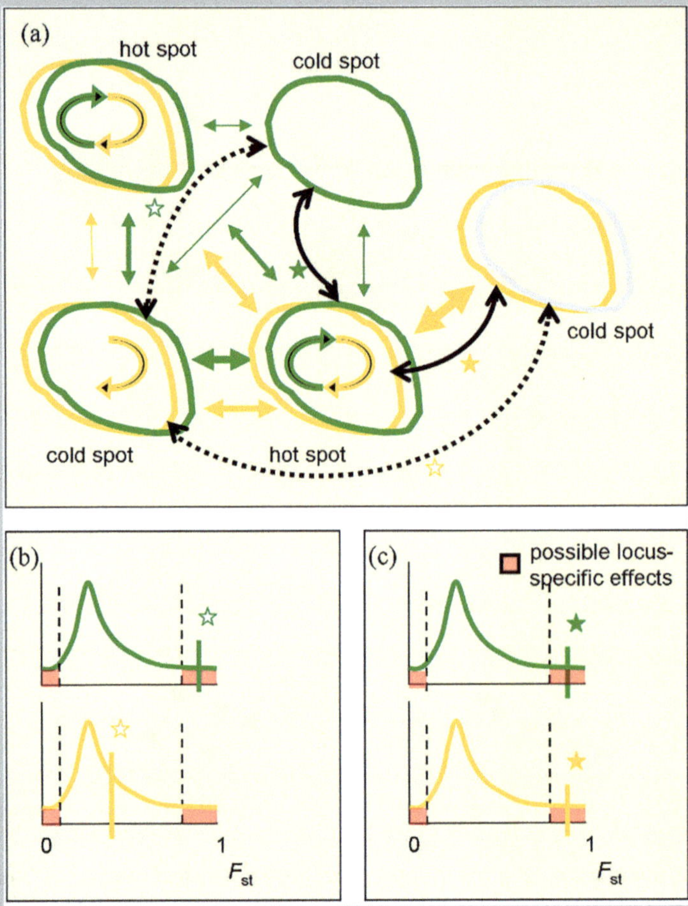

(continued)

Box 1 (continued)

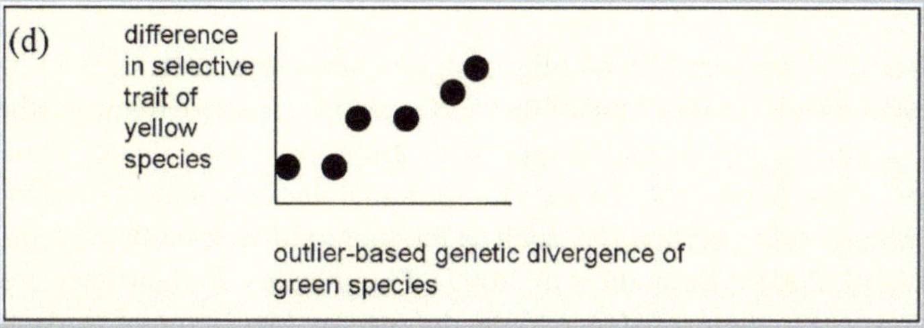

B. Implementation of a population genomics approach to analyse the GMC: using population genomics to detect selection and to identify hot spots and cold spots. (a) A hypothetical example has been drawn for a set of local populations of two interacting species (green and yellow patches, respectively). Green and yellow arrows between patches represent dispersal of the respective species, with thicker arrows indicating more dispersal between the local populations. Curved arrows within patches show selective interactions within local populations: a green arrow shows selection from the green species on the yellow species, and a yellow arrow vice versa. The occurrence of both a yellow and a green arrow within one patchindicates reciprocal selection, so that such a patch becomes a coevolutionary hot spot. In some patches, selection only acts from one species upon the other, and in other patches, no selection occurs between the two species of interest, e.g. when one of them is not present in that local patch. This is, for example, the case in the right-hand patch, where the yellow species is interacting with a third species (grey), and the green species is absent. (b, c) Examples of detection of selection within two local populations (the ones at the bottom of part a of the figure) by applying an Fst-based population genomics approach, comparing each of the two populations with local patches where only one of the two species is present (and hence no selection is exerted between them). These comparisons are indicated with curved, black (dashed and solid, respectively) double-headed arrows in part a, accompanied by green and yellow open, and closed asterisks, respectively, for the two different pairwise comparisons. The vertical bars in b and c show the Fst value for candidate loci, suspected to be under selection. In b, the Fst for the candidate locus for the green species is an outlier, but for the yellow species, it is not. This shows that the yellow species exerts selection on the green one, but not voce versa, and the local community is thus a coevolutionary cold spot. In c, the Fst-values for the candidate loci for both interacting species are outliers, showing that reciprocal selection may be occurring and thus indicating a coevolutionary hot spot. (d) A validation involves the correlation of outlier Fst values of one species with the hypothesized selective pressure imposed by the other species (the latter needs to be determined empirically). Reproduced with permission from Vermeer et al. (2011).

1.1 Barbarea vulgaris *G-type*—Phyllotreta nemorum *System*

The flea beetle *P. nemorum* is an oligophagous herbivore, attacking a limited range of brassicaceous host plants. One of these, *B. vulgaris*, has two distinguished types in Denmark: a chemically defended type with **g**labrous leaves (hence called G-type) and a type with **p**ubescent leaves (P-type) that lacks chemical defence against *P. nemorum* and other herbivores, such as the diamondback moth *Plutella xylostella* (Agerbirk et al. 2003a; Shinoda et al. 2002). Flea beetles, in their turn, are polymorphic with respect to their resistance to the defence of the G-type of *B. vulgaris*: some show virulence (or resistance) towards its chemical defence and some are susceptible to it. Crossing experiments demonstrated that major dominant resistance genes (R-genes) enable the flea beetles to overcome the plant defence (De Jong et al. 2000; Nielsen and de Jong 2005; Breuker et al. 2005). Different modes of inheritance of these genes (sex-linked and autosomal) present a puzzle with respect to the number of loci that may be responsible for the resistance of the flea beetles (Nielsen 1997b, 1999, 2012). However, it has been demonstrated that one dominant resistant allele at any of these putative loci is sufficient to enable the beetle to use *B. vulgaris* G-type as a host plant. Near Kværkeby, Denmark, the genetic basis of the resistance trait appears to be straightforward because a previous study showed that one dominant allele at a single major autosomal locus is sufficient to provide a beetle with resistance to the plant defence (De Jong et al. 2000).

The geographical distribution of resistance in the flea beetle is remarkable (De Jong and Nielsen 1999). Whereas virtually all flea beetles that are collected on *B. vulgaris* G-type are apparently homozygous resistant, the proportion of flea beetles using other host plants than this G-type that possess the resistance gene (s) is surprisingly low, and homozygous resistant individuals appear to be absent on these plants. This geographic distribution of adaptive genotypes may be explained in two different ways: (1) the flea beetle populations are highly genetically structured, and local selection favours resistance of the flea beetles on *B. vulgaris* G-type. Limited dispersal, a high frequency of local extinction and recolonization, and host fidelity may all contribute to a limited spread of flea beetle resistance to other host plants (Peterson and Denno 1998); (2) there is a cost associated with the resistance trait, so that selection acts against this specific genotype on other host plants than *B. vulgaris* G-type. This would also limit the spread of resistance, even when dispersal is not limited. Previous data show evidence that both of these mechanisms may operate in the field: allozyme analysis has shown a limited, though statistically significant, genetic differentiation between flea beetles collected on different plant patches, which was both influenced by the geographical distance between the collection sites, and the resistance phenotype of the beetles (De Jong et al. 2001, 2009). Under certain conditions, resistant flea beetles appeared to have a lower fitness than susceptible ones through a high pre-adult mortality, both of homozygous and heterozygous resistant individuals, especially after outcrossing with a susceptible line (De Jong and Nielsen 2000; Breuker et al. 2007). This effect has only been found for laboratory lines into which the autosomal R-allele was crossed. Sex-linked R-genes do not seem to have any negative fitness consequences on other host plants than *B. vulgaris* G-type (Nielsen 1996; Agerbirk et al. 2003b). In earlier work, it was proposed that the abundance of apparently homozygous resistant flea beetles on *B. vulgaris* G-type plants in the field, despite their relatively low fitness in the experiments mentioned above, may be explained by the presence of co-adapted gene complexes, consisting of the resistance gene and modifiers that counteract the negative pleiotropic effects (i.e. high mortality) of these R-genes (De Jong et al. 2000; Nielsen and de Jong 2005; De Jong and Nielsen 2002). Such co-adapted gene complexes are thought to break up after migrants from *B. vulgaris* outbreed with beetles living on other host plants, thus limiting the spread of resistance of the beetles (especially in genetically structured populations). At the same time, this mechanism would account for the relatively low levels of genetic differentiation that have been found between flea beetle populations, because the selection against resistance on other plants than *B. vulgaris* would take place only after outbreeding of resistant migrants with the populations on the non-*Barbarea* host plants, thus allowing for a large amount of genetic exchange.

The present study aims to explain the geographic distribution of resistance in natural flea beetle populations in terms of migration/dispersal versus selection. Based on our earlier results described above, we hypothesize that the observed geographical distribution of the resistance trait cannot solely be explained by limited dispersal, but also involves selection against resistant individuals on other host plants than *B. vulgaris* G-type. To test this hypothesis under natural conditions, it is necessary to employ a method to disentangle the roles of dispersal/migration and selection, i.e. a population genomics approach.

Flea beetle resistance is favoured by selection on the *B. vulgaris* G-type, as a logical result of only resistant beetles being able to survive on this plant. Our research interest, however, is to explain the relative rarity of the resistant flea beetles on other, non-defended host plants. The essential research question is therefore: Is there evidence for selection against resistance in beetles that use other host plants than *B. vulgaris* G-type? The specific hypothesis is: *If coadapted gene complexes are indeed present in resistant flea beetles and break up through recombination during outbreeding with other local flea beetle populations, then we expect that genetic differentiation between two flea beetle populations living on other host plants than the G-type, at different geographical distances from a G-type patch, is larger for the resistance trait than the expectation based on neutral markers.*

1.2 General Approach

Most of the earlier work on the flea beetle-host plant interactions in Denmark has focused on populations around Kværkeby (De Jong et al. 2000, 2001; Nielsen and de Jong 2005; Breuker et al. 2005). Even though allozyme studies have revealed limited, but significant, population structure for flea beetles using distinct plant patches around Kværkeby—even at a scale of ≈ 1 km between patches (De Jong et al. 2001)—we tested for the sampled locations if they harbour genetically differentiated subpopulations. Ten microsatellite markers had already been developed for the flea beetle and we chose to use these highly variable microsatellite loci and a further 11 newly discovered microsatellite loci for our analyses. Microsatellite polymorphism of these new loci was assessed per locus using 192 up to 288 individuals of *P. nemorum* collected nearby Kværkeby from different locations. Genotyping errors, such as stuttering and null alleles in microsatellites, can lead to apparent departure from the Hardy–Weinberg equilibrium (HWE). Therefore, we first checked if the microsatellite loci were in HWE across subpopulations. We then tested if the subpopulations were each in HWE and used STRUCTURE (Pritchard et al. 2000), a software program to cluster genetic data to check for population structure. The basic algorithm was described by Pritchard et al. (2000). Extensions to the method were published by Falush et al. (2003, 2007) and Hubisz et al. (2009). In addition, the differentiation between subpopulations was visualized by factorial correspondence analysis (FCA) using GENETIX version 4.03 (Belkhir et al. 1998) and principal coordinate analysis (PCA) using GENALEX version 6.13 (Peakall and Smouse 2006).

A population genomics approach was used to measure the variation in heterozygosity and allele frequencies within and between subpopulations. As in the classic population genomics approach (Black et al. 2001; Luikart et al. 2003), numerous individuals were sampled from each subpopulation and various loci were genotyped across the genome. When a locus shows a significantly deviating level of differentiation between the subpopulations compared to the majority of the loci (i.e. if the value of the measurement of differentiation at this locus forms an outlier), then this locus is a candidate locus for being under selection. Our specific aim was to test whether flea beetle resistance is likely to be under selection in subpopulations on other host plants than *B. vulgaris* G-type. Because the genetic basis of resistance has not been resolved at the DNA level, we estimated genotype frequencies at the resistance locus using the phenotype frequencies, assuming that the resistance is governed by one resistance gene in the Kværkeby subpopulations (as supported by data from de Jong et al. (2000)) and that the subpopulations are in HWE at this locus. These estimates were then used to calculate differentiation at this locus between subpopulations, which was subsequently compared to differentiation at the neutral microsatellite loci. Different measures for genetic differentiation were used, because different underlying models lack consensus about which one of these is correct to use for microsatellites (Jost 2008; Goldstein et al. 1995; Slatkin 1995; Leng and Zhang 2011; Takezaki and Nei 1996; Balloux and Lugon-Moulin 2002). By using F_{ST} (estimator θ as in Weir and Cockerham (1984)) and R_{ST} (Slatkin 1995), both the infinite allele model of mutation (IAM) and the stepwise-mutation model (SMM) have been considered. We used the program LOSITAN for the frequentist (fdist) (Antao et al. 2008; Beaumont and Nichols 1996) and ran it under the assumption of the IAM as well as the SMM. Moreover, we also used the program BayeScan for the Bayesian method (Foll and Gaggiotti 2008; Balding 2003; Beaumont and Balding 2004). Further details of this approach are described in the next section.

2 Methods

2.1 *Sample Collection*

For the present study, we sampled the same, and one new nearby, locations as in previous studies (Nielsen and de Jong 2005; Vermeer et al. 2012) focussing on the flea beetle populations around Kværkeby (Denmark). Both *B. vulgaris* G-type and other host plants, such as *Sinapis arvensis*, *B. vulgaris* P-type and *Capsella bursa-pastoris*, occur at this location. In 2009 on the 11th and 26th of June leaves with leaf mining, flea beetle larvae were collected on *Sinapis arvensis* and *B. vulgaris* G-type nearby Kværkeby from various plant patches (see Fig. 1 and Table 1 for exact locations). The area has been thoroughly searched for host plant- and flea beetle occurrence. Other patches with host plants were present, but not all were colonized by flea beetles. Patches of host plants that were colonized at least during the 2008 and 2009 sampling periods are shown in Fig. 1. The host plants grow in distinct patches surrounded by areas where no suitable host plants for *P. nemorum* occur. On the 11th of June, 75 leaves were collected from plant patch A (corresponding to site 4 in Nielsen and de Jong (2005)). On the 26th of June, an additional ±100 leaves were collected from plant patch A and ±200 leaves from each of the two other plant patches, B and C. Location C (corresponding with site 1 in Nielsen and de Jong (2005)) consists of a G-type plant patch and is located respectively 2.3 and 3.4 km away from locations A and B where *S. arvensis* is the host plant. Mined leaves were picked from various plant individuals in each patch at each location to avoid potential bias from family structuring across plants. Picked leaves usually contained 1–3 larvae, though some leaves (especially *S. arvensis*) occasionally contained more larvae. These leaves were transferred to 500 ml plastic vials containing a layer of moist peat and vermiculite (5:1 volume ratio). Vials were closed using a plastic lid with a central hole of 15 mm diameter, plugged with cotton wool, and kept at 20 ± 2 °C. Full-grown larvae pupated in the layer of soil, after which we removed plant material from the vials. After 2.5 weeks, adult beetles started to emerge from the soil. Individuals were sexed using a stereomicroscope (25x magnification) (De Jong and Nielsen 1999). The sex ratio for each subpopulation was approximately evenly balanced between males and females (Chi square test, $DF = 1$, $P > 0.05$).

2.2 *Bioassays*

After sexing, each beetle was used in a no-choice bioassay within 24 h of emergence to determine whether they were resistant to defences of B. vulgaris G-type. Beetles had no access to food prior to the bioassay. The bio-assays were set up as in Vermeer et al. (2012). Because 96-wells plates were used for the DNA analysis (see below), only 96 beetles per sample location were used for further analyses. Supernumerary

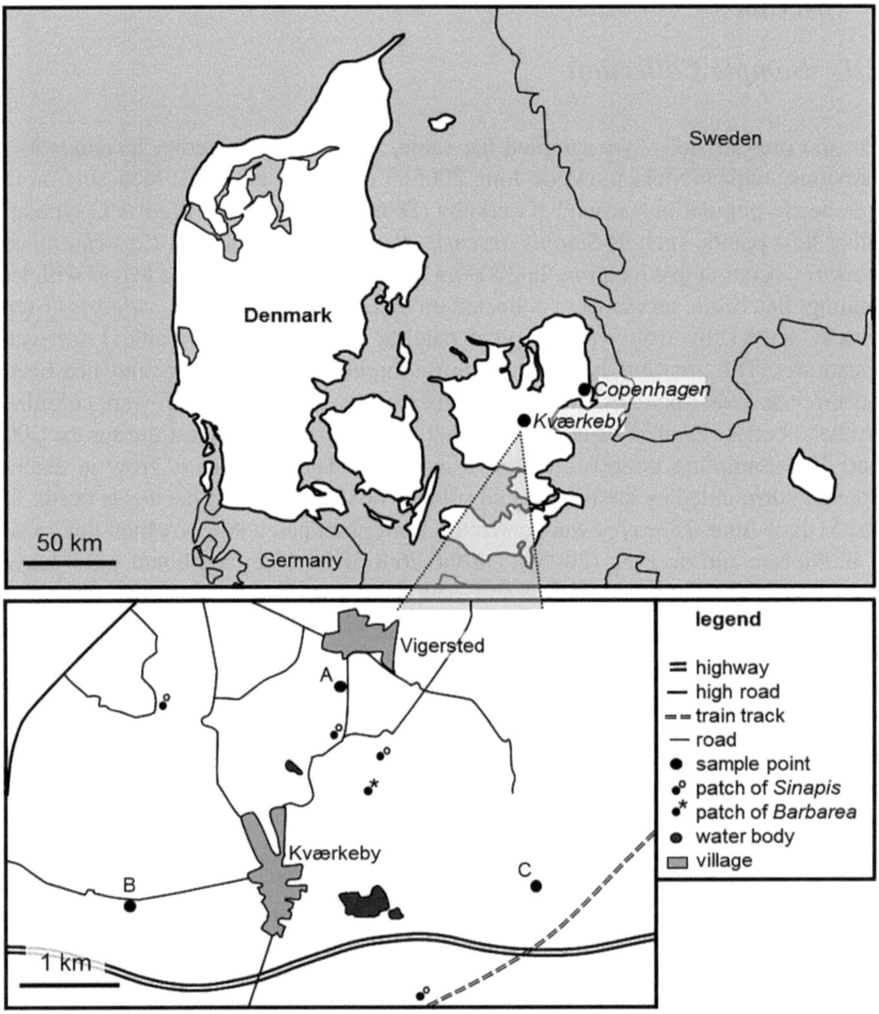

Fig. 1 Map of the sites at which *Phyllotreta nemorum* was sampled, and other plant patches in the vicinity that were not studied in detail for the population genomics approach. (A) Kværkeby, *Sinapis arvensis*; (B) Kværkeby, *Sinapis arvensis*; (C) Kværkeby, *Barbarea vulgaris* G-type (for more details, see Table 1). Only patches of *Sinapis arvensis* or *Barbarea vulgaris* G-type colonized by flea beetles are shown on the map

beetles were also bio-assayed but excluded from further analysis. As in Vermeer et al. (2012), non-feeders were classified as susceptible beetles and feeding beetles were classified as resistant beetles. Only $<1.5\%$ of the bioassays were inconclusive when using this method.

The bioassays provide information on the flea beetle's phenotype in terms of the resistance to the G-type of *B. vulgaris*. To be able to use statistics on genotypic

Table 1 Sampling sites and their coordinates in Denmark, near Kværkeby

Location	Host plant	Latitude	Longitude	*N*	Distance (km) from A	Distance (km) from B
A	*Sinapis arvensis*	+55° 28′ 47.24″	+11° 53′ 2.90″	162	0	2.5
B	*Sinapis arvensis*	+55° 27′ 32.67″	+11° 51′ 4.01″	129	2.5	0
C	*Barbarea vulgaris* G-type	+55° 27′ 40.82″	+11° 55′ 0.08″	113	2.3	3.4

N refers to the number of sampled adult beetles that were subsequently used in bio-assays

microsatellite data, the resistance genotype of each beetle was calculated assuming HWE within each sampled population. The frequency of the susceptible genotype is the same as for the susceptible phenotype (rr), because resistance is a dominant trait. The resistant phenotype can either be heterozygous (Rr) or homozygous (RR). We consider this assumption of HWE within samples justified, because (a) we are interested in outliers of measures that indicate differentiation between subpopulations, and (b) if directional selection at the resistance locus would lead to a deviation from HWE (with a deficiency of heterozygotes), our assumption of HWE actually represents a conservative approach with respect to the chance of finding the resistance to be an outlier relative to the neutral microsatellite loci.

2.3 DNA and Microsatellite Isolation and Amplification

Whole genomic DNA was isolated from each beetle using the DNeasy® Blood & Tissue Kit

(QIAGEN), following Purification of Total DNA from Animal Tissues (Spin-Column Protocol). Ten polymorphic microsatellite loci were already available for usage. The microsatellite primers for these loci have been characterized in Verbaarschot et al. (2007). New microsatellite markers were constructed by using di-, tri-, and tetranucleotide repeat-enriched libraries of *P. nemorum* genomic DNA constructed for previous microsatellite marker development using the same method as in Verbaarschot et al. (2007). The newly designed microsatellite primer pairs were evaluated via polymerase chain reaction (PCR) amplification using the isolated DNA of 96 beetles of each sampled plant patch. Together with the 10 microsatellites characterized by Verbaarschot et al. (2007), these newly designed microsatellite primer pairs were multiplexed (true multiplexing) for efficient amplification and scoring of alleles for the three locations near Kværkeby.

Amplification was performed in a 25 μl reaction volume using 15.875 μl nuclease-free water, 5 μl 5X colourless GoTaq® flexi buffer, 1.5 μl 25 mM MgC12, 0.5 μl 200 μM dNTPs, 0.5 μl 500 nM of each primer, 0.125 μl GoTaq® DNA polymerase and 1 μl of DNA (at least 4 ng of DNA in water). Each forward primer was 5′-fluorescently labelled with either 6-FAM, NED and HEX (Applied Biosystems). Annealing temperatures used for the PCR amplification are shown in Table 2 and (Verbaarschot et al. 2007). PCR conditions consisted of an initial cycle at 94 °C for 3 min, 30–35 repeat cycles at 94 °C for 30 s, the annealing temperature for 30 s, 72 °C for 1 min and a final extension at 72 °C for 5 min. For all markers, 35 repeat cycles were used except for markers PnD09, PnH09 and PnH12 (30 cycles) (see also Verbaarschot et al. (2007); coding of the markers was done as in Verbaarschot et al. (2007)).

The PCR products were then purified using Sephadex G-50 (Sigma) spin columns in a MultiScreen®-HV 96-well filter plate (Millipore) according to the Sequencing Reaction Purification Protocol (step 1–5) (Millipore 2012). The purified sequencing products were then run on an ABI PRISM 3700 sequencer (Applied Biosystems, Foster City, CA, at Greenomics, Wageningen) using GeneScan-LIZ 500 as an internal marker (Applied Biosystems, Foster City, CA).

2.4 Data Analysis

Genetic diversity Results were compiled and analysed with GeneMarker® V1.5 software (Softgenetics®). Allele evaluation and the stutter peak filter failed to detect some alleles that were noticed after visual inspection, yet the program summed up many alleles that did not seem legitimate. Adjusting the parameters of the program did not satisfactorily improve this and thus fragment scoring was done mainly by eye using the electrophoregrams visualized by GeneMarker. We made sure that scoring occurred 'blind', i.e. during scoring the sample location was not known to the person who did the scoring. Allelic size, total number of alleles (T_N), effective number of alleles (E_N), observed (H_o) and expected heterozygosity (H_e) were determined for each microsatellite locus. All these statistics were calculated using the GENALEX version 6.41 add-in (Peakall and Smouse 2006) in Excel and FSTAT version 2.9.3 (Goudet 1995, 2002). Parameters H_o and H_e were calculated using the estimate for genetic diversity as described by Nei (1973, 1987), and E_N was calculated as

$$E_N = 1/\sum p_i^2 \tag{1}$$

Presence of null alleles and scoring errors due to stuttering were checked with MICROCHECKER version 2.2.3 (Van Oosterhout et al. 2004). As in Verbaarschot et al. (2007), locus PnBB08 showed an excess of homozygotes for all sampled locations. Also newly developed primer pair PnG06 showed possible presence of null alleles due to a heterozygote deficit when tested for all sample locations. Primer pair PnC10 only showed an excess in homozygotes for sample location A and B, but not for location C. Primer pairs PnF03 and PnH09 only showed a heterozygote deficit for sample location C, and primer pair PnA06 only showed a heterozygote deficit for sample location A. There was no evidence for the presence of null alleles

Table 2 Summary of polymorphic microsatellite markers developed for *Phyllotreta nemorum*

Microsatellite locus	Primer sequence	Repeat sequence motif[a]	Allelic size range	Annealing temperature	Number of alleles (Effective number)
(GenBank ID)	5′—3′				
PnA06	F: FAM-CCAGAAATGTCATCGTACA	(AC)8	190–228	60	20
(FJ217683)	R: GTTTGGTCTTTGTGATGGACAAGG				(5.02)
PnA10	F: NED-ACTCACGCCCGAATCGCTTC	(CGA)5	126–141	57	6
(FJ217684)	R: GTTTAGAATGGACATGGTCGGCGG				(2.57)
PnB11.311	F: FAM-CTTCGAAATAATCGTCTTC	(CGT)4N22(TCG)4	133–154	57	8
(FJ217687)	R: GTTTAATCTGGAGACGATGATGAC				(1.41)
PnC05.11	F: HEX-TCACCAAATTGTGACATGTACC	(TAAA)5	147–163	57	5
(FJ217688)	R: GTTTACCATAAACGCACTGTTGA				(2.27)
PnC10	F: NED-GGGCAACGAAACAAACGACGGTACC	(GAC)5	198–213	50	6
(FJ217689)	R: CGTTTGAGTACCGCTGCCGG				(1.71)
PnD01	F: FAM-AACTTTGCATCGAATTGTGC	(GCC)5N85(TGC)8	347–398	50	18
(FJ217690)	R: GTTTGTCTTGGAACGCTAATCGC				(7.4)
PnD02	F: HEX-ATCAGCAGGCCATGCAGCAG	(CAG)5	135–162	57	10
(FJ217691)	R: GTTTGCGTTGTAGCCGGATTTTG				(5.74)
PnD06	F: NED-ATCACGTTCGGCACCACCTG	(CTG)7	171–186	53	6
(DQ507814)	R: GTTTCTTCAGCAGCCTGATGGGC				(4.28)
PnE08.61	F: FAM-GCAGCAGGTCGAGGCGACTG	(CAG)4	133–160	50	10
(FJ217694)	R: GTTTATTCGCCACCGTACCGTTCG				(4.33)
PnE10	F: NED-TTTGAAAACTATTGCCCATC	(TGC)7N4(GCT)4	184–214	50	11
(FJ217695)	R: GTTTATGTCTTGGAACGCTAATCG				(5.14)
PnF03	F: FAM-AAATCCTTCAAAGGCTAAGCCAGC	(TGC)11	327–360	55	12
(FJ217696)	R: CAACGGTTCAGCAGCAACG				(1.62)
PnG06	F: HEX-CGTGGAATTTGTCAACACTTG	(GCC)8	192–228	50	13
(FJ217697)	R: GTTTGAGGACTTAGGTTTTGGTATAG				(4.66)

Sequences are given for forward (F) and reverse (R) PCR primer
[a]The number after '*N*' denotes the number of nucleotides in between repeat sequences

Table 3 The Fixation Index (F_{IS}) is given for each sampled plant patch (i.e. A, B and C) and the mean for all samples together (All)

Fixation Index				
Microsatellite locus	A	B	C	All
PnA03	−0.008	−0.016	−0.137	−0.051
PnA04	0.185	−0.041	0.009	0.043
PnA06	0.090	−0.015	−0.112	−0.061
PnA10	0.133	0.098	−0.342[a]	−0.025
PnAB12	0.033	−0.109	0.098	0.016
PnBB08	0.154	0.447[a]	0.222[a]	0.288[b]
PnB11.311	−0.078	−0.034	0.087	−0.005
PnC05.11	0.019	0.183	−0.226	−0.087
PnC10	0.362[a]	0.346[a]	−0.143	0.232[b]
PnD01	0.037	−0.012	0.042	0.026
PnD02	−0.11	−0.076	−0.038	−0.069
PnD04	−0.161	−0.141	−0.18	−0.142
PnD06	−0.119	0.047	0.037	−0.018
PnD09	0.152	0.025	−0.04	0.013
PnE08.61	−0.191	−0.152	−0.091	−0.135
PnE10	−0.145	−0.061	−0.027	−0.073
PnE11	−0.064	0.015	−0.044	−0.02
PnF03	0.065	0.142	0.261	0.099
PnG06	0.162[a]	0.288[a]	0.27[a]	0.239[b]
PnH09	−0.085	−0.126	0.303	−0.102
PnH12	−0.348[a]	−0.254[a]	−0.252[a]	−0.268

Indicative adjusted nominal level (5%) per plant patch is: 0.00079, indicative adjusted nominal level (5%) for the whole population is: 0.00217. Asterisks indicate the proportion of randomizations for each microsatellite locus that gave a significantly larger or smaller F_{IS} than the observed based on 6300 randomizations for each sampled plant patch and 1000 randomizations for all

[a]Significance at $P < 0.0001$

[b]Significance at $P < 0.001$

for all other loci. Scoring errors due to stutter bands and large allele dropout were not detected either. FSTAT version 2.9.3.2 (Goudet 2002) was used to test deviation from HWE, inbreeding and linkage disequilibrium by calculating F_{IS} (Wright 1951) using the estimator f as in Weir and Cockerham (1984) and log-likelihood ratio G-statistics. Deviation from HWE was globally tested for each pair of loci across all three sample locations based on 1000 permutations. Consistent with MICROCHECKER, PnBB08, PnC10 and PnG06 were not in HWE (all P value <0.001), with an observed heterozygosity significantly lower than expected (Table 3).

HWE deviation was also tested for all pairs of loci per sampled plant patch based on 6300 randomizations with a 5% nominal level. Again PnBB08, PnC10 and PnG06 were not in HWE, with an observed homozygosity significantly higher than expected for some sampled locations (each P value <0.0001 for two sample locations, Table 3). Moreover, primer pair PnH12 showed an excess of heterozygotes for each sampled location (P value <0.0001) and across all locations (P value <0.001; data not shown). We ruled out a recent bottleneck in the population (assuming neutral theory) and a sudden significant increase in population size (Cornuet and Luikart 1996) by not finding a significant heterozygosity excess nor a significant lack of heterozygosity as in Nei (1978) for all markers when compared to expected heterozygosity (Table 3). We found a heterozygosity excess for some markers, however. Heterozygosity deficiency of some markers may represent a presence of null alleles or genetic hitchhiking (Barton 2000). Because markers with a heterozygosity deficiency are not regarded as 'neutral', PnBB08, PnG06 and PnC10 were discarded in subsequent analyses. Linkage disequilibrium was tested between all pairs of loci based on a confidence level of 1%, i.e. 21,000 permutations. Primer pairs PnE11, PnE10, PnD01 and PnD02 showed linkage amongst each other when considering a 1% nominal level after Bonferroni correction (P value <0.00005). All other loci showed no significant deviation from linkage equilibrium (all P values $>$ adjusted P value for a 5% nominal level) so we assumed that these loci were independent.

Analysis of population structure A Bayesian clustering method (see Shoemaker et al. (1999) for a good review about the use of this method in genetics) for inference of population structure based on unlinked (or at least not tightly linked) markers has been implemented in the computer program STRUCTURE version 2.3.3 (Pritchard et al. 2000, 2007; Falush et al. 2003; Hubisz et al. 2009). This program assigns individuals to clusters, giving the number of clusters as K, estimating for each individual the proportion (q) of the genome that originated from a given genetic cluster. This estimation is done by the Maximum Likelihood method (see e.g. Beebee and Rowe (2004)). STRUCTURE assumes HWE within subpopulations and performs best with unlinked loci. Therefore, we excluded from the analyses those markers that are in linkage disequilibrium, namely, PnE10, PnD01 and PnD02. Marker PnH12, which showed an excess of heterozygotes for all samples, was excluded from the analyses as well. Individuals were first clustered in STRUCTURE with no prior population information. Afterwards the data sets were analysed again, this time with prior location information in the form of sample location A, B and C to assist clustering. STRUCTURE v2.3.3 can detect structure at lower levels of divergence than the original STRUCTURE models by using prior sampling location information, without bias towards detecting structure when none is present (Hubisz et al. 2009). Sampling information is ignored by this version of the program when there is evidence of population structure uncorrelated with sampling locations. For all analyses, we set most of the parameters to their default values as advised in the user's manual of STRUCTURE 2.3.3 (Pritchard et al. 2010). Exploring runs with varying λ, however, showed that the estimations of λ were often much lower than the $\lambda = 1$ expected under the independent allele frequencies model (Falush et al. 2003) and dependent on the number of assumed genetic clusters (K). We therefore let λ vary during each run. Data sets were analysed under the admixture model and dependent allelic frequencies (Falush et al. 2003). The algorithm was run 10 times for each number of genetic clusters, which was set from 1 to 5, for 100,000 burn-in steps and 100,000 replications, respectively. Additionally, the overall mean values of the *r* parameter in STRUCTURE were calculated for each genetic cluster when sample location information was used to assist clustering. The *r* parameter estimates if the location data were informative in defining the clusters. Values close to and/or below 1 indicate that the location data were informative, while values much larger than 1 indicate that the location data were not informative for the structure found (Hubisz et al. 2009; Pritchard et al. 2010). The results were compiled with STRUCTURE HARVESTER version 0.6.92 (Earl and Vonholdt 2012). Output files from STRUCTURE HARVESTER were run in CLUMPP version 1.1.2 (Jakobsson and Rosenberg 2007) to get summarized *q*-matrices of the repeated runs for each *K* and visualized in Distruct 1.1 (Rosenberg 2004). Differences between the data Maximum Likelihood of successive *K* values were tested using the non-parametric Wilcoxon two-sample test for dependent samples and SPSS 20 software. The most likely *K* value was estimated based on: (1) the standard method for estimating *K* as described in the manual of STRUCTURE (Pritchard et al. 2010) by calculating the posterior probability of *K* and (2) the largest *K* value with significantly higher likelihood than that from $K - 1$ runs. Often the ΔK method of Evanno et al. (2005) is also used to estimate *K*, especially when the estimated probability of data for successive *K* values plateaus or continues to increase slightly after the presumed real *K* value is reached (Evanno et al. 2005; Campana et al. 2011). We chose not to include this method because the data Maximum Likelihood decreased after $K = 3$, and we only wanted to assure that our sampled plant patches could be considered genetically distinct subpopulations to at least some extent.

Finally, to corroborate clustering done by STRUCTURE, both a principal coordinate analysis (PCA) and a factorial corresponding analysis (FCA) were used to make a graphical representation of the repartition of the three individually sampled subpopulations. The PCA was done with data standardization after calculating the pairwise genetic distance between samples using the GENALEX add-in (Peakall and Smouse 2006) in Excel. This PCA provides major axes explaining the variation in the dataset. The FCA was carried out with notice of sample information with GENETIX software (Belkhir et al. 1998). This FCA plot identifies the major contributors to the overall genetic variation using a microsatellite individual-genotype matrix and puts higher priorities on finding relationships within a population than between populations.

Outlier detection Both the frequentist method (fdist) and the Bayesian method were used to identify outlier loci in the microsatellite dataset and test for outlier behaviour of the resistance trait and microsatellites (Beaumont and Nichols 1996; Beaumont and Balding 2004). Multiple analyses were done per program to get more robust results. LOSITAN (Antao et al. 2008), a user-friendly program in which simulations were done assuming either the infinite allele model (IAM) or the stepwise mutation model (SMM), was used for fdist. Fifty thousand simulations were run in LOSITAN with a confidence interval of 99% to test for patterns of possible selection by evaluating the relationship between the F_{ST}, using the estimator θ as in Weir and

Cockerham (Weir and Cockerham 1984) and expected H_e for each locus. A 'neutral' mean F_{ST} was computed for each pair of subpopulations and all three subpopulations together. To find the 'neutral' mean F_{ST}, LOSITAN removes potential outliers during the simulations when computing the F_{ST} value. We used the program BayeScan version 2.1 (Foll and Gaggiotti 2008) for the Bayesian method (Balding 2003; Beaumont and Balding 2004). A multinomial-Dirichlet distribution is used as prior distribution of the allele frequencies in the subpopulation (Foll and Gaggiotti 2008). The program performs a logistic regression over two different models: one model that includes a locus-specific component (α) that is shared by all populations, and a population-specific component (β) that is shared by all loci, and the other model has only the population-specific component β. BayeScan estimates a posterior probability of each one of these models using a reversible-jump MCMC algorithm. Because α is locus-specific, the posterior probability indicates for each locus if there is an effect of selection. One lack simulations were run in total, with 50,000 burn-in steps and 50,000 replications. Posterior probabilities were computed for each pair of subpopulations and all three subpopulations together. These posterior probabilities should not be seen as P-values. A posterior probability above 0.91 can already be considered strong support for a model including α. When for a locus the posterior probability exceeds 0.99, this provides decisive evidence for selection according to Jeffreys' scale of evidence for Bayes factors (Foll 2012).

Genetic differentiation between two subpopulations was further estimated by pairwise G_{ST} (Nei and Chesser 1983), G'_{ST} (Hedrick 2005) and D (Jost 2008) for each microsatellite and the resistance trait, using SMOGD (Crawford 2010). Microsatellites that showed an excess of natural selection against resistance of a flea beetle to host plant defences homozygotes in a subpopulation according to MICROCHECKER and FSTAT were left out of the analysis, because we are only interested in neutral microsatellite loci. G_{ST} is equivalent to Wright's F_{ST} (Wright 1951; Nei 1973) but also accounts for multiple loci with multiple alleles. In 1983, Nei and Chesser improved G_{ST} by taking into account deviation from HWE within subpopulations and sample size (Nei and Chesser 1983). Hedrick developed the standardized measurement G'_{ST}, because the outcome of G_{ST} is dependent on the level of heterozygosity. The value of G_{ST} will always be very small with high levels of heterozygosity, even if not many alleles are shared by the subpopulations compared. Jost's D is calculated with the effective number of alleles rather than heterozygosity, so D is independent of average within-subpopulation heterozygosity. Furthermore, the estimator D is also dependent on the mutation rate of the marker. Leng and Zhang (2011) advise to calculate both G'_{ST} and Jost's D and the relative levels of heterozygosity when using microsatellites. The estimator of G_{ST}, G'_{ST} and Jost's D are given by SMOGD as well as their 95% confidence interval after 500 bootstrap replicates done at the individual level.

3 Results

All newly found microsatellites were di-, tri- or tetra-nucleotide repeat motifs and polymorphic for the flea beetle population in Kværkeby; the number of alleles per marker ranged from 5 up to 20 (Table 2). The previously developed microsatellite markers were also polymorphic for this population (Table 4). For none of the sampled plant patches, the markers were fixed. For each microsatellite, the effective number of alleles (E_N) ranged between 1.42 and 6.42 per locus, with an average of 4.07, and expected heterozygosity (H_e) ranged from 0.29 to 0.89 (Table 4), with an average of 0.68 (average over all loci except PnBB08, PnC10, and PnG06, see Sect. 2). For each plant patch, the average H_e was between 0.67 and 0.70. The average F_{IS} value per locus ranged between −0.27 and 0.29 (Table 3) with an overall F_{IS} of −0.012, indicating that inbreeding was not likely because F_{IS} values are expected to be close to zero when there is no inbreeding. Only for one locus a strong deviation

Table 4 Descriptive statistics of the microsatellite loci and their allelic size range, total number of alleles (T_N), effective number of alleles (E_N), observed (H_o) and expected (H_e) heterozygosity per locus based on HWE

				Heterozygosity			
Microsatellite	Allelic	T_N	E_N	A	B	C	All
Locus	Size range			H_o / H_e	H_o / H_e	H_o / H_e	H_o / H_e
PnA03	197–225	15	5.24	0.82 / 0.81	0.81 / 0.79	0.94 / 0.83	0.86 / 0.81
PnA04	171–183	5	1.87	0.33 / 0.40	0.51 / 0.49	0.50 / 0.50	0.45 / 0.47
PnA06	190–228	16	6.37	0.81 / 0.89	0.84 / 0.84	0.85 / 0.76	0.84 / 0.83
PnA10	126–141	6	2.55	0.55 / 0.63	0.44 / 0.48	0.90 / 0.67	0.63 / 0.61
PnAB12	185–224	14	6.42	0.83 / 0.86	0.97 / 0.87	0.72 / 0.79	0.84 / 0.85
PnBB08	209–257	16	6.02	0.73 / 0.86	0.44 / 0.78	0.67 / 0.85	0.61 / 0.84
PnB11.311	133–154	8	1.42	0.46 / 0.42	0.24 / 0.23	0.18 / 0.2	0.29 / 0.29
PnC05.11	147–163	5	2.24	0.56 / 0.57	0.47 / 0.55	0.66 / 0.54	0.16 / 0.49
PnC10	198–213	6	1.94	0.22 / 0.35	0.20 / 0.30	0.75 / 0.65	0.39 / 0.47
PnD01	347–398	17	7.18	0.85 / 0.87	0.88 / 0.86	0.81 / 0.84	0.85 / 0.89
PnD02	135–162	10	5.58	0.93 / 0.83	0.89 / 0.82	0.94 / 0.86	0.89 / 0.83
PnD04	136–166	10	4.31	0.83 / 0.71	0.85 / 0.74	0.98 / 0.82	0.89 / 0.78
PnD06	171–186	5	4.37	0.79 / 0.76	0.74 / 0.75	0.72 / 0.79	0.75 / 0.78
PnD09	164–197	10	4.56	0.60 / 0.68	0.80 / 0.82	0.84 / 0.81	0.75 / 0.79
PnE08.61	133–160	10	4.29	0.86 / 0.72	0.88 / 0.76	0.88 / 0.8	0.88 / 0.77
PnE10	184–214	11	5.08	0.93 / 0.81	0.86 / 0.80	0.82 / 0.79	0.87 / 0.81
PnE11	208–235	10	5.37	0.90 / 0.84	0.80 / 0.81	0.83 / 0.79	0.84 / 0.82
PnF03	327–360	12	1.51	0.42 / 0.45	0.26 / 0.31	0.15 / 0.21	0.28 / 0.32
PnG06	192–228	12	4.58	0.67 / 0.80	0.56 / 0.77	0.56 / 0.77	0.60 / 0.79
PnH09	159–165	3	1.90	0.58 / 0.52	0.39 / 0.34	0.66 / 0.52	0.54 / 0.46
PnH12	143–164	8	2.44	0.88 / 0.65	0.80 / 0.63	0.28 / 0.43	0.65 / 0.57

H_o and H_e are given for each sampled plant patch (i.e. A, B and C) and all samples together (All)

from HWE was found. Hardy–Weinberg equilibrium for the whole population and over all loci was not rejected ($P > 0.99$).

3.1 Bioassays

All beetles from plant patch C (*B. vulgaris* G-type plants) were classified as being resistant. From plant patch A (*Sinapis arvensis* plants), nearly 70% of the beetles (66 of 95 individuals) were classified as resistant, and from plant patch B (*S. arvensis* plants), only 30% (28 of 93 individuals) were classified as resistant (Table 5). Of one beetle from plant patch A, we could not with certainty determine the phenotype, and for plant patch B, this was the case for 3 beetles. All other beetles were classified as susceptible beetles. Table 5 displays both the resistance phenotype and the inferred genotype when HWE is assumed within subpopulations.

3.2 Genetic Structure in the Flea Beetle Population Near Kværkeby

The genetic structure of all individuals was analysed in STRUCTURE with no prior location information using microsatellite markers that did not significantly deviate from HWE and did not show linkage disequilibrium. The data likelihood of successive K values (i.e. the number of genetic clusters assumed by the model) only decreased by increasing the K value and the posterior probability of $K = 1$ was the highest probability ($\Pr(K) = 0.996$) compared to the posterior probabilities found for the other K values, both suggesting $K = 1$ as the most likely K (Table 6). Additionally, the proportion of cluster membership (q) of each individual and each plant

Table 5 Resistance phenotype of the flea beetle and inferred genotype when HWE is assumed within each sampled population

Location	Plant type	Phenotype	#	Genotype	#
A	*Sinapis arvensis*	res	66	RR	19
				Rr	47
		sus	29	rr	29
B	*Sinapis arvensis*	res	28	RR	3
				Rr	25
		sus	65	rr	65
C	*Barbarea vulgaris* G-type	res	96	RR	96
				Rr	0
		sus	0	rr	0

'res' is a resistant phenotype and 'sus' a susceptible phenotype. R is the dominant allele for resistance, r the recessive allele

Table 6 Results of the population genetic analyses carried out with STRUCTURE and SPSS

K	Prior location data	Number of runs	Ln P(*K*)		Pr (*K*)	Wilcoxon significance	Mean *r*
			Mean	Standard dev.			
1	No	10	−8041	0.08	1	NA	NA
2	No	10	−8065	8.31	0	0.005	NA
3	No	10	−8115	38.02	0	0.005	NA
4	No	10	−8141	33.99	0	n.s.	NA
5	No	10	−8170	47.15	0	n.s.	NA
1	Yes	10	−8041	0.06	0	NA	8.88
2	Yes	10	−7972	1.52	0.004	0.005	0.99
3	Yes	10	−7967	7.66	0.996	n.s.	0.57
4	Yes	10	−7994	36.59	0	n.s.	0.76
5	Yes	10	−8004	32.06	0	n.s.	0.61

K is the number of genetic clusters assumed by the model, Ln P(*K*) is the log likelihood of the *K* value, and Pr(*K*) is the posterior probability of the *K* value. '*r*' is a parameter which estimates if the location data were informative in defining the clusters. Values below 1 indicate that the location data were informative and values much larger than 1 indicate that the location data were not informative for the structure found. 'n.s.' stands for *not significant* ($P > 0.05$), 'NA' stands for *not available*

patch sample was fairly close to 1/*K* for each *K* value, meaning that each individual could be assigned to any genetic cluster. All of this suggests little to no population structure or that too few microsatellites/individuals were used for the program to pick up structure (Pritchard et al. 2010).

When individuals were classified into clusters with prior location information to assist clustering, the program was able to detect structure. The data likelihood of *K* values larger than 1 was higher than found when no prior location information was used (Table 6). Also the mean of the *r* parameter shows that for *K* values 2–5, the location data were helpful in defining the clusters, since the mean value of *r* was always close to 1 or smaller than 1. The data likelihood for each *K* value increased slightly when prior location information was incorporated into the clustering method until a *K*-value of 3, though the increase from $K = 2$ to $K = 3$ was not significant (non-parametric Wilcoxon two-sample test). Bigger *K*-values produced lower data likelihoods.

Figure 2 shows for $K = 3$ for each plant patch sample the proportion of cluster membership (q) when prior location information is not used (Fig. 2c) and when it is used to assist clustering (Fig. 2d). Without prior location information, the proportions of cluster membership are equally divided over each plant patch sample as mentioned earlier. When prior location information is included, the proportions of cluster membership are not equally divided over each plant patch sample, suggesting that there is genetic structure. All samples show an average estimated major membership proportion for the same genetic cluster suggesting that we sampled admixed subpopulations in which all individuals inherited most ancestry from the same ancestral population. Despite that, we can still see that the genetic profiles of each

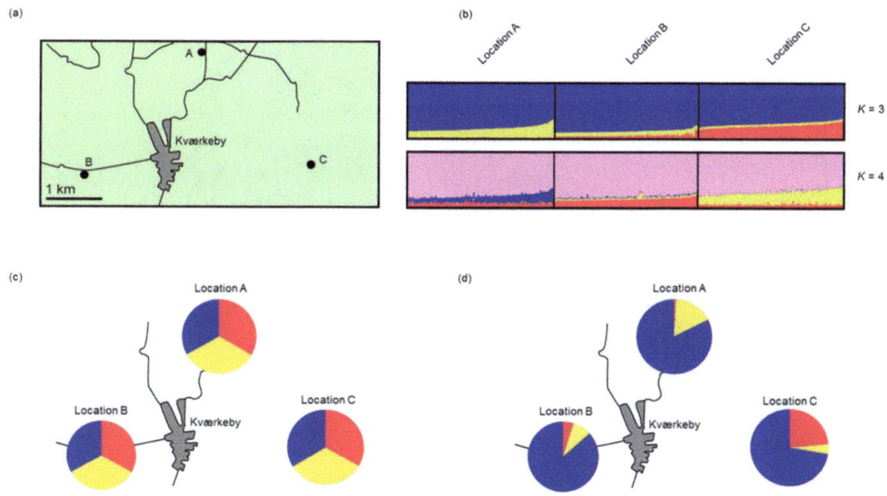

Fig. 2 STRUCTURE analysis of *Phyllotreta nemorum* subpopulations near Kværkeby (Denmark). (**a**) Geographic representation of the sample area. (**b**) Graphical representations of ancestry membership coefficients inferred from STRUCTURE analysis with prior location knowledge. Each individual is shown as a vertical line divided into coloured segments representing the estimated membership proportions in the three and four genetic clusters (K) inferred with STRUCTURE. Individuals are arranged according to each sampled plant patch at location A, B and C. Magenta is the fraction of membership (q) of cluster 1, yellow of cluster 2, purple of cluster 3 and pink of cluster 4. (**c**) Graphical representation of each sampled plant patch and the percentage of each cluster as an average over 10 repeats at $K = 3$ *without* prior location knowledge. Each sample is shown as a bar divided into segments representing the overall membership proportions in the three genetic clusters inferred with STRUCTURE. As for (**b**) magenta stands for the fraction of membership (q) of cluster 1, yellow of cluster 2 and purple of cluster 3. (**d**) Graphical representations of each sampled plant patch and the percentage of each cluster as an average over 10 repeats at $K = 3$ as in (**c**) but this time *with* prior location knowledge

sample exhibit a different composition of cluster membership. Moreover, looking at the cluster membership of each individual from a certain plant patch (Fig. 2b), we can see that the genetic profiles from individuals of one plant patch exhibit a different composition of subpopulation membership than genetic profiles from individuals of another plant patch, whether we assume three or four genetic clusters. Different compositions of genetic cluster membership of each plant patch sample at the subpopulation level and the individual level, both support the conclusion that the sampled plant patches harbour groups of flea beetles with distinct genetic divergence.

The partitioning of individuals in the FCA and PCA (Figs. 3 and 4) corresponded with the three samples as well. Factorial Correspondence axis (FC) 1 explained 70% of the variation in the FCA and FC 2 the remaining 30%. All three clouds of individuals from a location over-lap partially, though not completely, so that three different groups can be distinguished that each roughly resembles a group of individuals from a certain plant patch. The first principal component (PC) in the

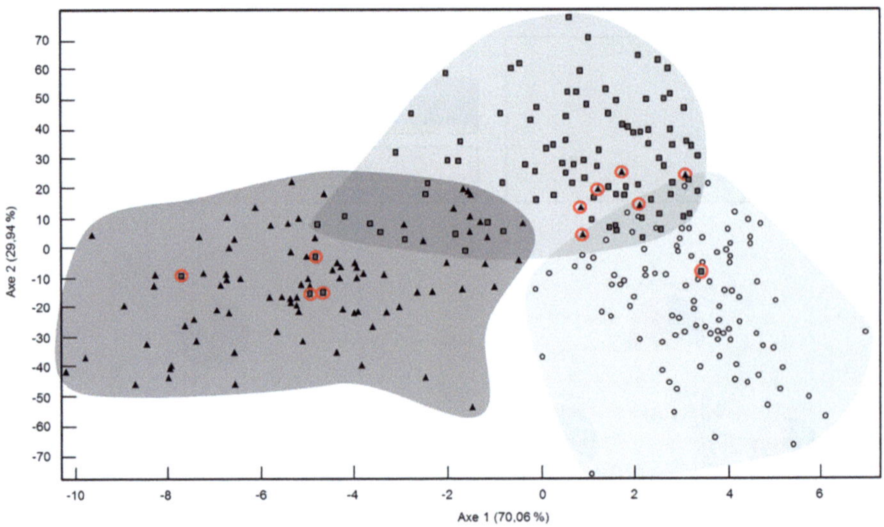

Fig. 3 Graphical representation of the repartition of the three sampled subpopulations made with GENETIX software version 4.05.2 (Belkhir et al. 1998) using factorial correspondence analysis (FCA). Each grey square depicts a sample/an individual from location B, each black triangle from location A and each white circle from location C. Individuals that are situated in another cloud than the other individuals from that sampled plant patch are highlighted with a red circle

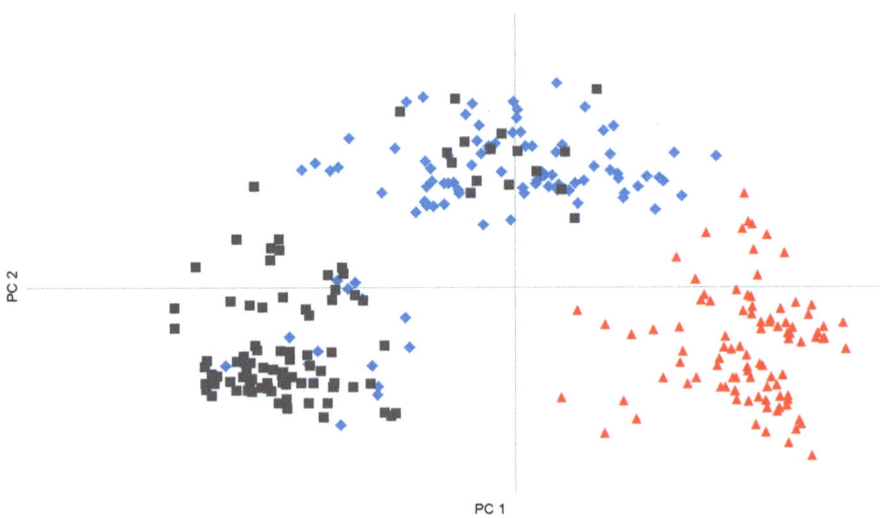

Fig. 4 Graphical representation of the repartition of the three sampled subpopulations made with GENALEX software version 6.41 (Peakall and Smouse 2006) using principal component analysis (PCA). Each grey square depicts a sample/an individual from location A, each blue diamond an individual from location B and each red triangle an individual from location C

PCA explained 33% of the variation and the second PC 26%. No individuals of locations A and B were partitioned in the cloud of individuals from location C. Neither did individuals from location C show up between individuals from locations A and B, while clouds of individuals from location A and B also contained individuals of the other sample location.

Thus, we have identified three distinct subpopulations which are roughly in agreement with the sampled plant patches. We therefore continued the analysis, while assuming that the three samples were genetically distinct subpopulations.

3.3 Genetic Differentiation Among Subpopulations

Outlier detection We used the program LOSITAN (Antao et al. 2008) for an F_{ST} outlier test, by comparing the 'neutral' mean F_{ST} with the heterozygosity in the loci. For both the infinite alleles model (IAM) and the stepwise mutation model (SMM), the resistance trait appeared as a potential outlier locus when using a 99% confidence interval with a probability of the actual F_{ST} being bigger than the simulated F_{ST} ($0.9998 < P < 1$) (see Table 7 and Fig. 5 for the results under the IAM; the results under SMM were congruent) (for the results under the IAM, the results under SMM

Table 7 LOSITAN results of the simulated F_{ST} values per locus and the probability (P) that the simulated F_{ST} < sample F_{ST}, using 50,000 simulations, a 99% confidence interval and the infinite allele model. Loci with a probability higher than 0.99 (in bold) are highlighted with grey shading

	Population A–B		Population A–C		Population B–C		All populations	
Locus	F_{ST}	P	F_{ST}	P	F_{ST}	P	F_{ST}	P
PnA03	0.0027	0.4765	0.0017	0.4039	0.0064	0.5302	0.0036	0.4106
PnA04	0.0041	0.6056	0.0068	0.6267	−0.0055	0.5278	0.0016	0.4847
PnA06	NA	NA	NA	NA	0.0117	0.633	NA	NA
PnA10	0.0261	0.7929	0.019	0.7134	0.0662	0.9498	0.037	0.9141
PnAB12	0.0005	0.3877	0.0171	0.7035	0.0361	0.9215	0.0179	0.8053
PnB11.311	0.0341	0.8476	0.0503	0.8845	−0.0002	0.5101	0.0313	0.8334
PnC05.11	0.0675	0.9557	0.0092	0.6189	0.023	0.7637	0.0342	0.8777
PnD01	0	0.3401	0.0131	0.633	0.0045	0.4728	0.0059	0.4639
PnD02	−0.0004	0.3975	0.0134	0.6432	0.0008	0.3984	0.0046	0.4438
PnD04	−0.0031	0.3867	0.0443	0.923	0.028	0.8471	0.0239	0.8448
PnD06	0.0050	0.5435	0.004	0.4748	0.0023	0.4811	0.0038	0.4421
PnD09	0.0191	0.7610	0.0185	0.721	−0.0008	0.3588	0.012	0.6402
PnE08.61	0.0099	0.6311	0.0202	0.7436	0.0043	0.4992	0.0114	0.6277
PnE10	−0.0038	0.2955	0.0032	0.4398	0.0009	0.4205	0.0001	0.2866
PnE11	0.0141	0.7039	0.0084	0.5527	0.0207	0.7706	0.0144	0.7128
PnF03	0.0088	0.6565	NA	NA	NA	NA	NA	NA
PnH09	0.0383	0.8676	NA	NA	NA	NA	NA	NA
Resistance	0.3218	**0.9998**	0.4021	**1**	0.8324	**1**	0.5760	**1**

'NA' stands for *not available*

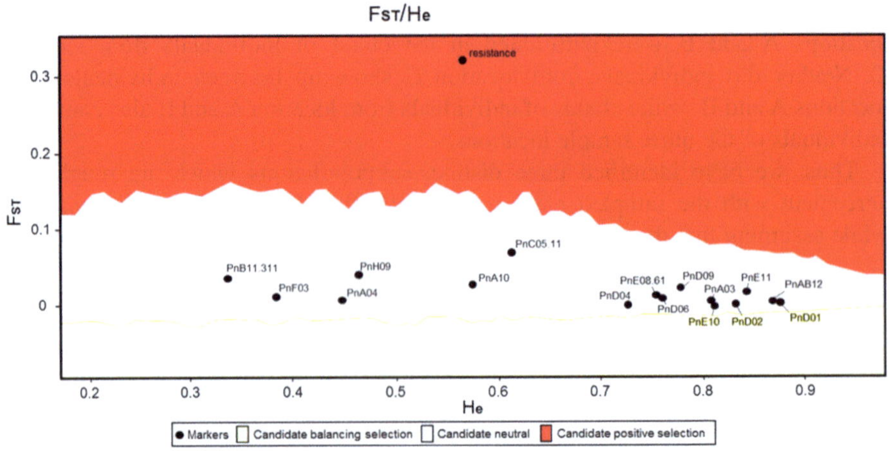

Fig. 5 A comparison between the F_{ST} value and the expected heterozygosity (H_e) of each microsatellite locus to identify outliers using LOSITAN (Antao et al. 2008). In this case, the F_{ST} values were simulated for the comparison between population A and population B (see text). Loci in the red area are candidates for directional selection and the ones in the white area are selectively neutral. None of the loci are positioned in the yellow area representing candidates for balancing selection

were congruent). For each pairwise population comparison and also for all subpopulations together, the resistance trait appeared as a potential outlier. In none of these comparisons, the microsatellite loci used in the simulations were a candidate for selection.

The Bayesian method was also used to test for outliers with the program BayeScan (Foll and Gaggiotti 2008). The posterior probability for each scenario did not exceed 0.91 for any microsatellite locus (see Table 8), which can be interpreted as no strong evidence for an effect of selection on those loci (Foll 2012). For the resistance trait, the posterior probability was 1 for the global analysis over all three populations and for the pairwise analysis of populations A and C and populations B and C. Only for the pairwise analysis of populations A and B, the posterior probability was not high enough to infer evidence of selection. The inclusion of alpha in the model did not matter for any microsatellite locus nor for any pairwise comparison between subpopulations. For the resistance trait, however, alpha was estimated to be 3.05 for all populations together and 2.86 for populations B and C and 2.41 for populations A and C, all indicating that directional selection might be present.

Additionally, genetic differentiation between each pair of subpopulations was estimated by calculating G_{ST} (Nei and Chesser 1983), G'_{ST} (Hedrick 2005) and D_{est} (Jost 2008) and their 95% Confidence Interval (CI) for each microsatellite and the resistance trait using the program SMOGD (Crawford 2010) (Table 9).

For the pairwise comparison between sampled subpopulations B and C (host plant *S. arvensis* versus *B. vulgaris* G-type, respectively), the resistance trait shows

Table 8 BayeScan results of each locus F_{ST} and the posterior probability (P), showing that a model including selection for that locus fits the data better than a model wihout selection

	Population A–B		Population A–C		Population B–C		All populations	
Locus	F_{ST}	P	F_{ST}	P	F_{ST}	P	F_{ST}	P
PnA03	0.0230	0.1600	0.0222	0.4713	0.0246	0.4139	0.0132	0.6145
PnA04	0.0259	0.0888	0.0327	0.0964	0.0317	0.1622	0.0189	0.2146
PnA06	NA	NA	NA	NA	0.0347	0.0456	NA	NA
PnA10	0.0255	0.0730	0.0333	0.0642	0.0367	0.0754	0.0225	0.0524
PnAB12	0.0143	0.5947	0.0251	0.3525	0.0337	0.0626	0.0168	0.3617
PnB11.311	0.0258	0.0718	0.0538	0.2281	0.0336	0.1008	0.0227	0.0632
PnC05.11	0.0288	0.0966	0.0327	0.0886	0.0361	0.0782	0.0229	0.0760
PnD01	0.0240	0.1178	0.0321	0.0840	0.0316	0.1532	0.0160	0.4109
PnD02	0.0137	0.6107	0.0326	0.0646	0.0273	0.2975	0.0187	0.2204
PnD04	0.0260	0.0888	0.0608	0.4101	0.0351	0.0618	0.0232	0.0766
PnD06	0.0245	0.0924	0.0286	0.2088	0.0254	0.3577	0.0165	0.3403
PnD09	0.0263	0.0788	0.0317	0.1056	0.0285	0.2549	0.0169	0.3243
PnE08.61	0.0241	0.1064	0.0318	0.1030	0.0253	0.3823	0.0169	0.3331
PnE10	0.0166	0.4543	0.0325	0.0766	0.0338	0.0694	0.0173	0.3063
PnE11	0.0255	0.0592	0.0313	0.1216	0.0344	0.0582	0.0220	0.0526
PnF03	0.0241	0.1218	NA	NA	NA	NA	NA	NA
PnH09	0.0278	0.0900	NA	NA	NA	NA	NA	NA
Resistance	0.0340	0.1714	0.2764	**1**	0.3808	**1**	0.3101	**1**

Loci with a posterior probability higher than 0.91 (in bold) for at least one scenario are highlighted with grey shading. 'NA' stands for *not available*

for all estimated parameters a substantially higher value ($G_{ST} = 0.717$, $G'_{ST} = 0.944$ and $D_{est} = 0.808$) than for each microsatellite marker ($G_{ST} = 0$–0.037, $G'_{ST} = 0$–0.225 and $D_{est} = 0$–0.209). In all cases, the estimated parameter of the resistance trait even exceeds the upper level of the 95% confidence interval of the microsatellite estimated parameter.

For the pairwise comparison between sampled subpopulations A and C (host plant *S. arvensis* versus *B. vulgaris* G-type), the difference between estimators of the resistance trait and the microsatellites is not as pronounced as above. For G_{ST}, the resistance trait shows an estimate substantially higher than the confidence interval of this estimator for each microsatellite marker. For G'_{ST} and D, the resistance trait shows an estimate higher than all estimates for the microsatellites, but their confidence interval does not always come up with an upper limit lower than the estimator D of the resistance trait.

For the pairwise comparison between sampled subpopulations A and B (both patches with host plant *S. arvensis*), the resistance trait shows for all estimators a higher value than the upper limit of the confidence interval of this estimator for each microsatellite marker. Estimators of the resistance trait's genetic differentiation ($G_{ST} = 0.196$, $G'_{ST} = 0.289$ and $D_{est} = 0.295$) are substantially higher than the estimators of the microsatellites' genetic differentiation ($G_{ST} = 0$–0.048, $G'_{ST} = 0$–0.167 and $D_{est} = 0$–0.127).

Table 9 Genetic differentiation estimators and 95% confidence interval (CI) per pair of subpopulations for each microsatellite and for the resistance phenotype

	Pairwise comparison between locations A and B																	
	Microsatellite marker																	
	PnA03	PnA04	PnA10	PnAB12	PnB11.311	PnC05.11	PnD01	PnD02	PnD04	PnD06	PnD09	PnE08.61	PnE10	PnE11	PnF03	PnH09	PnH12	Resistance
G_{ST}	0.004	0.005	0.017	0.003	0.021	0.048	0.003	0.003	0.001	0.006	0.020	0.005	0.001	0.010	0.007	0.032	0.008	0.196
CI min	0	−0.002	0.002	0	0.005	0.008	−0.001	−0.001	−0.002	−0.001	−0.002	0.001	−0.002	0.002	−0.001	−0.008	0.001	0.114
CI max	0.011	0.021	0.038	0.009	0.041	0.108	0.009	0.008	0.005	0.014	0.056	0.018	0.005	0.018	0.045	0.099	0.017	0.289
G'_{ST}	0.037	0.014	0.058	0.038	0.040	0.167	0.041	0.028	0.005	0.040	0.131	0.034	0.008	0.100	0.016	0.086	0.037	0.430
CI min	0.003	−0.005	0.008	−0.007	0.010	0.029	−0.014	−0.010	−0.012	−0.005	−0.008	0.004	−0.019	0.026	−0.001	−0.015	0.005	0.266
CI max	0.091	0.054	0.123	0.110	0.081	0.347	0.126	0.081	0.031	0.095	0.320	0.112	0.050	0.186	0.045	0.271	0.079	0.588
D_{est}	0.033	0.008	0.043	0.035	0.020	0.127	0.038	0.026	0.004	0.034	0.059	0.029	0.007	0.092	0.009	0.057	0.030	0.295
CI min	0.003	−0.003	0.005	−0.006	0.005	0.021	−0.013	−0.010	−0.010	−0.005	−0.007	0.003	−0.017	0.024	−0.001	−0.008	0.004	0.174
CI max	0.081	0.033	0.091	0.102	0.043	0.272	0.117	0.074	0.026	0.081	0.278	0.097	0.045	0.171	0.025	0.201	0.063	0.423
	Pairwise comparison between locations A and C																	
	Microsatellite marker																	
	PnA03	PnA04	PnA10	PnAB12	PnB11.311	PnC05.11	PnD01	PnD02	PnD04	PnD06	PnD09	PnE08.61	PnE10	PnE11	PnH12	Resistance		
G_{ST}	0.003	0.007	0.012	0.011	0.028	0.008	0.010	0.009	0.025	0.005	0.020	0.012	0.004	0.007	0.014	0.252		
CI min	0	−0.002	0.004	0.003	0.010	−0.003	0.003	0.004	0.016	−0.001	−0.002	0.005	0	0	0.003	0.197		
CI max	0.009	0.022	0.023	0.023	0.050	0.030	0.019	0.018	0.035	0.014	0.058	0.021	0.011	0.017	0.027	0.313		
G'_{ST}	0.034	0.017	0.058	0.117	0.053	0.028	0.122	0.094	0.187	0.036	0.127	0.092	0.039	0.070	0.054	0.413		
CI min	−0.001	−0.006	0.020	0.035	0.020	−0.009	0.039	0.036	0.123	−0.006	−0.015	0.035	−0.004	0.002	0.012	0.311		
CI max	0.087	0.058	0.107	0.221	0.099	0.103	0.236	0.172	0.257	0.102	0.332	0.158	0.102	0.151	0.104	0.522		
D_{est}	0.031	0.011	0.046	0.107	0.026	0.020	0.114	0.085	0.167	0.032	0.110	0.081	0.035	0.064	0.040	0.217		
CI min	−0.001	−0.004	0.016	0.032	0.010	−0.006	0.036	0.032	0.109	−0.005	−0.013	0.030	−0.003	0.002	0.009	0.142		
CI max	0.079	0.037	0.086	0.202	0.049	0.075	0.222	0.158	0.230	0.088	0.291	0.140	0.094	0.137	0.078	0.305		
	Pairwise comparison between locations B and C																	
	Microsatellite marker																	
	PnA03	PnA04	PnA06	PnA10	PnAB12	PnB11.311	PnC05.11	PnD01	PnD02	PnD04	PnD06	PnD09	PnE08.61	PnE10	PnE11	PnH12	Resistance	
G_{ST}	0.006	0	0.010	0.037	0.021	0.002	0.023	0.005	0.003	0.016	0.004	0.003	0.005	0.003	0.013	0.007	0.717	
CI min	0.001	−0.003	0.003	0.018	0.010	−0.002	−0.002	0	−0.001	0.009	−0.002	−0.002	0	−0.001	0.004	0	0.643	

CI max	0.013	0.009	0.018	0.060	0.035	0.010	0.059	0.012	0.009	0.027	0.016	0.011	0.013	0.009	0.025	0.018	0.793
G'_{ST}	0.055	0.001	0.087	0.135	0.225	0.003	0.077	0.063	0.031	0.135	0.029	0.031	0.043	0.029	0.116	0.027	0.944
CI min	0.008	−0.008	0.027	0.073	0.120	−0.003	−0.007	0.001	−0.009	0.074	−0.014	−0.017	−0.004	−0.007	0.031	0.001	0.907
CI max	0.123	0.025	0.160	0.207	0.343	0.015	0.197	0.149	0.085	0.211	0.106	0.110	0.104	0.079	0.215	0.056	0.974
D_{est}	0.050	0.001	0.078	0.103	0.209	0.001	0.056	0.058	0.028	0.121	0.025	0.028	0.038	0.026	0.105	0.020	0.808
CI min	0.007	−0.006	0.024	0.054	0.110	−0.001	−0.005	0.001	−0.008	0.066	−0.012	−0.015	−0.003	−0.006	0.028	0.001	0.740
CI max	0.110	0.018	0.145	0.159	0.320	0.005	0.152	0.138	0.077	0.190	0.091	0.099	0.092	0.071	0.195	0.050	0.872

4 Discussion

The objective of this study was to provide support for an explanation for the geographical distribution of the resistant flea beetle phenotype. This distribution may theoretically be explained by either high population structure with little dispersal or by a cost to resistance on other host plants than *B. vulgaris* G-type. Because resistance is highly favoured on *B. vulgaris* G-type, we expect that for comparisons between populations on *B. vulgaris* G-type and another host plant, genetic differentiation of the resistance trait is larger than for neutral markers if coadapted gene complexes are indeed present in resistant beetles and break up through recombination during outbreeding with other local flea beetle populations, then we expect a larger genetic differentiation for the resistance trait than expected for neutral markers, not only for comparisons between *B. vulgaris* G-type and another host plant but also between other host plants at different geographical distances from a G-type patch.

The outlier detection methods showed in all comparisons between populations on G-type and on *S. arvensis* that resistance emerged as a candidate for a locus that is subject to selection. This was expected because there is strong selection favouring resistance on the *B. vulgaris* G-type. To also find an indication that resistance is an outlier in the comparison between two *P. nemorum* populations on *S. arvensis* is more surprising and indicates that also differences in resistance allele frequencies between populations on *S. arvensis* are influenced by local selection, rather than a restriction to dispersal. Whereas the genetic differentiation of microsatellites between the two *S. arvensis* patches is low ($D_{est} < 0.1$), nearly 70% of the flea beetles on plant patch A were resistant compared to only 30% on plant patch B. So there seems to be directional selection on the resistance trait even on other host plants than *B. vulgaris* G-type.

Earlier work with neutral allozymes had already shown a weak, though significant, genetic differentiation between the samples under the influence of geographical distance (De Jong et al. 2001). In addition, it was shown that the differentiation was not only influenced by geographical distance, but also by the resistance phenotype of the beetles (De Jong et al. 2009). Our results are consistent with the allozyme analysis by de Jong et al. (2009), for weak but significant differentiation between flea beetles collected on different plant patches was found as well as a high level of gene flow between the plant patches. The populations on different plant patches do not seem sufficiently differentiated to explain the distribution of the resistance trait, because the microsatellites show only weak differentiation, whereas differences in percentage of resistant flea beetles between populations on different plant patches are substantial. The level of genetic differentiation of neutral loci may be accounted for by genome-wide effects such as random drift and gene flow. When the level of genetic differentiation is substantially larger for the resistance trait than for neutral loci, then this could reflect locus-specific effects such as directional selection or a difference in mutation variation (Balloux and Lugon-Moulin 2002; Hedrick 2005; Noor and Feder 2006). However, with a high rate of gene flow between the subpopulations, there is not a strong effect of mutation rate (Balloux and Lugon-Moulin 2002; Beaumont and Nichols 1996; Hedrick 2005). Certain measurements for differentiation, such as G'_{ST}, also take a possible effect of different mutation rates into account. As high mutation rates would lead to a more different allele composition between subpopulations, differentiation measurements like Jost's D (Jost 2008) would be higher for markers with a high mutation rate compared to markers with a lower mutation rate (in contrast to estimators such as G_{ST} that decrease with high mutation (Balloux and Lugon-Moulin 2002; Hedrick 1999)). Because of the presumably high mutation rate of microsatellites, it could very well differ from the mutation rate of the genetic region causing resistance in the flea beetle. So a difference in mutation rate of the neutral loci compared to the candidate gene may have affected our results. It has not caused an underestimation of genetic differentiation between subpopulations though. In this case, taking different mutation rates into account or not leads to the same conclusion: the resistant trait behaves as an outlier. By using multiple differentiation measurements (including mutation models or not) and acknowledging the substantial amount of gene flow between subpopulations, we believe that differences in mutation rate did not influence our findings in any substantial way.

Selection on the resistance trait seems to differ in strength as well as direction between subpopulations. While selection strongly favours resistance on *B. vulgaris* G-type, the ratio of resistant beetles decreases substantially with distance from the *B. vulgaris* G-type patches and the resistance trait behaves as an outlier for other host plant comparisons. Differences in percentage of resistant flea beetles between populations on the same host plant patches are substantial. It seems that there is a cost associated with the resistance trait, so that selection acts against this specific genotype on other host plants than the *B. vulgaris* G-type. This would limit the spread of resistance, even when dispersal is not limited. The difference in strength and direction of selection fits with the hypothesis of the breaking up of coadapted gene complexes (De Jong and Nielsen 2002). When gene complexes break up, the resistance trait becomes less favourable and, therefore, selection will act against it. This is likely to occur at different rates in different subpopulations, leading to a stronger signal of negative selection in subpopulations further away from *B. vulgaris* G-type. That is why we would find a larger genetic differentiation for the resistance trait than expected for neutral markers for comparisons between other host plants at different geographical distances from a G-type patch.

Furthermore, we found higher values of the estimated parameters G_{ST}, G'_{ST} and D for the resistance trait than for the microsatellites. When we looked at the subpopulations on *S. arvensis*, none of the 95% confidence intervals constructed for the estimators of differentiation amongst neutral loci had an upper level exceeding the estimator of differentiation for the resistance trait. We cannot conclude from this that the resistance trait is under selection, because these estimators each have their drawbacks. G'_{ST} and D are not comparable between loci when mutation rates differ substantially. G_{ST} is influenced by high within population heterozygosity (H_s). When H_s is high (as is often the case with microsatellites because they may contain many alleles), G_{ST} estimators can only produce low values, whether there is high differentiation between subpopulations or not.

There are some limitations of this study that should be taken into account. One of the limitations is that the genotype of the resistant and susceptible beetles was not known because resistance is a dominant marker. Therefore, homozygous and heterozygous genotypes cannot be distinguished by our bioassays. When heterozygosity is used as a measure for differentiation, the ratio of heterozygotes and homozygotes is needed. To estimate the genotype of the collected beetles, we assumed this locus to be in HWE like most of the sampled markers in the subpopulations. Of course we know that loci under selection—whether diversifying or balancing selection—are not in HWE according to the neutral theory. We believe that using estimated genotypes for resistance does not affect our conclusions. If the resistant beetles would include more homozygous beetles than estimated using HWE, then differences between the resistant trait and neutral loci would only become larger. In addition, the fact that the resistance trait is an outlier even while HWE was used to estimate genotype frequencies shows that between the subpopulations the allele distribution is not what one would expect for neutral loci. The estimated genotypes for beetles from location C (*B. vulgaris* G-type) are all homozygous, even under HWE, because no susceptible beetles were found at the site. This is supported by data from previous studies where nearly all beetles were found to be homozygous for the resistance trait in specific crosses with susceptible beetles to determine their genotype (De Jong et al. 2000).

Different assumptions are made when applying the population genomics approach. One is assuming the neutral theory of molecular evolution (Kimura 1983) to find selection signatures. This approach is widely used (see e.g. review (Butlin 2010)) but should be done taking into account other evolutionary processes that can cause a departure from predictions (Charlesworth et al. 2003). In our case, we tested for neutrality, recent bottlenecks, inbreeding and linkage before using STRUCTURE and looking for outliers, and we used a stringent criterion for the comparisons in LOSITAN and BayeScan. In the population genomic approach, multiple comparisons are made which can more easily lead to false positives than a single comparison. Therefore, usually different outlier detection methods are used as well as corrections such as the Bonferroni method (e.g. the false discovery rate correction (FDR) of P-values by Benjamini and Hochberg (1995)). Our findings are robust against variation in the methods.

The present study forms an example of the use of the population genomics approach as proposed in Vermeer et al. (2011), i.e. rather than using it to identify parts of the genome that are likely to be under selection without a priori knowledge of function of these genomic regions, now we used this approach with a candidate for a genetic factor under selection to answer ecological questions (in this case, about the cause of the geographic distribution of resistance).

This appears to be a suitable method to eventually identify coevolutionary hotspots in a geographic context, as proposed by Vermeer et al. (2011). The results warrant further work on the detection of the postulated coadapted gene complexes in the flea beetles. This local genetic coadaptation might be a more general mechanism underlying local ecological adaptation. A potentially fruitful direction for future research is to identify the detailed molecular mechanism of resistance. This might lead to an understanding of why this resistance trait experiences negative selection on other host plants than the *B. vulgaris* G-type.

5 Conclusions

A population genomics analysis was successfully used to answer a population-ecological question: Is selection involved in the geographical distribution of a trait influencing host plant use in a flea beetle, or can this distribution be explained by (limitations to) dispersal? The outlier detection methods showed that the trait 'resistance' (to defence of one of the host plants) emerged as an outlier, and we conclude that its geographical distribution is therefore most likely influenced by not only dispersal, but also selection. Surprisingly, this apparent selection did not only operate on flea beetle populations using different host plants but also must be invoked to explain differences in frequencies of the resistance trait between populations using the same host plant.

Acknowledgements We would like to thank Patrik Nosil for critically reading of, and constructively commenting on, an earlier version of this manuscript. We would also like to thank John Thompson for inspiring us to work with population genomics/genetics to unravel some aspects of his comprehensive geographic mosaic theory of coevolution. We also would like to show our gratitude to Ulrich B. Madsen, Erikholm Skovdistrikt (Kværkeby) for permission to work on the property.

Availability of Supporting Data DNA sequences: Genbank accessions FJ217683-FJ217691, DQ507814, FJ217694-FJ217697.

The other data will be made available on our institutional data repository: (W:\PROJECTS \entomology_Research\deJong_Peter).

Competing Interests The authors declare that they have no competing interests.

Ethics No ethical legislation was applicable to this study since it involved work on native insects.

Authors' Contributions KV carried out the sampling and molecular work and drafted the manuscript, including statistical analysis. PV participated in designing and carrying out the molecular analysis. MD provided a major contribution in writing the paper. PJ originally conceived the idea of this study, participated in the sampling and analysis and helped draft the manuscript. All authors read and approved the final manuscript.

Authors' Information KV performed this work as part of her PhD thesis and has obtained her PhD successfully. PV is head of the molecular laboratory at the Entomology laboratory of Wageningen University. MD is head of the department of Entomology, and PJ is assistant professor, was co-promotor of KV and is the principal investigator of the line of research to which this work belongs.

References

Agerbirk N, Olsen CE, Bibby BM, Frandsen HO, Brown LD, Nielsen JK, et al. A saponin correlated with variable resistance of Barbarea vulgaris to the diamondback moth Plutella xylostella. J Chem Ecol. 2003a;29(6):1417–33.

Agerbirk N, Orgaard M, Nielsen JK. Glucosinolates, flea beetle resistance, and leaf pubescence as taxonomic characters in the genus Barbarea (Brassicaceae). Phytochemistry. 2003b;63(1):69–80.

Agrawal AF, Brodie ED 3rd, Wade MJ. On indirect genetic effects in structured populations. Am Nat. 2001;158(3):308–23.

Antao T, Lopes A, Lopes RJ, Beja-Pereira A, Luikart G. LOSITAN: a workbench to detect molecular adaptation based on a F(st)-outlier method. BMC Bioinformatics. 2008;9:323.

Avise JC. Molecular markers, natural history, and evolution. Sunderland: Sinauer Associates; 2004. 684 p.

Balding DJ. Likelihood-based inference for genetic correlation coefficients. Theor Popul Biol. 2003;63(3):221–30.

Balloux F, Lugon-Moulin N. The estimation of population differentiation with microsatellite markers. Mol Ecol. 2002;11(2):155–65.

Barton NH. Genetic hitchhiking. Philos Trans R Soc Lond Ser B Biol Sci. 2000;355(1403):1553–62.

Beaumont MA, Balding DJ. Identifying adaptive genetic divergence among populations from genome scans. Mol Ecol. 2004;13(4):969–80.

Beaumont MA, Nichols RA. Evaluating loci for use in the genetic analysis of population structure. Proc Roy Soc London Ser B Biol Sci. 1996;263(1377):1619–26.

Beebee TJC, Rowe G. Introduction to molecular ecology, vol. XII. Oxford: Oxford University Press; 2004. 346 p.

Belkhir K, Borsa P, Chikhi L, Raufaste N, Bonhomme F. GENETIX 4.05, logiciel sous Windows TM pour la génétique des populations. Laboratoire Génome, Populations, Interactions, CNRS UMR 5000. 4.05.2 ed. Montpellier: Université de Montpellier II; 1998.

Benjamini Y, Hochberg Y. Controlling the false discovery rate – a practical and powerful approach to multiple testing. J R Stat Soc Ser B Stat Method. 1995;57(1):289–300.

Benkman CW, Holimon WC, Smith JW. The influence of a competitor on the geographic mosaic of coevolution between crossbills and lodgepole pine. Evolution. 2001;55(2):282–94.

Black WC, Baer CF, Antolin MF, DuTeau NM. Population genomics: genome-wide sampling of insect populations. Annu Rev Entomol. 2001;46:441–69.

Breuker CJ, Victoir K, De Jong PW, van der Meijden E, Brakefield PM, Vrieling K. AFLP markers for the R-gene in the flea beetle, Phyllotreta nemorum, conferring resistance to defenses in Barbarea vulgaris. J Insect Sci. 2005;5:38.

Breuker CJ, de Jong PW, Victoir K, Vrieling K, Brakefield PM. Pleiotropic effects associated with an allele enabling the flea beetle Phyllotreta nemorum to use Barbarea vulgaris as a host plant. Evol Ecol. 2007;21(1):13–26.

Butlin RK. Population genomics and speciation. Genetica. 2010;138(4):409–18.

Campana MG, Hunt HV, Jones H, White J. CorrSieve: software for summarizing and evaluating structure output. Mol Ecol Resour. 2011;11(2):349–52.

Charlesworth B, Charlesworth D, Barton NH. The effects of genetic and geographic structure on neutral variation. Annu Rev Ecol Evol Syst. 2003;34:99–125.
Cornell HV, Hawkins BA. Herbivore responses to plant secondary compounds: a test of phytochemical coevolution theory. Am Nat. 2003;161(4):507–22.
Cornuet JM, Luikart G. Description and power analysis of two tests for detecting recent population bottlenecks from allele frequency data. Genetics. 1996;144(4):2001–14.
Crawford NG. smogd: software for the measurement of genetic diversity. Mol Ecol Resour. 2010;10(3):556–7.
De Jong PW, Nielsen JK. Polymorphism in a flea beetle for the ability to use an atypical host plant. Proc Roy Soc London Ser B Biol Sci. 1999;266(1414):103–11.
De Jong PW, Nielsen JK. Reduction in fitness of flea beetles which are homozygous for an autosomal gene conferring resistance to defences in Barbarea vulgaris. Heredity. 2000;84 (Pt 1):20–8.
De Jong PW, Nielsen JK. Host plant use of Phyllotreta nemorum: do coadapted gene complexes play a role? Entomol Exp Appl. 2002;104:207–15.
De Jong PW, Frandsen HO, Rasmussen L, Nielsen JK. Genetics of resistance against defences of the host plant Barbarea vulgaris in a Danish flea beetle population. Proc Biol Sci. 2000;267 (1453):1663–70.
De Jong PW, De Vos H, Nielsen JK. Demic structure and its relation with the distribution of an adaptive trait in Danish flea beetles. Mol Ecol. 2001;10(5):1323–32.
De Jong PW, Breuker CJ, De Vos H, Vermeer KMCA, Oku K, Verbaarschot P, et al. Genetic differentiation between resistance phenotypes in the phytophagous flea beetle, Phyllotreta nemorum. J Insect Sci. 2009;9:1–8.
Earl DA, Vonholdt BM. STRUCTURE HARVESTER: a website and program for visualizing STRUCTURE output and implementing the Evanno method. Conserv Genet Resour. 2012;4(2): 359–61.
Ehrlich PR, Raven PH. Butterflies and plants – a study in coevolution. Evolution. 1964;18(4): 586–608.
Evanno G, Regnaut S, Goudet J. Detecting the number of clusters of individuals using the software STRUCTURE: a simulation study. Mol Ecol. 2005;14(8):2611–20.
Falush D, Stephens M, Pritchard JK. Inference of population structure using multilocus genotype data: linked loci and correlated allele frequencies. Genetics. 2003;164(4):1567–87.
Falush D, Stephens M, Pritchard JK. Inference of population structure using multilocus genotype data: dominant markers and null alleles. Mol Ecol Notes. 2007;7(4):574–8.
Foll M. BayeScan v2.1 User Manual. 2012. Swiss Institute of Bioinformatics, Bern.
Foll M, Gaggiotti O. A genome-scan method to identify selected loci appropriate for both dominant and codominant markers: a Bayesian perspective. Genetics. 2008;180(2):977–93.
Forde SE, Thompson JN, Bohannan BJ. Adaptation varies through space and time in a coevolving host-parasitoid interaction. Nature. 2004;431(7010):841–4.
Fox LR. Defense and dynamics in plant-herbivore systems. Am Zool. 1981;21(4):853–64.
Goldstein DB, Linares AR, Cavallisforza LL, Feldman MW. An evaluation of genetic distances for use with microsatellite loci. Genetics. 1995;139(1):463–71.
Gomulkiewicz R, Drown DM, Dybdahl MF, Godsoe W, Nuismer SL, Pepin KM, et al. Dos and don'ts of testing the geographic mosaic theory of coevolution. Heredity. 2007;98(5):249–58.
Goudet J. FSTAT (Version 1.2): a computer program to calculate F-statistics. J Hered. 1995;86(6): 485–6.
Goudet J. FSTAT: a program to estimate and test gene diversities and fixation indices (version 2.9.3, February 2002). Available from http://www2.unil.ch/popgen/softwares/fstat.htm.
Hedrick PW. Perspective: highly variable loci and their interpretation in evolution and conservation. Evolution. 1999;53(2):313–8.
Hedrick PW. A standardized genetic differentiation measure. Evolution. 2005;59(8):1633–8.
Hu S, Dilcher DL, Jarzen DM, Winship TD. Early steps of angiosperm pollinator coevolution. Proc Natl Acad Sci USA. 2008;105(1):240–5.

Hubisz MJ, Falush D, Stephens M, Pritchard JK. Inferring weak population structure with the assistance of sample group information. Mol Ecol Resour. 2009;9(5):1322–32.

Jakobsson M, Rosenberg NA. CLUMPP: a cluster matching and permutation program for dealing with label switching and multimodality in analysis of population structure. Bioinformatics. 2007;23(14):1801–6.

Jost L. G(ST) and its relatives do not measure differentiation. Mol Ecol. 2008;17(18):4015–26.

Kimura M. The neutral theory of molecular evolution. Cambridge: Cambridge University Press; 1983.

Laine AL. Role of coevolution in generating biological diversity: spatially divergent selection trajectories. J Exp Bot. 2009;60(11):2957–70.

Leng L, Zhang DE. Measuring population differentiation using GST or D? A simulation study with microsatellite DNA markers under a finite Island model and nonequilibrium conditions. Mol Ecol. 2011;20(12):2494–509.

Luikart G, England PR, Tallmon D, Jordan S, Taberlet P. The power and promise of population genomics: from genotyping to genome typing. Nat Rev Genet. 2003;4(12):981–94.

Millipore. Sequencing Reaction Purification Protocol 2000 [cited 2012 19-04-2012]. Protocol. Available from: http://www.millipore.com/publications.nsf/a73664f9f981af8c852569b9005b4eee/d447617b3cc8f5bb852568f5006b2ab4/$FILE/tn053.pdf.

Mopper S, Strauss SY. Genetic structure and local adaptation in natural insect populations: effects of ecology, life history, and behavior. New York [etc.]: Chapman & Hall [etc.]; 1998. 449 p.

Nei M. Analysis of gene diversity in subdivided populations. Proc Natl Acad Sci USA. 1973;70(12):3321–3.

Nei M. Estimation of average heterozygosity and genetic distance from a small number of individuals. Genetics. 1978;89(3):583–90.

Nei M. Molecular evolutionary genetics. New York: Colombia University Press; 1987.

Nei M, Chesser RK. Estimation of fixation indexes and gene diversities. Ann Hum Genet. 1983;47: 253–9.

Nielsen JK. Intraspecific variability in adult flea beetle behaviour and larval performance on an atypical host plant. Entomol Exp Appl. 1996;80(1):160–2.

Nielsen JK. Variation in defences of the plant Barbarea vulgaris and in counter adaptations by the flea beetle Phyllotreta nemorum. Entomol Exp Appl. 1997a;82(1):25–35.

Nielsen JK. Genetics of the ability of Phyllotreta nemorum larvae to survive in an atypical host plant, Barbarea vulgaris ssp arcuata. Entomol Exp Appl. 1997b;82(1):37–44.

Nielsen JK. Specificity of a Y-linked gene in the flea beetle Phyllotreta nemorum for defences in Barbarea vulgaris. Entomol Exp Appl. 1999;91:359–68.

Nielsen JK. Non-random segregation of an autosomal gene in males of the flea beetle, Phyllotreta nemorum: implications for colonization of a novel host plant. Entomol Exp Appl. 2012;143: 301–12.

Nielsen JK, de Jong PW. Temporal and host-related variation in frequencies of genes that enable Phyllotreta nemorum to utilize a novel host plant, Barbarea vulgaris. Entomol Exp Appl. 2005;115:265–70.

Noor MA, Feder JL. Speciation genetics: evolving approaches. Nat Rev Genet. 2006;7(11):851–61.

Peakall R, Smouse PE. GENALEX 6: genetic analysis in excel. Population genetic software for teaching and research. Mol Ecol Notes. 2006;6(1):288–95.

Pellmyr O. Yuccas, yucca moths, and coevolution: a review. Ann Mo Bot Gard. 2003;90(1):35–55.

Peterson MA, Denno RF. The influence of dispersal and diet breadth on patterns of genetic isolation by distance in phytophagous insects. Am Nat. 1998;152(3):428–46.

Pritchard JK, Stephens M, Donnelly P. Inference of population structure using multilocus genotype data. Genetics. 2000;155(2):945–59.

Pritchard VL, Jones K, Metcalf JL, Martin AP, Wilkinson P, Cowley DE. Characterization of tetranucleotide microsatellites for Rio Grande cutthroat trout and rainbow trout, and their cross-amplification in other cutthroat trout subspecies. Mol Ecol Notes. 2007;7(4):594–6.

Pritchard JK, Wen W, Falush D. Documentation for structure software: Version 2.32010 [cited]. 2010. Available from http://pritch.bsd.uchicago.edu.
Rosenberg NA. DISTRUCT: a program for the graphical display of population structure. Mol Ecol Notes. 2004;4(1):137–8.
Sagan L. On the origin of mitosing cells. J Theor Biol. 1966;14(3):225–74.
Shinoda T, Nagao T, Nakayama M, Serizawa H, Koshioka M, Okabe H, et al. Identification of a triterpenoid saponin from a crucifer, Barbarea vulgaris, as a feeding deterrent to the diamondback moth, Plutella xylostella. J Chem Ecol. 2002;28(3):587–99.
Shoemaker JS, Painter IS, Weir BS. Bayesian statistics in genetics – a guide for the uninitiated. Trends Genet. 1999;15(9):354–8.
Slatkin M. A measure of population subdivision based on microsatellite allele frequencies. Genetics. 1995;139(1):457–62.
Strong DR, Lawton JH, Southwood R. Insects on plants: community patterns and mechanisms. Oxford [etc.]: Blackwell; 1984. 313 p.
Takezaki N, Nei M. Genetic distances and reconstruction of phylogenetic trees from microsatellite DNA. Genetics. 1996;144(1):389–99.
Thompson JN. Specific hypotheses on the geographic mosaic of coevolution. Am Nat. 1999;153: S1–S14.
Thompson JN. The geographic mosaic of coevolution, vol. XII. Chicago [etc.]: University of Chicago Press; 2005. 443 p.
Thompson JN. The coevolving web of life. Am Nat. 2009;173(2):125–40.
Thompson JN, Cunningham BM. Geographic structure and dynamics of coevolutionary selection. Nature. 2002;417(6890):735–8.
Van Oosterhout C, Hutchinson WF, Wills DPM, Shipley P. MICRO-CHECKER: software for identifying and correcting genotyping errors in microsatellite data. Mol Ecol Notes. 2004;4(3): 535–8.
Verbaarschot P, Calvo D, Esselink D, Molina JM, Vrieling K, De Jong PW. Isolation of polymorphic microsatellite loci from the flea beetle Phyllotreta nemorum L. (Coleoptera: Chrysomelidae). Mol Ecol Notes. 2007;7:60–2.
Vermeer KMCA, Dicke M, de Jong PW. The potential of a population genomics approach to analyse geographic mosaics of plant—insect coevolution. Evol Ecol. 2011;25(5):977–92. https://doi.org/10.1007/s10682-010-9452-8.
Vermeer KMCA, Verbaarschot P, de Jong PW. Changes in frequencies of genes that enable Phyllotreta nemorum to utilize its host plant, Barbarea vulgaris, vary in magnitude and direction, as much within as between seasons. Entomol Exp Appl. 2012;144(1):37–44.
Weir BS, Cockerham CC. Estimating F-statistics for the analysis of population-structure. Evolution. 1984;38(6):1358–70.
Wheat CW, Vogel H, Wittstock U, Braby MF, Underwood D, Mitchell-Olds T. The genetic basis of a plant-insect coevolutionary key innovation. Proc Natl Acad Sci USA. 2007;104(51): 20427–31.
Wright S. Size of population and breeding structure in relation to evolution. Science. 1938;87:430–1.
Wright S. The genetical structure of populations. Ann Eugenics. 1951;15(4):323–54.
Wright S. Evolution and the genetics of populations: a treatise in four volumes. Chicago [etc.]: University of Chicago Press; 1968. 4 dl; p.

Population Genomic Insights into Insecticide Resistance in the Colorado Potato Beetle

Sean D. Schoville, Zachary P. Cohen, and Michael S. Crossley

Abstract The ability of pests to develop resistance is still poorly understood, particularly regarding the process of adaptive change that leads to resistance. Understanding the mechanisms involved in pesticide adaptation is important because common methods of field-level resistance management assume that simple (single-gene) mutations cause resistance, but may not perform effectively when multiple loci and several molecular pathways confer resistance. Emerging experimental and genomic evidence suggests the single-gene model may be insufficient to explain many cases of rapid pest evolution. Here we review such evidence in the Colorado potato beetle, *Leptinotarsa decemlineata*, widely considered to be a super-pest due to its rapid evolution to most classes of insecticides. A combination of studies suggests that polygenic resistance drawn from standing variation best explains regional differences in patterns of insecticide resistance. However, these studies have lacked a temporal framework to demonstrate polygenic evolution unequivocally, as well as the ability to estimate key evolutionary parameters to model processes of rapid evolution. We suggest a temporal framework to empirically test for polygenic insecticide resistance evolution, as well as determine the proximate genetic and evolutionary mechanisms contributing to resistance. Emerging data from population genomics promise to increase our knowledge of insect pest molecular genetics by identifying gene networks involved in pesticide resistance, thereby providing a means to monitor the spread of resistance, develop management approaches that increase pesticide longevity, and identify new methods of gene-targeted pest control.

Keywords Genomic variation · Insect pest · Insecticide resistance · *Leptinotarsa decemlineata* · Pesticide treadmill · RAPID evolution

S. D. Schoville (✉) · Z. P. Cohen
Department of Entomology, University of Wisconsin-Madison, Madison, WI, USA
e-mail: sean.schoville@wisc.edu

M. S. Crossley
Department of Entomology, University of Wisconsin-Madison, Madison, WI, USA

Department of Entomology, University of Georgia, Athens, GA, USA

Julian R. Dupuis and Om P. Rajora (eds.), *Population Genomics: Insects*, Population Genomics [Om P. Rajora (Editor-in-Chief)], https://doi.org/10.1007/13836_2021_91,

1 Introduction

Pesticide resistance in agricultural pests remains a major concern for the global food supply, as it limits productivity, incurs financial costs, and requires unsustainable levels of chemical inputs. Pests evolve rapidly and can be devastating if they are unmanaged or allowed to reach outbreak conditions (Mallet 1989; Gould 1991). Despite considerable research effort over the last 70 years, efforts to manage pesticide resistance continue to fail, and observed patterns of resistance evolution challenge the assumptions of the evolutionary models employed in pest management (Hoy 1998). While resistance management models have been highly effective in certain scenarios (Tabashnik et al. 2013), emerging genomic evidence now raises questions about whether simple genetic changes (i.e., single-gene mutations) adequately explain most cases of insecticide resistance (Ffrench-Constant 2013). Despite early studies arguing that polygenic (multiple-gene) evolution from standing (ancestral) variation was likely to be the norm (Crow 1957), most research of the last few decades focused on de novo mutational change as the source of adaptive variation (Feyereisen et al. 2015). The importance of new mutations has been demonstrated, for example, in episodes of insecticide resistance in *Anopheles funestus* (Wondji et al. 2011; Riveron et al. 2014) and *Drosophila melanogaster* (Karasov et al. 2010). Yet, the question of what mechanisms underlie rapid genetic adaptation remains a highly debated topic in evolutionary biology (Orr 2005; Barrett and Schluter 2008). If species rapidly evolve by drawing upon standing genetic variation, this has implications for resistance management models and future control methods in integrated pest management. Such evolutionary change is evident, for example, in plant species that have colonized heavy mineral-contaminated mine tailings (Baker 1987) or evolved herbicide resistance (Kreiner et al. 2018), and in marine fish that have colonized highly polluted superfund sites (Whitehead et al. 2017). To be plausible, this standing variation model of resistance evolution requires certain evolutionary conditions: large and stable population sizes, high mutation rates, and/or sufficient spatial gene flow. Otherwise, standing variation should be rapidly removed from pest populations experiencing high mortality and undergoing genetic drift. However, most evolutionary biologists would agree that a combination of de novo mutational change and selection on standing genetic diversity probably occurs in many cases of adaptation, as it is evident that phenotypic traits are typically quantitative in nature (influenced by multiple loci) and polygenic adaptation frequently emerges in theoretical simulations (Wellenreuther and Hansson 2016). Studies of pest species have generally lacked the genomic, spatial, and temporal framework to properly test these mechanisms (Pélissié et al. 2018). Greater effort is needed to empirically assess the relative importance of these modes of adaptive evolution in cases of insecticide resistance.

2 Population Genetic Models of Insecticide Resistance

Research on insecticide resistance has been of interest to population geneticists for more than 60 years (Crow 1957), and is frequently used in textbook examples for understanding population genetic responses to selective pressures (Hartl and Clark 1997). This interest has led, over time, to the development of resistance management models to inform the field deployment of insecticides and prolong their efficacy. Resistance management strategies assume that simple (single-gene) mutations of large effect cause resistance and should be rare (Roush 1989), while writing off the importance of alternative mechanisms that involve multiple loci (Denholm and Rowland 1992; Haridas and Tenhumberg 2018). They also assume that resistance mutations might be present at low frequency prior to insecticide exposure, but would need to be recessive in order to be maintained in the population (Georghiou and Taylor 1986). This is because resistance mutations are presumed to be deleterious in natural populations compared to the wild-type allele, related to their strong phenotypic effects. The fitness difference among alleles is an important assumption affecting evolutionary outcomes, as lower fitness acts to slow or even prevent the mutation from going to fixation (Georghiou and Taylor 1986), and has the implication that resistance mutations will gradually disappear in large populations when selective pressures are removed. Furthermore, spatial resistance management models, which are premised on gene flow from susceptible "refuge" populations carrying higher fitness wild-type alleles (Tabashnik 1990), only prevent the rapid fixation of resistance mutations when fitness tradeoffs exist. While fitness costs associated with resistance have been found in many insects (Crow 1957; Kliot and Ghanim 2012), mutations that ameliorate these fitness costs may be present in some populations or may arise over time.

Collectively, these underlying assumptions may provide unrealistic expectations for many cases of insecticide adaptation (Hoy 1998; Bourguet et al. 2000), although more recent efforts to model the rate of pesticide resistance evolution are exploring polygenic models (Haridas and Tenhumberg 2018). Understanding the evolutionary mechanisms underlying pesticide adaptation has become increasingly important because of the widespread occurrence of the "pesticide treadmill" phenomenon in agricultural pests (Luck et al. 1977). Repeated and escalating use of pesticides has led to a need to more quickly develop new pesticide chemistries in order to keep pace with pest evolution (Alyokhin et al. 2015). Furthermore, it is of fundamental interest to understand the evolutionary genetics of insecticide resistance, as these real-world experiments provide some of the best cases for understanding how organisms successfully respond to rapid environmental change (Mallet 1989).

3 The Colorado Potato Beetle as a Model for Insecticide Resistance

The Colorado potato beetle (CPB), *Leptinotarsa decemlineata* (Fig. 1), is among the most notorious top insect pests on the Insecticide Resistance Action Committee database (IRAC 2016), with documented resistance to 56 insecticides spanning all modes of action (Whalon et al. 2016). From early insecticides such as Paris Green, to modern insecticides such as to DDT, pyrethroids and neonicotinoids, CPB populations have amassed resistance over impressively short periods of time, even within 1 year of the introduction of novel chemicals (Forgash 1985; Ioannidis et al. 1991; Alyokhin et al. 2008). At a continental scale, CPB exhibits regional differences in rates of adaptation to insecticides (Fig. 2). Populations in the Eastern U.S. typically evolve resistance to novel insecticides within a few generations (Forgash 1985; Olson et al. 2000; Alyokhin et al. 2008, 2015), while Midwestern U.S. populations have evolved resistance to a similar breadth of insecticides but at a slower rate (Szendrei et al. 2012; Huseth and Groves 2013; Clements et al. 2017b). In contrast, CPB populations in the Northwestern U.S. exhibit limited insecticide resistance (Alyokhin et al. 2008; Crossley et al. 2018; Dively et al. 2020). The cause of this geographic variation in adaptation to insecticides is unknown, though some authors have argued that regional differences in pesticide use are important (Clements et al. 2018), while others have argued that levels of standing genetic variation may play a role (Crossley et al. 2019a).

A fundamental question with important implications for resistance management is whether the same genes and molecular pathways repeatedly underlie CPB's ability to adapt to a broad range of insecticides across its range. While there is widespread evidence of insecticide resistance evolving in different geographical populations, there is relatively limited evidence that resistance evolves according to the assumptions of most pest management models. Traditionally, these models used classical population genetic approaches to model a single, recessive locus under selection,

Fig. 1 Adult Colorado potato beetles, *Leptinotarsa decemlineata*, on a potato plant

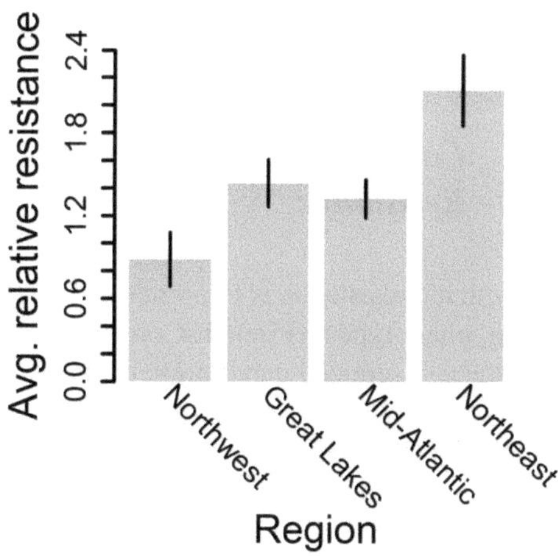

Fig. 2 Regional variation in relative rates of insecticide resistance in Colorado potato beetle. Data span 15 years of bioassays (1994–2009) and are summarized from Dively et al. (2020)

often with fitness tradeoffs in its homozygous form (Georghiou and Taylor 1986). Studies have shown that insecticide resistance traits in CPB range from recessive (or partially recessive) to dominant (or partially dominant), and in some cases are inherited in a sex-linked manner (Argentine et al. 1989a; Heim et al. 1992; Ioannidis et al. 1992; Hawthorne 2003; Alyokhin et al. 2008). Furthermore, to date there is limited evidence for fitness tradeoffs. A study of an organophosphate (azinophosmethyl) resistant CPB population in Massachusetts showed reduced larval and adult viability compared to a susceptible strain (Argentine et al. 1989a; Zhu et al. 1996). Crosses to the susceptible population showed no reduction in fitness in F1 heterozygotes, suggesting the resistance mutations were recessive. Resistance was associated with a point mutation causing an amino acid substitution in acetylcholinesterase. The same population was also resistant to pyrethroid insecticides (permethrin), but showed no fitness disadvantage relative to the susceptible strain. In a laboratory population from Michigan, 35 generations of selection to *Bacillus thuringiensis* (Bt) crystal inclusion (Cry) toxins (Trisyono and Whalon 1997; Alyokhin and Ferro 1999) led to fitness costs in terms of longer development time, reduced overwintering success, and reduced female fecundity. While Bt Cry toxins have been widely used in transgenic corn production, there has been comparatively limited use in potatoes and no known cases of CPB populations developing Bt resistance in nature. Resistance to neonicotinoids (imidacloprid) is more equivocal, with a few studies showing reduced viability in larvae and overwintering adults (Baker et al. 2007; Alyokhin et al. 2008), although the effects may not persist across generations. A lack of persistent fitness costs might be explained by emerging evidence that insecticide exposure can lead to epigenetic changes that have transgenerational effects on fitness (Margus et al. 2019; Brevik et al. 2020), but don't alter the genetic inheritance of traits.

4 Insights from Candidate Gene and Transcriptomics Research

4.1 Candidate Genes and the Genetic Basis of Insecticide Resistance

Insecticide resistance is hypothesized to occur through evolutionary changes in four main phenotypes (Ffrench-Constant 2013): target-site insensitivity (knockdown resistance), reduced penetration (thicker cuticle), metabolic detoxification and transport (e.g., cellular processing by ABC transporters, cytochrome P450s, glutathione S-transferases, and carboxylesterases), and behavioral avoidance of toxic compounds (Mamidala et al. 2011). Genetically, resistance phenotypes involve point mutations to structural or regulatory genes, duplication or gene conversion, or loss-of-function mutations (Feyereisen et al. 2015).

Early studies of the genetics of CPB resistance to insecticides employed an array of approaches, including insecticide bioassays with synergists, insecticide metabolite profiling, detoxification enzyme activity assays, experimental crosses to ascertain inheritance, and candidate gene sequencing. Notably, none adopted a spatial population genetic perspective, but instead drew conclusions from comparisons of resistant and susceptible reference populations, with 10 out of the 25 studies reviewed here using the same resistant reference population (Long Island, NY). This population is well known for its rapid development of insecticide resistance, with rates of evolution averaging 2.1 years to multiple modes of action (Georghiou 1986; Alyokhin et al. 2015). Experimental crosses revealed that some resistance traits were incompletely recessive (avermectin, neonicotinoid, and pyrethroid resistance), while others were incompletely dominant (*Bt,* carbamate, and organophosphate resistance) (Rose and Brindley 1985; Silcox et al. 1985; Argentine et al. 1989b, 1992, 1994, 1995; Ioannidis et al. 1992; Wierenga and Hollingworth 1993; Anspaugh et al. 1995; Rahardja and Whalon 1995; Zhu and Clark 1997; Miyo et al. 1999; Zhao et al. 2000; Hawthorne 2003; Mota-Sanchez et al. 2006; Baker et al. 2007; Tan et al. 2008; Jiang et al. 2011; Piiroinen et al. 2013; Wan et al. 2013; Qu et al. 2016). Many of these studies confirmed the importance of target-site genes, or metabolic detoxification genes, but the identities of these genes or further evidence of genetic changes were not always established.

Few studies have employed standard quantitative trait mapping experiments, such as crossing inbred lines or bulk segregant analysis, to identify and estimate the number of loci contributing to resistance in CPB. In one such experiment, Hawthorne (2003) mapped a pyrethroid-resistance mutation to the voltage-sensitive sodium channel (*Vssc*), located on the X-chromosome. The observed knockdown resistance (kdr) mutation in *Vssc* (allele L1014F) is associated with reduced sensitivity to pyrethroids and was also found to confer pyrethroid resistance in house flies, *Musca domestica* (Williamson et al. 1996). A review of U.S. CPB populations suggests multiple origins and several alleles conferring kdr mutations (Scott 2019), highlighting the inherent complexity of linking candidate genes to resistance phenotypes. Furthermore, a national monitoring effort in Germany determined that the resistance-conferring mutation in *Vssc* was widespread among CPB, but that field-relevant doses of the pyrethroid lambda-cyhalothrin still achieved economic control (Tebbe et al. 2016), suggesting that candidate gene approaches focusing on reference populations miss other sources of genetic variation that contribute to resistance in different populations.

4.2 Transcriptomic Studies of Insecticide Resistance

Metabolic detoxification enzymes, including the cytochrome P450s, glutathione S-transferases, esterases, and ABC transporters, can become constitutively upregulated in resistant insect pests (Alyokhin and Chen 2017; Dermauw et al. 2018), or rapidly induced when insects are exposed to insecticides (Bass et al. 2013; Wybouw et al. 2015; Zhu et al. 2016). Across major potato growing regions in the U.S., metabolic detoxification via constitutive changes in gene regulation is well supported in gene expression studies (Clements et al. 2016; Zhu et al. 2016; Kaplanoglu et al. 2017; Dively et al. 2020). Different pesticides can drive similar transcriptomic responses (expression of the same candidate genes) in resistant populations (Clements et al. 2018). Recent research on the master regulatory transcription factor *Cap-n-collar* suggests trans-regulatory control of multiple-gene networks in the metabolic resistance pathway in CPB (Gaddelapati et al. 2018). However, the genes involved in metabolic resistance vary regionally (Table 1), suggesting more complex quantitative variation in this trait in CPB. The role of different genes in metabolic resistance has been demonstrated by RNA interference experiments where knockdown of different upregulated CYP450 genes restores susceptibility in different pesticide-resistant populations (Clements et al. 2017b; Kalsi and Palli 2017). This suggests a simple upstream shift in *Cap-n-collar* expression is not sufficient to explain all cases of metabolic resistance in CPB and instead additional cis-regulatory changes to metabolic genes are required (Wilding 2018). Additionally, there is evidence for neonicotinoid target-site insensitivity in Wisconsin (Clements et al. 2017a), which provides evidence for multiple pathways of resistance in CPB.

5 Insights from Population Genomic Approaches

5.1 Patterns of Resistance at the Landscape-Scale

Landscape-level studies have found high spatial heterogeneity in levels of insecticide resistance (Ioannidis et al. 1991; Stewart et al. 1997; Kim et al. 2005; Hitchner et al. 2012; Szendrei et al. 2012; Huseth and Groves 2013; Huseth et al. 2015; Scott et al. 2015; Tebbe et al. 2016). Field populations separated by as little as two

Table 1 Statistically significant differentially expressed genes (with a known role in pesticide resistance) vary by potato growing region and experiment in neonicotinoid resistant CPB. Genes shared across experiments shown in bold type

Mechanisms	Categories	Long Island, NY (Hawthorne et al. Accepted 2020)	Long Island Lab selected strain (Kaplanoglu et al. 2017)	Hancock, WI (Clements et al. 2016)
Metabolic detoxification	**ABC transporters**			LDEC003183 LDEC004153 LDEC004861 LDEC005086 LDEC008896 LDEC009711 LDEC013093 LDEC017372 LDEC017376 LDEC017378 LDEC020972 LDEC020973
	CYP450s	LDEC006940 LDEC013538 **LDEC018533** **LDEC020421** **LDEC020527** LDEC021333 LDEC021334 LDEC022309 LDEC022856 LDEC023239 **LDEC023780** LDEC024396	**LDEC007599** LDEC012478	LDEC004541 LDEC005459 LDEC007597 LDEC007598 **LDEC007599** LDEC008257 LDEC011286 LDEC013790 LDEC015922 LDEC016278 LDEC016357 LDEC018119 **LDEC018533** LDEC020420 **LDEC020421** **LDEC020527** **LDEC023780**
	Esterases	LDEC022531	LDEC020362	LDEC004000 LDEC006847 LDEC011645 LDEC013801 LDEC018985 LDEC019607 LDEC021268
	GSTs	LDEC001123 LDEC023323	LDEC020265	LDEC003607
	MFS transporters	LDEC019606		LDEC013314

(continued)

Table 1 (continued)

Mechanisms	Categories	Long Island, NY (Hawthorne et al. Accepted 2020)	Long Island Lab selected strain (Kaplanoglu et al. 2017)	Hancock, WI (Clements et al. 2016)
Target-site genes	**Glutamate receptor pathways**	LDEC005342	LDEC015313 LDEC006139 LDEC014124	LDEC002149 LDEC015534 LDEC010832
	Voltage-dependent channels	LDEC013670 LDEC017037	LDEC019444 LDEC005749 LDEC020439	LDEC010130 LDEC000112
	Known insecticide resistance genes			LDEC016741

kilometers can have very different sensitivities to the same insecticide (Crossley et al. 2017), and even a single field population can be comprised of individuals with varying levels of insecticide resistance (Heim et al. 1990; Crossley et al. 2017, 2018). This high spatial and temporal variability presents a challenge to pest management, but an opportunity to identify genomic variation responding to selection from insecticides. We explore how one case study leveraged this high spatial genomic variation to understand the evolution of insecticide resistance (Box 1).

Box 1

Crossley et al. (2017) employed tools from population genomics and landscape genomics to study insecticide resistance patterns in Wisconsin. In particular, they employed genome-environment association tests to detect statistical associations between single nucleotide polymorphism (SNP) allele frequencies and the intensity of potato production (Fig. 3). After screening more than 30,000 SNPs, 207 exhibited significant correlations, and were found in or near 462 genes. Forty-two of these genes could easily be related to insecticide resistance, encoding: CYPs, glutathione-*S*-transferases, ATP-binding cassette transporters, insecticide target sites (e.g., glutamate receptor), and cuticle proteins. This suggests that the genetic variation underlying resistance is likely polygenic, and most notably, it was distributed among populations at a relatively small spatial scale.

A key limitation of the environmental-association approach is that it is not able to determine causal linkages between selection events and a particular episode of pesticide resistance evolution. Crossley et al. (2017) used a landscape metric for potato land cover intensity as a surrogate for insecticide use, but selection has been exerted on CPB by many insecticides for over a century. Historical insecticide use may have selected for several genetic changes

(continued)

Box 1 (continued)

(Fig. 4), which were followed by other selection events from more recent insecticides (Ioannidis et al. 1992). Thus, the result supporting a polygenic basis of resistance is likely biased by a historical legacy of cross-resistance to multiple insecticides, as well as the independent trajectories taken by each sampled population. Decoupling resistance loci and the insecticidal selective agent in multi-resistant strains can be approached through backcross breeding designs to a susceptible strain (Cattel et al. 2020), or alternatively through the use of historical specimens that span the history of insecticide use and population genetic approaches that model selection for individual alleles (Malaspinas et al. 2012).

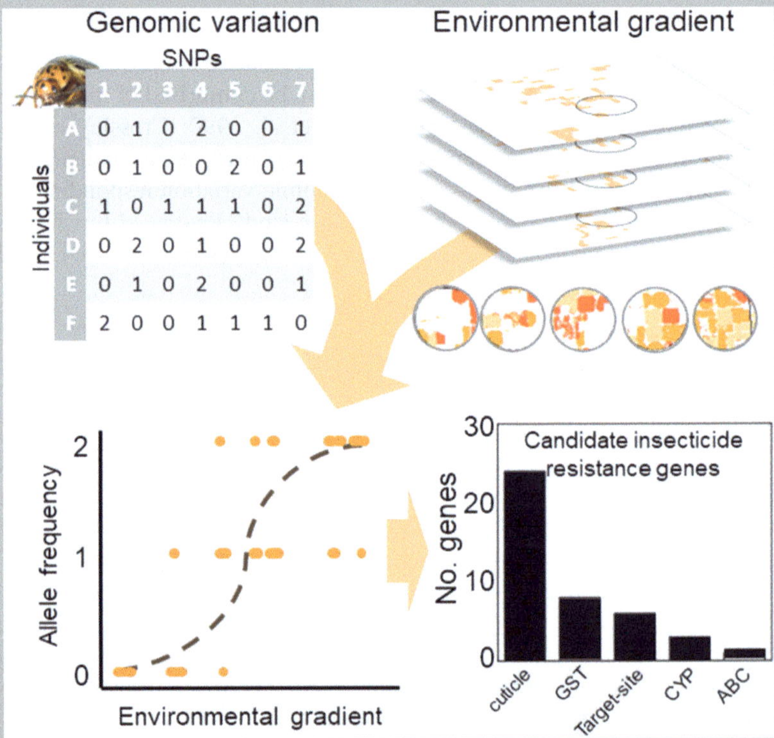

Fig. 3 Schematic showing genome environmental-association analysis employed by Crossley et al. (2017) to detect polygenic adaptation to insecticide use in Colorado potato beetle

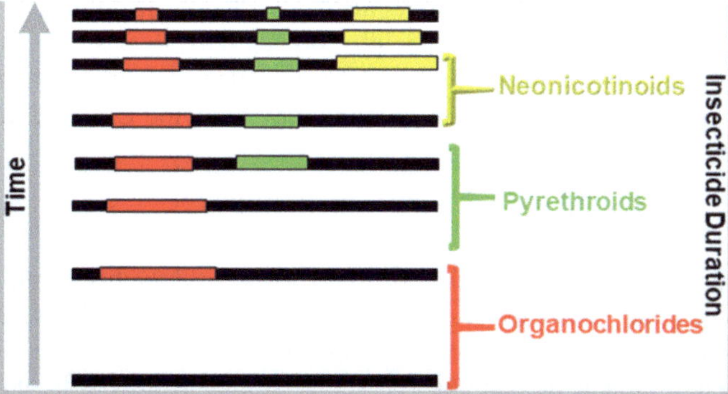

Fig. 4 Graphical depiction of how the legacy of pesticide use leads to multiple episodes of selection that complicate population genetic inferences of insecticide resistance phenotypes. Here, resistance alleles are shown as colored regions of a black chromosome accumulating over time as hard selective sweeps. Hard sweeps reduce genetic variation around the selected mutation as the resistance allele becomes fixed under strong selection. Resistance may arise at a faster rate with subsequent pesticides, reducing the duration of those pesticides, especially if the mechanism of resistance co-opts previous resistance alleles (i.e., metabolic detoxification). We show the length of the haplotype block declining over time due to mutation and recombination, but also note that drift or purifying selection can drive the resistance alleles to extinction in the absence of the specific mode of action. While population genetic methods can be used to estimate the best-fit models of these dynamics, including the "allele-age" of a particular resistance mutation, in practice the best opportunities to reconstruct resistance evolution may come from leveraging temporal samples

5.2 *Regional Patterns of Resistance in Pest and Non-pest Lineages*

Interestingly, the geography of first-reports of CPB outbreaks suggests a distinct pattern of invasion of potato agroecosystems that may also be relevant to geographic variation in insecticide resistance. The first report of CPB feeding on potato was in 1859 in central Nebraska (Walsh 1866). Prior to that, CPB was only known to feed primarily on buffalo bur, *Solanum rostratum* (Tower 1906). The beetle subsequently spread east and north, occupying much of the Midwestern and Eastern U.S. by the 1870s (Walsh 1866; Riley 1869). CPB outbreaks were common in Western Canada by the early 1900s (Foster 1876), but did not affect the potato production centers of the Western U.S. until the 1930s (Haegele and Wakeland 1932). Recent genetic analysis (Izzo et al. 2018) has shown that patterns of population divergence mirror this expansion history from the Great Plains. Genetic diversity is also reduced in the invasive pest lineages and shows a stronger reduction in the Northeast and Northwest U.S., where CPB colonized later in its expansion history. Thus, geographic variation in standing genetic diversity may partly explain differential rates of resistance evolution at larger regional scales (Crossley et al. 2019a). More recent transcriptomic work shows that gene regulatory evolution may underlie differences in Northwestern and Eastern resistance phenotypes (Dively et al. 2020). However, emerging whole-genome resequencing data (see Box 2) suggest that resistance evolution is occuring in most potato growing regions in the U.S., including Northwestern populations.

Box 2

Whole-genome analyses of selection have only recently been employed to study insecticide resistance in CPB (Pélissié et al. 2021). These analyses, based on 85 resequenced genomes, employed SNP-based outlier detection, genome-environment association (GEA) tests, and haplotype-based tests to identify genes under selection. Although there is discordance among methods in terms of the candidate selected loci, Pélissié et al. (2021) show that signatures of selection frequently involve insecticide resistance genes. Within individual populations, multiple candidate genes are implicated, including both target-site mutations and metabolic detoxification genes. Furthermore, there is evidence that selection has affected the same set of candidate genes independently in different populations. Thus, whole-genome resequencing data suggest that resistance evolves rapidly and repeatedly and is likely to be polygenic.

It should be noted that a large number of SNPs identified as putatively under selection occur in or near genes whose functions are unknown. These SNPs could reflect false positives or selection by other environmental factors, but alternatively may be linked to insecticide resistance in yet unknown ways. As genome assemblies and annotations continue to improve with insights gleaned from functional genomics studies, these knowledge gaps will be filled and understanding of how these genes contribute to CPB adaptation to insecticides will improve.

The main limitation in this study is that sampling was limited, and power in genome scan approaches scales with the number of individuals sampled per population (Lotterhos and Whitlock 2015). Discordance among selection tests could be an artifact of the different assumptions of underlying models and in how they balance the occurrence of false positives and false negatives (François et al. 2016). On the other hand, differences among the methods potentially reflect true biological variation in how different CPB populations respond to selection exerted by insecticides. Indeed, this is a core strength of alternative population genomic approaches, in that they permit the discovery of a diversity of ways in which insects adapt to insecticides. Other population genomics approaches, such as genome wide association (GWA) mapping, have yet to be leveraged for CPB, but as costs for whole-genome sequencing drop, these will undoubtedly provide greater resolution of causative insecticide resistance genes.

6 Insights from Comparative Genomic Approaches

In his seminal paper on the population genetic basis of insecticide resistance, Crow (1957) argued that standing genetic variation was a form of "pre-adaptation" that might be drawn upon to confer insecticide resistance. More recently, large-scale analyses of insect pests have suggested that generalist-feeding (polyphagous) insects have demonstrated greater evolutionary propensity, relative to specialists, in their ability to develop resistance to insecticides (Hardy et al. 2018). Thus, one might broaden the "pre-adaptation" hypothesis to include generalist herbivores that have a diverse host breadth, because they are exposed to more and varied plant secondary metabolites than specialists (Gordon 1961), and insect herbivores employ detoxification enzymes to counteract these diverse plant xenobiotics (Feyereisen et al. 2015; Rane et al. 2016; Alyokhin and Chen 2017; Calla et al. 2017). These same enzymes, including the cytochrome P450s, glutathione S-transferases, esterases, and ABC transporters, also detoxify insecticides in many resistant insect populations (Alyokhin and Chen 2017; Dermauw et al. 2018). This molecular arsenal, colloquially referred to as the "defensome," could be associated with greater evolutionary potential (therefore, be "pre-adaptive") in pest lineages through a variety of genomic mechanisms conferring greater capacity to overcome pesticides. Species with greater host breadth might possess structural genomic features such as greater gene family diversity (Hahn et al. 2007), altered mutability/regulatory control via a propensity for gene duplication (copy number variation) (Cattel et al. 2020), or transposable element mobilization (Wilson 1993). However, a recent review of pest genomes has largely discounted the theory that generalists demonstrate rapid insecticide resistance evolution compared to specialists (Dermauw et al. 2018). Instead, fundamental properties of pest lineages, such as effective population size, mutation rate, and mode of reproduction (Chen and Schoville 2018), might provide more predictive power, as these factors are associated with greater rates of positive selection (Whitehead et al. 2017). Indeed, recent comparative genomic work of *Leptinotarsa* species (Cohen et al. 2021) suggests that rates of evolution are coupled with greater ancestral standing genetic diversity (see Box 3).

Box 3

The Colorado potato beetle is a specialist on plants within the *Solanaceae* family and an internationally notorious pest on potato and other *Solanum* crops. However, relative to other beetles in the *Leptinotarsa* genus, it has by far the largest host breadth and is considered an oligophage (Jacques 1988). In order to investigate the pre-adaptation hypothesis, as well as assess rates of positive selection, Cohen et al. (2021) compared pest and non-pest CPB individuals and the genomes of nine additional *Leptinotarsa* species using a phylogenomic framework. This analysis provides strong support for standing genetic diversity (as a product of large effective population size) and high rates

(continued)

Box 3 (continued)
of positive selection underlying rapid adaptation. In a direct comparison of genomic variation among *Leptinotarsa* species, there were significant changes in coding regions for genes associated with insecticide resistance, including cytochrome P450s, esterases, ABC transporters, ion channel receptors, and cuticle proteins. In contrast, other features that could explain rapid genome evolution were not significantly different between non-pest CPB and non-pest *Leptinotarsa* species. Gene family turnover was limited, with only one gene family significantly changing (contracting) in the CPB lineage, a cysteine-type peptidase that is most likely associated with host plant digestion. Similarly, the transposable element (TE) diversity and abundance was similar among *Leptinotarsa* species, including within coding regions.

One caveat of the above study is that the other *Leptinotarsa* species' genomes were highly fragmented due to relatively low coverage (20x) sequencing. Thus, missing sequence data, or poor coverage in orthologs, could influence these results. As long-read whole-genome sequencing becomes more inexpensive, comparative genomics will provide greater power to reexamine the pre-adaptive hypothesis and also consider the role of structural genomic variation (inversions, insertions, and deletions) in explaining patterns of rapid evolution.

7 Perspectives: Major Challenges in Understanding Insecticide Resistance Evolution

7.1 *The Legacy of Selection and the Complexity of Phenotypes*

Population genomics studies on the molecular basis of insecticide resistance evolution have only emerged since the 2010s, with the first CPB transcriptome being sequenced in 2014 (Kumar et al. 2014), and the first genome assembly reported in 2018 (Schoville et al. 2018). Insecticide use, in contrast, became widespread with the advent of DDT in the 1940s, while arsenical chemicals were in use as early as the mid-1800s (Casagrande 1987). This presents a serious challenge to population genomics studies attempting to link allele frequency differences with selection exerted by insecticides (see Fig. 4). Some of the most important questions include: which insecticide caused a given genomic signature of selection? Are changes across multiple genes required for a given resistance phenotype (or to balance fitness effects of major resistance mutations)? Efforts to unravel the origins of multi-resistance phenotypes will need to move beyond standard population sampling by incorporating temporal sampling of museum specimens. Museum collections provide a unique opportunity to directly observe allele frequency change through time and can close the gap between genomic DNA sampling and the onset of selection by insecticides.

Alleles linked to pesticide resistance should increase in frequency as a result of selection, so a comparison of modern beetles and beetles that predate pesticide use should exhibit allele frequency differences (Holmes et al. 2016). Furthermore, if standing genetic variation plays a role in insecticide resistance evolution, alleles associated with the resistance phenotype should be present preceding pesticide usage. Thus, museum samples provide an opportunity to not only detect selection, but a means to directly assess the role of standing variation in resistance evolution. Efforts are currently underway to examine changes in CPB populations from the 1900s to present. Additionally, these and future data will also be useful in determining fundamental population genetic parameters, such as mutation, effective population size change, recombination rate, and migration, thus enabling more precise modeling of rapid adaption to insecticides.

7.2 *Unknown Sources of Variation: Genotype by Environment Interactions and Microbiomes*

Additional mechanisms of variation in insecticide resistance traits, both among and within populations, include the effects of microbiome biology (Pietri and Liang 2018) and genotype by environment interactions (Bourguet et al. 2000). Oral and gut microbes have been implicated in reducing plant defensive chemistry in the Colorado potato beetle (Chung et al. 2013, 2017) and mitigating (via degradation) synthetic insecticides in another insect pest (Cheng et al. 2017). The interactions of these resident microbes with plant performance and insecticide resistance may vary geographically and should be considered when tailoring control strategies. Dominance and pleiotropy of insecticide resistance genes have been considered in CPB populations (Alyokhin et al. 2008), but much less is known of phenotypic plasticity as it might relate to insecticide resistance. Existing studies that have examined resistance in unrelated populations to multiple active ingredients have illustrated the common difference between lab-selected (monogenic) and field-evolved (polygenic) insecticide resistance as well as dominance. However, these studies failed to account for variation across life stages, as well as appropriately incorporate the significant interaction of environment and genotype, as suggested by Bourguet et al. (2000). Recent work has shown that sub-lethal pesticide exposure and transgenerational epigenetic changes in DNA methylation may explain some of the plasticity observed in Colorado potato beetle insecticide resistance phenotypes (Margus et al. 2019; Brevik et al. 2020). With whole-genome, methylome, and transcriptome sequencing becoming readily available tools, these features can be accounted for in future studies.

7.3 Functional Validation

Existing population genetics datasets in CPB (Piiroinen et al. 2013; Crossley et al. 2017, 2019b; Izzo et al. 2017), as well as gene expression studies (Clements et al. 2016; Kaplanoglu et al. 2017; Hawthorne et al. 2020), provide a number of strong candidate insecticide resistance genes. These genes have been functionally validated in only a few cases (Wan et al. 2014; Clements et al. 2017b; Gaddelapati et al. 2018). Thus, fundamental questions about the nature of insecticide resistance in CPB will remain as long as genetic evidence is only putatively associated with insecticide resistance. Fortunately, functional genomics tools, such as RNAi and CRISPR/Cas9, are feasible in CPB (Zhu et al. 2011; Gui et al. 2020), making it possible to directly link nucleotide variation to specific mechanisms of insecticide resistance. At present, however, significant challenges remain in scaling such studies to allow for screening a large number of loci, or appropriately quantifying phenotypic effects. There is considerable promise in pursuing functional genetics research, as the identification of genes involved in pesticide resistance could be developed as gene targets for novel insecticides (Clements et al. 2017b). Such efforts are already underway to utilize RNAi for novel insecticidal products. Such products could be designed to restore pest susceptibility to existing chemical control products, or be tiered to target multiple metabolic genes simultaneously in order to delay resistance, thereby facilitating pest management efforts in the future.

8 Future Perspectives and Conclusions

In order to obtain greater sustainability in food production, fundamental knowledge of insect population and molecular genetics is needed, particularly how and under what conditions pest resistance evolves to control methods. Knowledge of the genetic loci and gene networks involved in resistance will help management of resistance, potentially mitigating its spread, as well as refine IPM strategies in the use of novel gene-targeted pesticides. As we have shown here, emerging genomic evidence suggests polygenic resistance drawn from standing variation best explains cases of rapid evolution in the Colorado potato beetle. However, despite recent genome resequencing and comparative genomic research, studies on population genetic adaptation to insecticides lack a temporal framework to demonstrate polygenic evolution unequivocally. Temporal genomic analyses, for example drawing on museum samples, will uniquely provide insight into evolutionary mechanisms and parameters that underlie rapid adaptation in this agricultural insect pest. Such knowledge would provide a means to model the spread of resistance, reevaluate economic thresholds for interventions (such as pesticide applications), and perhaps to develop a predictive framework for improving pest management outcomes.

Acknowledgements The authors wish to thank Yolanda Chen, Sylvia Rondon, Justin Clements, David Hawthorne, Andrei Alyokhin, and Russell Groves for discussions related to this research. Funding for this research was provided by a USDA NIFA Exploratory grant (2015-67030-23495), USDA Hatch grant (WIS02004), USDA Potato Research Service grant (58-5090-7-073), a USDA AFRI-ELI doctoral fellowship (2018-67011-28058), and grants from the Wisconsin Potato Industry Board.

References

Alyokhin A, Chen YH. Adaptation to toxic hosts as a factor in the evolution of insecticide resistance. Curr Opin Insect Sci. 2017;21:33–8.

Alyokhin AV, Ferro DN. Relative fitness of Colorado potato beetle (Coleoptera: Chrysomelidae) resistant and susceptible to the *Bacillus thuringiensis* Cry3A toxin. J Econ Entomol. 1999;92:510–5.

Alyokhin A, Baker M, Mota-Sanchez D, Dively G, Grafius E. Colorado potato beetle resistance to insecticides. Am J Potato Res. 2008;85:395–413.

Alyokhin A, Mota-Sanchez D, Baker M, Snyder WE, Menasha S, Whalon M, et al. The red queen in a potato field: integrated pest management versus chemical dependency in Colorado potato beetle control. Pest Manag Sci. 2015;71:343–56.

Anspaugh DD, Kennedy GG, Roe RM. Purification and characterization of a resistance-associated esterase from the Colorado potato beetle, *Leptinotarsa decemlineata* (Say). Pestic Biochem Physiol. 1995;53:84–96.

Argentine J, Clark JM, Ferro D. Relative fitness of insecticide-resistant Colorado potato beetle strains (Coleoptera: Chrysomelidae). Environ Entomol. 1989a;18:705–10.

Argentine J, Clark MJ, Ferro D. Genetics and synergism of resistance to azinphosmethyl and permethrin in the Colorado potato beetle (Coleoptera: Chrysomelidae). J Econ Entomol. 1989b;82:698–705.

Argentine JA, Clark JM, Lin H. Genetics and biochemical mechanisms of abamectin resistance in two isogenic strains of Colorado potato beetle. Pestic Biochem Physiol. 1992;44:191–207.

Argentine JA, Zhu K, Lee S, Clark JM. Biochemical mechanisms of azinphosmethyl resistance in isogenic strains of Colorado potato beetle. Pestic Biochem Physiol. 1994;48:63–78.

Argentine JA, Lee SH, Sos MA, Barry SR, Clark JM. Permethrin resistance in a near isogenic strain of Colorado potato beetle. Pestic Biochem Physiol. 1995;53:97–115.

Baker A. Metal tolerance. New Phytol. 1987;106:93–111.

Baker MB, Alyokhin A, Porter AH, Ferro DN, Dastur SR, Galal N. Persistence and inheritance of costs of resistance to imidacloprid in Colorado potato beetle. J Econ Entomol. 2007;100:1871–9.

Barrett RDH, Schluter D. Adaptation from standing genetic variation. Trends Ecol Evol. 2008;23:38–44.

Bass C, Zimmer CT, Riveron JM, Wilding CS, Wondji CS, Kaussmann M, Field LM, Williamson MS, Nauen R. Gene amplification and microsatellite polymorphism underlie a recent insect host shift. Proc Natl Acad Sci. 2013;110(48):19460–5.

Bourguet D, Genissel A, Raymond M. Insecticide resistance and dominance levels. J Econ Entomol. 2000;93:1588–95.

Brevik K, Bueno EM, McKay S, Schoville SD, Chen YH. Insecticide exposure affects intergenerational patterns of DNA methylation in the Colorado potato beetle, *Leptinotarsa decemlineata*. Evol Appl. 2020; https://doi.org/10.1111/eva.13153.

Calla B, Noble K, Johnson RM, Walden KK, Schuler MA, Robertson HM, et al. Cytochrome P450 diversification and hostplant utilization patterns in specialist and generalist moths: birth, death and adaptation. Mol Ecol. 2017;26:6021–35.

Casagrande R. The Colorado potato beetle: 125 years of mismanagement. Bull Entomol Soc Am. 1987;33:142–50.

Cattel J, Faucon F, Le Péron B, Sherpa S, Monchal M, Grillet L, et al. Combining genetic crosses and pool targeted DNA-seq for untangling genomic variations associated with resistance to multiple insecticides in the mosquito *Aedes aegypti*. Evol Appl. 2020;13:303–17.

Chen YH, Schoville SD. Editorial overview: ecology: ecological adaptation in agroecosystems: novel opportunities to integrate evolutionary biology and agricultural entomology. Curr Opin Insect Sci. 2018;26:iv–viii.

Cheng D, Guo Z, Riegler M, Xi Z, Liang G, Xu Y. Gut symbiont enhances insecticide resistance in a significant pest, the oriental fruit fly *Bactrocera dorsalis* (Hendel). Microbiome. 2017;5:13.

Chung SH, Rosa C, Hoover K, Luthe DS, Felton GW. Colorado potato beetle manipulates plant defenses in local and systemic leaves. Plant Signal Behav. 2013;8:e27592.

Chung SH, Scully ED, Peiffer M, Geib SM, Rosa C, Hoover K, et al. Host plant species determines symbiotic bacterial community mediating suppression of plant defenses. Sci Rep. 2017;7:1–13.

Clements J, Schoville S, Peterson N, Lan Q, Groves RL. Characterizing molecular mechanisms of imidacloprid resistance in select populations of *Leptinotarsa decemlineata* in the Central Sands region of Wisconsin. PLoS One. 2016;11:e0147844.

Clements J, Schoville S, Clements N, Chapman S, Groves RL. Temporal patterns of imidacloprid resistance throughout a growing season in *Leptinotarsa decemlineata* populations. Pest Manag Sci. 2017a;73:641–50.

Clements J, Schoville S, Peterson N, Huseth AS, Lan Q, Groves RL. RNA interference of three upregulated transcripts associated with insecticide resistance in an imidacloprid resistant population of *Leptinotarsa decemlineata*. Pestic Biochem Physiol. 2017b;135, 35–40.

Clements J, Sanchez-Sedillo B, Bradfield CA, Groves RL. Transcriptomic analysis reveals similarities in genetic activation of detoxification mechanisms resulting from imidacloprid and chlorothalonil exposure. PLoS One. 2018;13:e0205881.

Cohen ZP, Brevik K, Chen YH, Hawthorne DJ, Weibel BD, Schoville SD. Elevated rates of positive selection drive the evolution of pestiferousness in the Colorado potato beetle (*Leptinotarsa decemlineata*, Say). Mol Ecol. 2021;30:237–54.

Crossley MS, Chen YH, Groves RL, Schoville SD. Landscape genomics of Colorado potato beetle provides evidence of polygenic adaptation to insecticides. Mol Ecol. 2017;26:6284–300.

Crossley MS, Rondon SI, Schoville SD. A comparison of resistance to imidacloprid in Colorado potato beetle (*Leptinotarsa decemlineata* Say) populations collected in the Northwest and Midwest US. Am J Potato Res. 2018:1–9.

Crossley MS, Rondon SI, Schoville SD. Effects of contemporary agricultural land cover on Colorado potato beetle genetic differentiation in the Columbia Basin and Central Sands. Ecol Evol. 2019a;9:9385–94.

Crossley MS, Rondon SI, Schoville SD. Patterns of genetic differentiation in Colorado potato beetle correlate with contemporary, not historic, potato land cover. Evol Appl. 2019b;12:804–14.

Crow JF. Genetics of insect resistance to chemicals. Annu Rev Entomol. 1957;2:227–46.

Denholm I, Rowland M. Tactics for managing pesticide resistance in arthropods: theory and practice. Annu Rev Entomol. 1992;37:91–112.

Dermauw W, Pym A, Bass C, Van Leeuwen T, Feyereisen R. Does host plant adaptation lead to pesticide resistance in generalist herbivores? Curr Opin Insect Sci. 2018;26:25–33.

Dively GP, Crossley MS, Schoville SD, Steinhauer N, Hawthorne DJ. Regional differences in gene regulation may underlie patterns of sensitivity to novel insecticides in *Leptinotarsa decemlineata*. Pest Manag Sci. 2020;76:4278–85.

Feyereisen R, Dermauw W, Van Leeuwen T. Genotype to phenotype, the molecular and physiological dimensions of resistance in arthropods. Pestic Biochem Physiol. 2015;121:61–77.

Ffrench-Constant RH. The molecular genetics of insecticide resistance. Genetics. 2013;194:807.

Forgash AJ. Insecticide resistance in the Colorado potato beetle. Research Bulletin, Massachusetts Agricultural Experiment Station. 1985. p. 33–52.

Foster GE. The Colorado potato beetle. Concord, NH. 1876.

François O, Martins H, Caye K, Schoville SD. Controlling false discoveries in genome scans for selection. Mol Ecol. 2016;25:454–69.

Gaddelapati SC, Kalsi M, Roy A, Palli SR. Cap'n'collar C regulates genes responsible for imidacloprid resistance in the Colorado potato beetle, *Leptinotarsa decemlineata*. Insect Biochem Mol Biol. 2018;99:54–62.

Georghiou G. The magnitude of the resistance problem. In: National Research Council, editor. Pesticide resistance: strategies and tactics for management. Washington, DC: National Academy Press; 1986. p. 14–43.

Georghiou GP, Taylor CE. Factors influencing the evolution of resistance. In: National Research Council, editor. Pesticide resistance: strategies and tactics for management. Washington, DC: National Academy Press; 1986. p. 157–69.

Gordon H. Nutritional factors in insect resistance to chemicals. Annu Rev Entomol. 1961;6:27–54.

Gould F. The evolutionary potential of crop pests. Am Sci. 1991;79:496–507.

Gui S, Taning CNT, Wei D, Smagghe G. First report on CRISPR/Cas9-targeted mutagenesis in the Colorado potato beetle, *Leptinotarsa decemlineata*. J Insect Physiol. 2020:104013.

Haegele RW, Wakeland C. Control of the Colorado potato beetle. Moscow, ID: University of Idaho, College of Agriculture, Extension Division; 1932.

Hahn MW, Han MV, Han S-G. Gene family evolution across 12 *Drosophila* genomes. PLoS Genet. 2007;3:e197.

Hardy NB, Peterson DA, Ross L, Rosenheim JA. Does a plant-eating insect's diet govern the evolution of insecticide resistance? Comparative tests of the pre-adaptation hypothesis. Evol Appl. 2018;11:739–47.

Haridas CV, Tenhumberg B. Modeling effects of ecological factors on evolution of polygenic pesticide resistance. J Theor Biol. 2018;456:224–32.

Hartl DL, Clark AG. Principles of population genetics. 3rd ed. Sunderland, MA: Sinauer Associates; 1997. p. 542.

Hawthorne DJ. Quantitative trait locus mapping of pyrethroid resistance in Colorado potato beetle, *Leptinotarsa decemlineata* (Say) (Coleoptera: Chrysomelidae). J Econ Entomol. 2003;96:1021–30.

Hawthorne DJ, Crossley MS, Schoville SD, Steinhauer N, Dively GP. Regional differences in gene regulation may underlie patterns of sensitivity to novel insecticides in *Leptinotarsa decemlineata*. Pest Manag Sci. 2020:794446.

Heim D, Kennedy G, Vanduyn J. Survey of insecticide resistance among North Carolina Colorado potato beetle (Coleoptera: Chrysomelidae) populations. J Econ Entomol. 1990;83:1229–35.

Heim DC, Kennedy GG, Gould FL, Van Duyn JW. Inheritance of fenvalerate and carbofuran resistance in Colorado beetles – *Leptinotarsa decemlineata* (Say) – from North Carolina. Pestic Sci. 1992;34:303–11.

Hitchner EM, Kuhar TP, Dively GP, Youngman RR, Philips CR, Anderson TD. Baseline toxicity and field efficacy of metaflumizone on Colorado potato beetle (Coleoptera: Chrysomelidae). J Econ Entomol. 2012;105:207–13.

Holmes MW, Hammond TT, Wogan GO, Walsh RE, LaBarbera K, Wommack EA, et al. Natural history collections as windows on evolutionary processes. Mol Ecol. 2016;25:864–81.

Hoy MA. Myths, models and mitigation of resistance to pesticides. Philos Trans R Soc London B Biol Sci. 1998;353:1787–95.

Huseth A, Groves R. Effect of insecticide management history on emergence phenology and neonicotinoid resistance in *Leptinotarsa decemlineata* (Coleoptera: Chrysomelidae). J Econ Entomol. 2013;106:2491–505.

Huseth AS, Petersen JD, Poveda K, Szendrei Z, Nault BA, Kennedy GG, et al. Spatial and temporal potato intensification drives insecticide resistance in the specialist herbivore, *Leptinotarsa decemlineata*. PLoS One. 2015;10:e0127576.

Ioannidis P, Grafius E, Whalon M. Patterns of insecticide resistance to azinphosmethyl, carbofuran, and permethrin in the Colorado potato beetle (Coleoptera: Chrysomelidae). J Econ Entomol. 1991;84:1417–23.

Ioannidis PM, Grafius EJ, Wierenga JM, Whalon ME, Hollingworth RM. Selection, inheritance and characterization of carbofuran resistance in the Colorado potato beetle (Coleoptera: Chrysomelidae). Pestic Sci. 1992;35:215–22.

IRAC. Insecticide Resistance Action Committee database. 2016. http://www.irac-online.org/.

Izzo V, Chen YH, Schoville SD, Wang C, Hawthorne DJ. Origin of pest lineages of the Colorado potato beetle, *Leptinotarsa decemlineata*. bioRxiv. 2017:156612.

Izzo VM, Chen YH, Schoville SD, Wang C, Hawthorne DJ. Origin of pest lineages of the Colorado potato beetle (Coleoptera: Chrysomelidae). J Econ Entomol. 2018;111:868–78.

Jacques RL. The potato beetles. Abingdon: Taylor & Francis; 1988.

Jiang W-H, Guo W-C, Lu W-P, Shi X-Q, Xiong M-H, Wang Z-T, et al. Target site insensitivity mutations in the AChE and LdVssc1 confer resistance to pyrethroids and carbamates in *Leptinotarsa decemlineata* in northern Xinjiang Uygur autonomous region. Pestic Biochem Physiol. 2011;100:74–81.

Kalsi M, Palli SR. Transcription factor cap n collar C regulates multiple cytochrome P450 genes conferring adaptation to potato plant allelochemicals and resistance to imidacloprid in Leptinotarsa decemlineata (Say). Insect Biochem Mol Biol. 2017;83:1–12.

Kaplanoglu E, Chapman P, Scott IM, Donly C. Overexpression of a cytochrome P450 and a UDP-glycosyltransferase is associated with imidacloprid resistance in the Colorado potato beetle, *Leptinotarsa decemlineata*. Sci Rep. 2017;7:1762.

Karasov T, Messer PW, Petrov DA. Evidence that adaptation in *Drosophila* is not limited by mutation at single sites. PLoS Genet. 2010;6:e1000924.

Kim H-J, Hawthorne DJ, Peters T, Dively G, Clark JM. Application of DNA-based genotyping techniques for the detection of *kdr*-like pyrethroid resistance in field populations of Colorado potato beetle. Pestic Biochem Physiol. 2005;81:85–96.

Kliot A, Ghanim M. Fitness costs associated with insecticide resistance. Pest Manag Sci. 2012;68:1431–7.

Kreiner JM, Stinchcombe JR, Wright SI. Population genomics of herbicide resistance: adaptation via evolutionary rescue. Annu Rev Plant Biol. 2018;69:611–35.

Kumar A, Congiu L, Lindström L, Piiroinen S, Vidotto M, Grapputo A. Sequencing, *de novo* assembly and annotation of the Colorado potato beetle, *Leptinotarsa decemlineata*, transcriptome. PLoS One. 2014;9:e86012.

Lotterhos KE, Whitlock MC. The relative power of genome scans to detect local adaptation depends on sampling design and statistical method. Mol Ecol. 2015;24:1031–46.

Luck RF, van den Bosch R, Garcia R. Chemical insect control – a troubled pest management strategy. Bioscience. 1977;27:606–11.

Malaspinas A-S, Malaspinas O, Evans SN, Slatkin M. Estimating allele age and selection coefficient from time-serial data. Genetics. 2012;192:599–607.

Mallet J. The evolution of insecticide resistance: have the insects won? Trends Ecol Evol. 1989;4:336–40.

Mamidala P, Jones SC, Mittapalli O. Metabolic resistance in bed bugs. Insects. 2011;2:36–48.

Margus A, Piiroinen S, Lehmann P, Tikka S, Karvanen J, Lindström L. Sublethal pyrethroid insecticide exposure carries positive fitness effects over generations in a pest insect. Sci Rep. 2019;9:1–10.

Miyo T, Keil C, Hough-Goldstein J, Oguma Y. Inheritance of resistance to esfenvalerate in Colorado potato beetles (Coleoptera: Chrysomelidae). J Econ Entomol. 1999;92:1031–8.

Mota-Sanchez D, Hollingworth RM, Grafius EJ, Moyer DD. Resistance and cross-resistance to neonicotinoid insecticides and spinosad in the Colorado potato beetle, *Leptinotarsa decemlineata* (Say)(Coleoptera: Chrysomelidae). Pest Manag Sci. 2006;62:30–7.

Olson E, Dively G, Nelson J. Baseline susceptibility to imidacloprid and cross resistance patterns in Colorado potato beetle (Coleoptera: Chrysomelidae) populations. J Econ Entomol. 2000;93 (2):447–458.

Orr HA. The genetic theory of adaptation: a brief history. Nat Rev Genet. 2005;6:119–27.

Pélissié B, Crossley MS, Cohen Z, Schoville SD. Rapid evolution in insect pests: the importance of space and time in population genomics studies. Curr Opin Insect Sci. 2018;26:8–16.

Pélissié B, Chen YH, Cohen ZP, Crossley MS, Hawthorne DJ, Victor I, et al. Genome resequencing reveals rapid, repeated evolution in the Colorado potato beetle, *Leptinotarsa decemlineata*. bioRxiv. 2021:430453.

Pietri JE, Liang D. The links between insect symbionts and insecticide resistance: causal relationships and physiological tradeoffs. Ann Entomol Soc Am. 2018;111:92–7.

Piiroinen S, Lindström L, Lyytinen A, Mappes J, Chen YH, Izzo V, et al. Pre-invasion history and demography shape the genetic variation in the insecticide resistance-related acetylcholinesterase 2 gene in the invasive Colorado potato beetle. BMC Evol Biol. 2013;13:13.

Qu Y, Chen J, Li C, Wang Q, Guo W, Han Z, et al. The subunit gene Ldα1 of nicotinic acetylcholine receptors plays important roles in the toxicity of imidacloprid and thiamethoxam against *Leptinotarsa decemlineata*. Pestic Biochem Physiol. 2016;127:51–8.

Rahardja U, Whalon ME. Inheritance of resistance to *Bacillus thuringiensis* subsp. *tenebrionis* CryIIIA δ-endotoxin in Colorado potato beetle (Coleoptera: Chrysomelidae). J Econ Entomol. 1995;88:21–6.

Rane RV, Walsh TK, Pearce SL, Jermiin LS, Gordon KH, Richards S, et al. Are feeding preferences and insecticide resistance associated with the size of detoxifying enzyme families in insect herbivores? Curr Opin Insect Sci. 2016;13:70–6.

Riley CV. First annual report of the noxious, beneficial and other insects of the state of Missouri. First annual report of the noxious, beneficial and other insects of the State of Missouri. 1869.

Riveron JM, Yunta C, Ibrahim SS, Djouaka R, Irving H, Menze BD, et al. A single mutation in the GSTe2 gene allows tracking of metabolically based insecticide resistance in a major malaria vector. Genome Biol. 2014;15:R27.

Rose RL, Brindley WA. An evaluation of the role of oxidative enzymes in Colorado potato beetle resistance to carbamate insecticides. Pestic Biochem Physiol. 1985;23:74–84.

Roush RT. Designing resistance management programs: how can you choose? Pestic Sci. 1989;26:423–41.

Schoville SD, Chen YH, Andersson MN, Benoit JB, Bhandari A, Bowsher JH, et al. A model species for agricultural pest genomics: the genome of the Colorado potato beetle, *Leptinotarsa decemlineata* (Coleoptera: Chrysomelidae). Sci Rep. 2018;8:1931.

Scott JG. Life and death at the voltage-sensitive sodium channel: evolution in response to insecticide use. Annu Rev Entomol. 2019;64:243–57.

Scott IM, Tolman JH, MacArthur DC. Insecticide resistance and cross-resistance development in Colorado potato beetle *Leptinotarsa decemlineata* Say (Coleoptera: Chrysomelidae) populations in Canada 2008–2011. Pest Manag Sci. 2015;71:712–21.

Silcox CA, Ghidiu GM, Forgash AJ. Laboratory and field evaluation of piperonyl butoxide as a pyrethroid synergist against the Colorado potato beetle (Coleoptera: Chrysomelidae). J Econ Entomol. 1985;78:1399–405.

Stewart JG, Kennedy GG, Sturz AV. Incidence of insecticide resistance in populations of Colorado potato beetle, *Leptinotarsa decemlineata* (Say) (Coleoptera: Chrysomelidae), on Prince Edward Island. Can Entomol. 1997;129:21–6.

Szendrei Z, Grafius E, Byrne A, Ziegler A. Resistance to neonicotinoid insecticides in field populations of the Colorado potato beetle (Coleoptera: Chrysomelidae). Pest Manag Sci. 2012;68:941–6.

Tabashnik BE. Modeling and evaluation of resistance management tactics. In: Roush RT, Tabashnik BE, editors. Pesticide resistance in arthropods. New York: Chapman & Hall; 1990. p. 153–82.

Tabashnik BE, Brevault T, Carriere Y. Insect resistance to Bt crops: lessons from the first billion acres. Nat Biotechnol. 2013;31:510–21.

Tan J, Salgado VL, Hollingworth RM. Neural actions of imidacloprid and their involvement in resistance in the Colorado potato beetle, *Leptinotarsa decemlineata* (Say). Pest Manag Sci. 2008;64:37–47.

Tebbe C, Breckheimer B, Racca P, Schorn C, Kleinhenz B, Nauen R. Incidence and spread of knockdown resistance (kdr) in German Colorado potato beetle (*Leptinotarsa decemlineata* Say) populations. EPPO Bull. 2016;46:129–38.

Tower WL. An investigation of evolution in chrysomelid beetles of the genus *Leptinotarsa*. Washington, DC: Carnegie Institution of Washington; 1906.

Trisyono A, Whalon ME. Fitness costs of resistance to *Bacillus thuringiensis* in Colorado potato beetle (Coleoptera: Chrysomelidae). J Econ Entomol. 1997;90:267–71.

Walsh BD. The new potato bug. Pract Entomol. 1866;2:13–6.

Wan P-J, Shi X-Q, Kong Y, Zhou L-T, Guo W-C, Ahmat T, et al. Identification of cytochrome P450 monooxygenase genes and their expression profiles in cyhalothrin-treated Colorado potato beetle, *Leptinotarsa decemlineata*. Pestic Biochem Physiol. 2013;107:360–8.

Wan P-J, Guo W-Y, Yang Y, Lü F-G, Lu W-P, Li G-Q. RNAi suppression of the ryanodine receptor gene results in decreased susceptibility to chlorantraniliprole in Colorado potato beetle *Leptinotarsa decemlineata*. J Insect Physiol. 2014;63:48–55.

Wellenreuther M, Hansson B. Detecting polygenic evolution: problems, pitfalls, and promises. Trends Genet. 2016;32:155–64.

Whalon ME, Mota-Sanchez D, Hollingworth RM. Arthropod pesticide resistance database. East Lansing, MI: Michigan State University; 2016. http://www.pesticideresistance.org/

Whitehead A, Clark BW, Reid NM, Hahn ME, Nacci D. When evolution is the solution to pollution: key principles, and lessons from rapid repeated adaptation of killifish (*Fundulus heteroclitus*) populations. Evol Appl. 2017;10:762–83.

Wierenga JM, Hollingworth RM. Inhibition of altered acetylcholinesterases from insecticide-resistant Colorado potato beetles (Coleoptera: Chrysomelidae). J Econ Entomol. 1993;86:673–9.

Wilding CS. Regulating resistance: CncC: Maf, antioxidant response elements and the overexpression of detoxification genes in insecticide resistance. Curr Opin Insect Sci. 2018;27:89–96.

Williamson MS, Martinez-Torres D, Hick CA, Devonshire AL. Identification of mutations in the housefly para-type sodium channel gene associated with knockdown resistance (kdr) to pyrethroid insecticides. Mol Gen Genet. 1996;252:51–60.

Wilson TG. Transposable elements as initiators of insecticide resistance. J Econ Entomol. 1993;86:645–51.

Wondji CS, Dabire RK, Tukur Z, Irving H, Djouaka R, Morgan JC. Identification and distribution of a GABA receptor mutation conferring dieldrin resistance in the malaria vector Anopheles funestus in Africa. Insect Biochem Mol Biol. 2011;41:484–91.

Wybouw N, Zhurov V, Martel C, Bruinsma KA, Hendrickx F, Grbić V, Van Leeuwen T. Adaptation of a polyphagous herbivore to a novel host plant extensively shapes the transcriptome of herbivore and host. Mol Ecol. 2015;24(18), 4647–4663.

Zhao J-Z, Bishop BA, Grafius EJ. Inheritance and synergism of resistance to imidacloprid in the Colorado potato beetle (Coleoptera: Chrysomelidae). J Econ Entomol. 2000;93:1508–14.

Zhu KY, Clark JM. Validation of a point mutation of acetylcholinesterase in Colorado potato beetle by polymerase chain reaction coupled to enzyme inhibition assay. Pestic Biochem Physiol. 1997;57:28–35.

Zhu KY, Lee SH, Clark JM. A point mutation of acetylcholinesterase associated with azinphosmethyl resistance and reduced fitness in Colorado potato beetle. Pestic Biochem Physiol. 1996;55:100–8.

Zhu F, Xu J, Palli R, Ferguson J, Palli SR. Ingested RNA interference for managing the populations of the Colorado potato beetle, *Leptinotarsa decemlineata*. Pest Manag Sci. 2011;67:175–82.

Zhu F, Moural TW, Nelson DR, Palli SR. A specialist herbivore pest adaptation to xenobiotics through up-regulation of multiple cytochrome P450s. Sci Rep. 2016;6

Part III
Population Genomics Case Studies of Insect Species

The Population Genomics of *Aedes aegypti*: Progress and Prospects

Yoosook Lee, Karla Saavedra-Rodriguez, Tse-Yu Chen, Lindsay P. Campbell, and Chelsea T. Smartt

Abstract Despite its significance in public health, the population genomics of *Aedes aegypti* is in its infancy. We suspect this dichotomy is largely driven by its relatively large genome size affecting cost of sequencing. Efforts to capture the subset of genomic markers have been conducted using exome capture and microarray technologies. The focus of the population genomic studies has been largely on identifying population structure and investigating insecticide resistance. Recent advances in library preparation technology and availability of a chromosome-level reference assembly allowed whole genome sequence-based study. These developments open new questions that can be addressed with further population genomic data. As with *Anopheles gambiae,* interests in advancing gene drive-based mosquito control methods are propelling additional collections of *Ae. aegypti* datasets aimed at characterizing dispersal and population size. With increasing interest in novel genetic control and their application in the field, we expect to accelerate research in characterizing field populations using genomics approaches in the near future.

Keywords *Aedes aegypti* · High-throughput sequencing analysis · Microarray · Population genomics · SNP chip · Transcriptomics · Yellow fever mosquito

1 Introduction

Aedes aegypti Linnaeus is a mosquito species serving as the main vector of various arthropod-borne viruses such as yellow fever, dengue, chikungunya, and Zika. Currently, this is one of the most widespread mosquito species globally (Leta et al.

Y. Lee (✉) · T.-Y. Chen · L. P. Campbell · C. T. Smartt
Florida Medical Entomology Laboratory, Institute of Food and Agricultural Sciences, University of Florida, Vero Beach, FL, USA
e-mail: yoslee@ucdavis.edu; papilioninae@ufl.edu; lcampbell2@ufl.edu; ctsmart@ufl.edu

K. Saavedra-Rodriguez
Colorado State University, Fort Collins, CO, USA
e-mail: Karla.Saavedra_Rodriguez@colostate.edu

Julian R. Dupuis and Om P. Rajora (eds.), *Population Genomics: Insects*, Population Genomics [Om P. Rajora (Editor-in-Chief)], https://doi.org/10.1007/13836_2021_93,

2018). Consequently, the diseases transmitted by *Ae. aegypti* have become the fastest spreading mosquito-borne viral diseases (WHO 2020). It is estimated that about 50% of world population is presently living under the risk of contracting dengue. Naturally, *Ae. Aegypti* is one of the most highly studied mosquito species comprising roughly 17% of the mosquito related literature as of January, 2021, based on PubMed data.

Despite its significance in public health, the population genomics of *Ae. aegypti* is in its infancy. However, there have been significant advances in the population genomics over the last decade. In this chapter, we will introduce the genomic characteristics of *Ae. aegypti* and review the work investigating population genomics aspects of *Ae. aegypti* completed thus far. Due to high interest in genetic control methods and their applications in field settings, we covered the topics that have relevance in mosquito control research. These topics include (1) sex determination, (2) gene flow/dispersal, (3) insecticide resistance, (4) immune response, and (5) genetic variability that may impact the rise of resistance mechanisms in natural populations.

2 Genome Assembly

The first genome assembly of *Aedes aegypti* was published in 2007 (Nene et al. 2007) at about 1.3 Gbp. This is slightly less than half the size of the human genome size but greater than 5–6 times the size of *Anopheles* mosquitoes. The *Ae. aegypti* genome contains ~15,000 protein-coding genes. About half of its genome consists of transposable elements (Nene et al. 2007). The presence of repetitive DNA, including transposable elements, was a hindrance in constructing a chromosome-level assembly and thus the initial version of the *Ae. aegypti* genome (AaegL1) had 4,578 scaffolds.

The major breakthrough in chromosome-level assembly occurred in 2017 (Dudchenko et al. 2017) using the Hi-C sequencing approach (AaegL4). Soon after, the revised assembly AaegL5 became available after combining Hi-C and PacBio data (Matthews et al. 2018). The final assembly includes three major chromosomes, named 1, 2, and 3, a mitochondrial genome, and 2,307 scaffolds that are not incorporated into chromosome groups. Chromosome 1 is 310.8 Mbp long and contains 3,204 coding genes and 1,239 non-coding genes or pseudogenes. It contains a male determiner protein-encoding gene called Nix (AAEL022912; Chromosome 1:152,616,641-152,718,167) (Hall et al. 2015; Matthews et al. 2018) around the centromeric region. Chromosome 2 is 474.4 Mbp long and contains 5,471 coding genes and 1,814 non-coding/pseudogenes. Chromosome 3 is 409.8 Mbp long and contains 4,925 coding genes and 1,527 non-coding/ pseudogenes.

3 Sex Determination

Sex determination mechanisms are of particular interest in mosquito biology because some novel genetic control strategies utilize these processes to distort sex ratio by suppressing the production of adult female offspring. Sex of *Ae. aegypti* is decided by a dominant male-determining factor (M factor) located within the M locus of the Chromosome 1. The exact location of the M factor has recently been discovered (Hall et al. 2015; Matthews et al. 2018). Due to the highly repetitive nature of *Ae. aegypti*'s genome, a team of researchers first compared the genomes of males and females to identify the regions over-represented in male groups (males >5× females), which filtered 164 candidate M locus contigs. Then, based on RNAseq data, 140 of 164 contigs were removed due to their absence in the transcriptome or presence in female RNAseq samples. A subsequent homolog search identified one transcript that is a distant homolog of *transformer-2,* which is involved in the splicing of *doublesex* and *fruitless* that are known to regulate sex determination in fruit flies. This gene is now known as *Nix* (AAEL022912) and *AaegL5* placed it in the centromere of Chromosome 1 (1:152,616,641-152,718,167).

The sex-determining region (SDR) in *Ae. aegypti* is thought to be occupying a small region of Chromosome 1. Further studies using double-digest restriction-site associated DNA sequencing (ddRADseq) by Fontaine et al. (2017) further illuminated the extent of the SDR. They discovered a 123 Mbp-long region (from 87 to 211 Mbp) lacking recombination events in their laboratory crosses. By comparing *Ae. aegypti* populations from Rio de Janeiro, Brazil and Queensland, Australia, authors uncovered unusually high levels of divergence around the centromeric region of Chromosome 1 beyond just the region surrounding *Nix* (Fontaine et al. 2017). This study provides the first chromosome level view of the elevated divergence around SDR.

4 Population Genomic Studies

Due to the massive size of the *Ae. aegypti* genome, most early population genomic studies focused on techniques based on a fraction of the genome such as microarrays (Strode et al. 2008), ddRADseq (Rašić et al. 2014; Schmidt et al. 2019), SNP chips (Evans et al. 2015; Pless et al. 2017), and exome capture libraries (Campbell et al. 2017; Crawford et al. 2017; Dickson et al. 2017). Only recently has whole genome sequencing (WGS) been used for population genomic studies (Lee et al. 2019; Rose et al. 2020). The first WGS study published 46 individual-level whole genome sequence datasets of samples originating from California, Florida, and 3 regions in South Africa (Lee et al. 2017). A later study by Schmidt et al. (2020) published a population genomic dataset that consists of 132 *Ae. aegypti* whole genome sequences including the 46 from Lee et al. (2019). Since then, additional studies have produced with whole genome sequencing data for population assessments (Rose et al. 2020).

Most studies mentioned above commonly show strong population subdivision within *Ae. aegypti* based on their geographic origin. Rašić et al. (2014) showed clearer separation of SNP genotype profiles among Australia, Brazil, Vietnam, and Indonesia populations as compared to microsatellites generated from the same samples. SNP chip data utilizing 50K SNPs were able to separate worldwide samples roughly by continents and island groups, although separation within American continents was less clear (Evans et al. 2015). The same SNP chip was used to investigate invasive California *Aedes aegypti.* Results revealed multiple genetic clusters indicating that multiple populations of distinct genetic background were introduced in California (Pless et al. 2017).

This hypothesis of multiple introductions was corroborated by the later study utilizing whole genome sequences (Lee et al. 2019). Crawford et al. (2017) used an exome capture approach and identified 1.3 M polymorphic sites from a dataset that spans Mexico, multiple locations in Africa, and Sri Lanka. Again, populations were all segregated by their geographic origin with the possibility of multiple populations within Uganda.

The latest paper by Rose et al. (2020) published the largest population genomic dataset thus far with 375 individual mosquito genomes mostly originating from Africa. This is a unique study that links host choice phenotype, a morphological feature that is known to distinguish between *Ae. aegypti* formosus and the *Ae. aegypti* aegypti domestic forms, as well as genome sequencing data showing some chromosomal regions largely in chromosome 2 and chromosome 1 associated with human biting preference. The distribution of the subpopulation which prefers to bite humans is confined to West African countries, for example Senegal and Burkina Faso (Fig. 1). Results also confirmed that the domestic form of the Rabai (RAB in Fig. 1) population in Kenya which was known to be more anthropophilic is likely a non-African population introduced to the region. Fig. 1 shows RAB domestic (collected in 2009) clusters together with Asian and American *Ae. aegypti* populations with potential hybridization with the local RAB population. The investigators could not find this domestic form in their 2017 collection attempt. It seems plausible that there have been opportunistic introductions to Africa from other continents given the history of *Anopheles stephensi* introduction from Asia to Africa (Tadesse et al. 2021).

The high informativeness of SNPs for discriminating population origin compared to conventional microsatellite markers (Rašić et al. 2014) spurred further research into workflows for identifying the source population of invasive *Ae. aegypti* (Lee et al. 2019; Schmidt et al. 2019). Discriminant analysis of principal components on SNP profiles identified via ddRADseq shows very encouraging performance of source identification (Fig. 2). Use of genome-wide SNPs combined with insecticide resistance mutations by Schmidt et al. (2019) was able to track multiple routes of dispersal in Australia and neighboring regions. It also revealed potential challenges with source identification if genetic similarity in the examined area is minute. For example, the identified genetic clusters overlap in 4 locations around the Gulf of

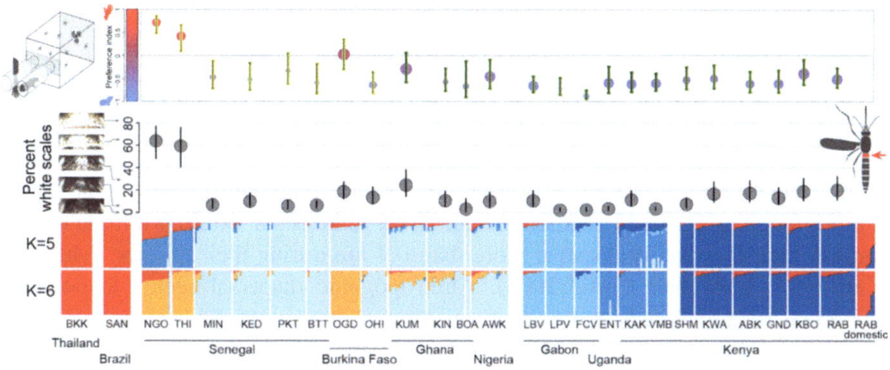

Fig. 1 Host choice, white scale proportions, and genetic clustering results point to a subpopulation of Ae. aegypti preferring to bite humans as blood source. Figure adapted from Rose et al. (2020)

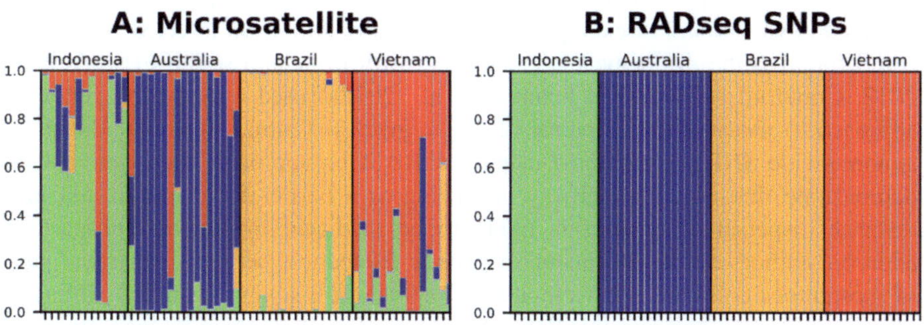

Fig. 2 Genotype clustering performance comparison between (**a**) microsatellite and (**b**) RADseq SNPs. The figure is adapted from Rašić et al. (2014)

Thailand (Bangkok, Singapore), South China Sea (Viet Nam), and Java Sea (Bali) (Schmidt et al. 2019). This may be a reflection of genetic relatedness correlated with their geographic proximity and frequent human-mediated traffic among those sites. In such regions, more advanced techniques tracking identity-by-descent segments or haplotypes may be required.

Some effort was made to identify chromosome inversions that may play a role in population differentiation (Redmond et al. 2020). Chromosome inversions have played important functions in adaptation and population differentiation in *Anopheles* mosquitoes (Kreutzer et al. 1972; Petrarca and Beier 1992; Touré et al. 1998; Black 4th and Lanzaro 2001; Cornel et al. 2016a). Redmond et al. (2020) used linked-read technology of 10× Genomics to study eight colonies of *Ae. aegypti* representing the extreme latitudes of each population. Although they did not find fixed inversion differences between populations, they found many microinversions that may impact behavior.

5 Dispersal

Accurately quantifying patterns of dispersal via direct observation is challenging, even at the micro-spatial scale, for small flying insects like *Ae. aegypti*. There have been many studies describing *Ae. aegypti* dispersal on a very small spatial scale using conventional mark-release-recapture (MRR) methods (Guerra et al. 2014; Harrington et al. 2014; Marcantonio et al. 2019; Juarez et al. 2020). Multiple studies indicate that they rarely disperse more than 0.4 km during their lifetime (Muir and Kay 1998; Liew and Curtis 2005). Anthropogenic dispersal seems to be more important for broad-scale patterns of *Ae. aegypti* movement. For example, eggs of *Ae. aegypti* laid in containers can be cryptically moved among cities miles apart by an unsuspecting person. *Aedes aegypti*, most likely using this hitchhiking strategy, was able to colonize urban areas over 200 miles apart in a time period of just 4 years (Lee et al. 2019).

Dispersal is of particular interest in genetic control strategy planning, deployment, and evaluation. For example, release of male *Wolbachia* infected *Ae. aegypti* could suppress the *Ae. aegypti* population growth in natural settings (Harris et al. 2012; Crawford et al. 2020). Schmidt et al. (2018a) used a landscape genomics approach to measure the dispersal of *Ae. aegypti* in Cairns, Australia. Using the genome-wide SNP set generated using a ddRADSeq approach, they were able to demonstrate that a major highway is serving as a barrier for dispersal and long distance dispersal of females is likely to be mediated by human transport. In addition, authors estimated kinship or genetic relatedness between individual pairs of mosquitoes collected and were able to separate full-sibling from half-sibling or other groups. Synthesizing this data and presence of *Wolbachia* in each mosquito specimen, they were able to confirm intergenerational loss of *Wolbachia* in the field. A similar approach was taken by other researchers to study dispersal of *Ae. aegypti* in Singapore (Filipović et al. 2020). Filipović et al. (2020) also introduced isolation by distance methods to estimate dispersal. With improved estimates of effective population size, this method could be widely used when resources are limited (Filipović et al. 2020).

6 Insecticide Resistance

Searching for insecticide-resistance genes was one of the early applications of *Ae. aegypti* reference genomes (Strode et al. 2008; Marcombe et al. 2009; Lertkiatmongkol et al. 2010), and efforts to this end continue today (Lien et al. 2019; Saavedra-Rodriguez et al. 2019). Major mechanisms of insecticide resistance involve target site insensitivity and enhanced metabolism. The target site insensitivity mechanism involves amino acid replacements at the major target sites of insecticides, such as the voltage-gated sodium channel (pyrethroid and DDT target site) and the acetylcholinesterase (organophosphate and carbamate target site). These replacements confer conformational changes that prevent the binding of insecticide chemicals and thus make mosquitoes "insensitive" to insecticides. Around 12 nonsynonymous mutations in the voltage-gated sodium channel gene have been associated with pyrethroid resistance and current research focuses on screening the frequency of these mutations in *Ae. aegypti* populations.

In contrast, enhanced metabolism involves either qualitative or quantitative changes in the expression of detoxification enzymes. Since detoxification is performed by a large group of P450 monooxygenases, glutathione transferases, and esterases, transcriptome technologies that compare the mRNA transcription between resistant and susceptible strains have greatly improved the knowledge of this mechanism. Early studies employed a cDNA microarray that included 206 detoxification genes (Strode et al. 2008; Saavedra-Rodriguez et al. 2012) which was subsequently replaced by a whole transcriptome microarray. Using this technology, over-expression of genes in temephos (Grisales et al. 2013; Poupardin et al. 2014) and pyrethroid resistance (Lertkiatmongkol et al. 2010; Kasai et al. 2014; Ishak et al. 2017; Lien et al. 2019) have been identified in *Ae. aegypti* populations. Furthermore, the metabolic properties of the enzymes encoded by these genes continue to be validated (Stevenson et al. 2012; Kasai et al. 2014; Ishak et al. 2017).

In the last 5 years, next generation sequencing technologies have improved the comparison of whole genomes or transcriptomes between insecticide resistant and susceptible strains from different regions of the world. For example, RNA-seq of permethrin-, imidacloprid-, and propoxur-selected strains allowed the identification of 413 differentially transcribed genes associated with resistance (David et al. 2014). Additionally, transcription patterns responding to different insecticides showed that permethrin was mostly associated with polymorphisms at cytochrome P450s, while other insecticides caused over-transcription of cuticle proteins, transporters, and metabolic genes (David et al. 2014).

Because >50% of the *Aedes* genome is constituted by repetitive DNA, different approaches to identify genomic variants associated with resistance have been tested. For example, using a target capture of a subset of 760 genes associated with resistance, 41 gene amplifications mostly affecting cytochrome P450s and 55 nonsynonymous variants were associated with deltamethrin resistance in *Ae. aegypti* populations from Thailand and French Guiana (Faucon et al. 2015). Subsequently, integration of DNA-seq and RNA-seq datasets identified variants at promoter regions associated with transcription. Additionally, polymorphism patterns in P450s were conserved within geographical regions but differed among continents (Faucon et al. 2015).

At a local scale, Saavedra-Rodriguez et al. (2019) compared exome-targeted sequencing of pools of resistant and susceptible mosquitoes from Yucatan Mexico. This effort identified novel mutations in the target site of pyrethroids – the voltage-gated sodium channel (vgsc) as well as potential clusters of esterases in chromosome 2 and P450 in chromosome 3. Furthermore, with the assistance of the recently available chromosome-level assembly of AaegL5, Matthews et al. (2018) were able to identify a genetic sweep in chromosome 3 caused by strong positive selection in favor of two replacements in the *vgsc* that confer resistance to pyrethroids (Matthews et al. 2018).

Transcriptome sequencing was used to identify mutations and copy number variations between pyrethroid-resistant strains originated from Khanh Hoa Province of Vietnam and Bora7 lab strain (Lien et al. 2019). Highly upregulated genes in resistant mosquitoes include P450s, glutathione S transferase (GST1), binding cassette transporter, and ABC transporter families.

Overall, P450 genes such as CYP6 and CYP9 consistently show up in these studies. Integration of these expression studies in response to insecticides has not yet been initiated. However, such an integration is feasible, as demonstrated in other fields. For instance, Jung et al. (2008) refined their light response gene expression analysis in rice by mining gene expression data published by different groups to exclude genes that had potential functional redundancy due to co-expressed family members and genes that had inconsistent light responses across publicly available microarray datasets. With careful records of metadata associated with each of these expression studies, it should be feasible to narrow down candidate insecticide-resistance genes that are consistently expressed across different experiments. Further development in correcting for population structure and uneven sample size (Sul et al. 2018) in genome association studies may also help integrate these datasets and discern mutations associated with their geographic origin vs. those truly associated with insecticide-resistance response.

7 Immune Response

Aedes aegypti transcriptome data became available long before genome data using the Liverpool lab strain of *Ae. aegypti* exposed to bacterial infection (Bartholomay et al. 2004). Studies of immune response against various viral strains have been predominately carried out on lab strains (Bartholomay et al. 2004; Batool et al. 2018; Wang et al. 2019), while studies related to immune response involving a specific geographic origin started to emerge in 2014 (Campbell et al. 2014; Etebari et al. 2017; Zhao et al. 2018). The most common subject of immune function studies have revolved around transcriptomic responses in geographically distinct populations following arbovirus infection. Most studies revealed the largest proportion of differentially expressed genes were of unknown function. Expression of immune response genes, including cysteine rich venom proteins, clip-domain serine proteases, C-type lectins, and serine proteases, was altered in response to dengue, Zika, and chikungunya viruses and varied between populations (Etebari et al. 2017; Zhao et al. 2018, 2019; Caicedo et al. 2019). Deep sRNA library mining revealed involvement of miRNA in antiviral responses in mosquito populations, where 35 modulated miRNA were involved in processes like transport, transcription/translation, and cytoskeleton/structural components (Campbell et al. 2014; Etebari et al. 2017). Similarly, studies investigating gene expression involved in interactions between Dengue virus (DENV) and distinct extrinsic incubation periods (EIP), viral load, and mosquito genetic lines in *Ae. aegypti* populations from Australia revealed involvement of processes like defense response, ion transport, and translation (Koh et al. 2018; Terradas and McGraw 2019).

Studies conducted with DENV-infected mosquitoes but analyzed for transcriptomic changes hours post infection show that autophagy/apoptosis, DNA replication, and fatty acid biosynthesis were enriched functions in infected mosquitoes from Trinidad and their responses were implicated in susceptibility (Kang et al. 2018). In contrast, DENV susceptibility in a population from Colombia was shown to involve Gram Negative Bacteria binding Protein, Cathepsin-b, and Keratinocyte lectin during early hours post infection; however, refractoriness, due to a midgut infection barrier, was controlled by expression of Niemann Pick Type-C2 (Caicedo et al. 2019). Koh et al. (2018) investigated the importance of the extrinsic incubation period (EIP) and measured gene expression changes in mosquitoes collected in Australia with different EIP for DENV, measured as viral particles in saliva. Mosquitoes with shorter EIPs had lower expression of immune response genes but higher levels of genes involved in translation and calcium ion transport while mosquitoes with longer EIP had lower expression of genes involved in metabolism (Koh et al. 2018). The role of mosquito genetic diversity in competence to transmit virus is also well established. A study exploring the role of expressed genes in individual mosquito lines with variation in viral load from dissemination of non-orally administered DENV found little variation in the genes analyzed across the lines and thus differences in viral load were not explained by variation in gene expression (Terradas and McGraw 2019).

With the availability of cheaper and more powerful tools, further studies on antiviral responses and variability in EIP in field mosquitoes to discover the mechanism that could be deployed for mosquito control efforts could be expanded. It would also be interesting to study variability of EIP in populations of mosquitoes as it relates to refractoriness especially if this phenotype can be used to control spread of disease-causing viruses.

8 Gene Drive System

The first gene drive system of *Ae. aegypti* was announced in 2020 (Li et al. 2020). High prevalence of insecticide resistance, difficulty in reaching cryptic urban environments, and high cost of intervention measures have made the control effort of this invasive species challenging (Vontas et al. 2012; Cornel et al. 2016b; Brelsfoard et al. 2019). Population suppression strategies using *Wolbachia* or the release of dominant lethal gene (RIDL) have been tried in the field as alternatives (Harris et al. 2012). However, these methods have not been widely adapted due to the high cost involved in continuous inundation. The discovery of the CRISPR-based gene drive system opened the possibility of developing a sustainable method to achieve population suppression without inundative releases and thus reducing the cost of genetic intervention methods.

The *Ae. aegypti* gene drive system developed by Li et al. (2020) is particularly notable compared to the *Anopheles* gene drive system published thus far (Gantz et al. 2015; Carballar-Lejarazu et al. 2020). It introduces a notion of a "split drive" where the endonuclease element and guide RNA components are separated. Subsequently, another alternative gene drive system based on chromosome translocation was introduced as a threshold-dependent drive system in *Ae. aegypti* (Sanchez et al. 2020). This threshold-dependent system opens a new possibility where a population could be reversed into its original state and remain confined to a target field site with much lower chance to spread to non-target regions (Li et al. 2020; Sanchez et al. 2020).

During the course of the gene drive system development in mosquitoes, a concern about the drive resistance has emerged as a key issue to address the development of a successful genetically engineered mosquito (GEM) (Marshall et al. 2017, 2019; Raban et al. 2020). This is of particular concern in *Ae. aegypti* as most GEM strategies are aimed at population suppression. Schmidt et al. (2020) explored a population genomic dataset of *Ae. aegypti* ($N = 133$ individuals) to identify conserved guided RNA target site polymorphisms throughout the genome. Although the total number of targets greatly diminishes after considering standing genetic variations in *Ae. aegypti* populations (33% of potential targets passed as "good" target site which do not contain any variation), 87.5% of protein-coding genes have at least one good target site for the CRISPR system. Of note, the geographic area captured in this genome dataset covers the states of California and Florida in the USA, Mexico, and South Africa. As the dataset expands to cover more geographic regions, we expect the proportion of good targets may decrease at a global level. But due to the localized nature of *Ae. aegypti* dispersal, one may find a higher number of good CRISPR targets that are conserved within the intended region for the genetic control method of choice.

Multiple species have shown reduced nucleotide diversity around the centromere, which experiences reduced recombination and may be responsible for reduced gene flow between populations of mosquitoes (Cruickshank and Hahn 2014). Similar patterns of reduced nucleotide diversity was observed in *Ae. aegypti* (Matthews et al. 2018; Lee et al. 2019). For the application of genetic control strategies, this may influence the likelihood of developing drive resistance depending on the genomic location of the genetic control target. Alternatively, one may find more suitable targets in those genomics regions with reduced nucleotide diversity.

9 Challenges in Data Integration

It is a well-known consensus that microsatellite data is difficult to integrate across different labs and times of data acquisition. Different fluorescent dyes attached to the same primer show different mobility in capillary sequencers and thus require careful control of instrument, protocol, and oligo design. Thus, when SNP genotyping platforms were introduced, it was widely accepted as a preferred choice because

Table 1 Genome properties of AaegL5 reference for chromosome (Chr) 1, 2, and 3

AaegL5	Chr. 1	Chr. 2	Chr 3	Sum
Length (bp)	310,827,022	474,425,716	409,777,670	1,195,030,408
Protein-coding genes	3,204	5,471	4,925	13,600
Non-coding genes/pseudogenes	1,239	1,814	1,527	4,580

scoring genotypes would be clearer-cut than microsatellites and thus data integration would be greatly improved. However, recent effort in trying to integrate SNPchip data from Evans et al. 2015 and SNP genotypes from WGS revealed that such integration is not straightforward (Lee et al. 2019).

The 71 bp-long probe sequences in SNPchip (Evans et al. 2015) were based on the AaegL1 reference sequence. The final genotype data used for data analysis in Evans et al. (2015) contained 25,589 SNPs out of 50,000 total SNPs. When these probe sequences were blasted to the latest AaegL5 reference, 80% of them remained unchanged in its sequence. The other 20% (4,981 probes) had either (1) more than 1 variant within the probe, (2) had gaps in the AaegL5 sequence, (3) probe length shorter than 71 bp (indicating too much divergence in probe sequence or large indel mutations), or (4) unable to be located in one of three chromosomes. Of the remaining 20,317 probe sequences, 191 had multiple hits in AaegL5 with identical match scores indicating duplication segments unfit for many population genetic analyses assuming a single copy gene in diploid organism.

After filtering out problematic probe sequences, comparison of WGS data with SNPchip data still showed mismatched genotypes similar to those in Table 1. Additional examples of discrepancy can be found in Lee et al. (2019). It would take a systematic approach to compare both systems with identical samples to test bias in genotype calls before datasets from different platforms could be integrated. However, due to the high upfront cost of SNPchip purchasing as well as high cost per individual for WGS (<$500 USD per sample for whole genome sequencing using HiSeq 4000 approach), we expect progress in SNP data integration efforts will remain slow-moving in the immediate future.

10 Special Handling of Mitochondrial Sequence During Mapping

The AaegL5 reference genome includes a mitogenome contig labelled "MT." Therefore mapping the whole genome to AaegL5 should provide variants information from both the nuclear genome and mitogenome. However, mitochondrial pseudogenes are known to be prevalent in the *Ae. aegypti* nuclear genome (Hlaing et al. 2009), which influence mapping performance (Schmidt et al. 2018b). Competitive mapping together with nuclear genome in AaegL5 resulted in much lower coverage in the mitogenome (mean = 853.3X; Fig. 3a) than non-competitive mapping (mapping to mitogenome only; mean 1,486.2X; Fig. 3b). This implies

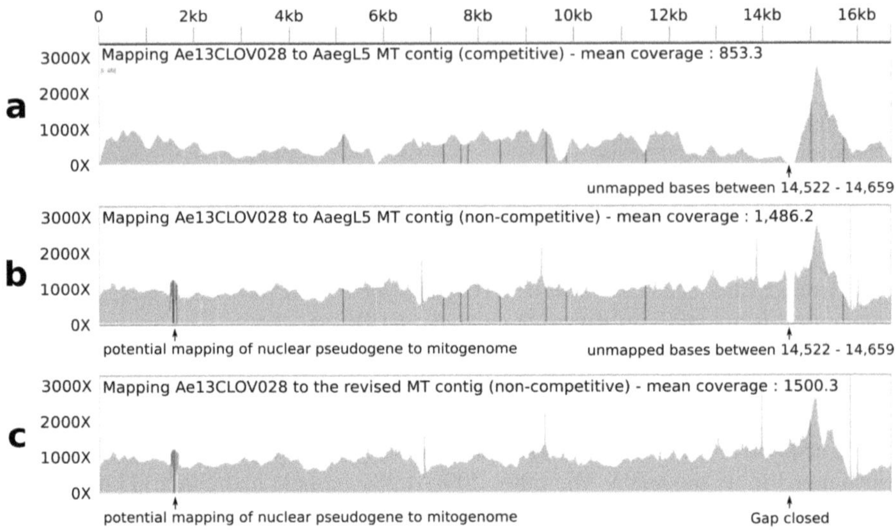

Fig. 3 Coverage data comparison. (**a**) a sample mapped to AaegL5 MT contig along with nuclear genome (competitive mapping) with mean coverage of 852.3X. (**b**) the non-competitive mapping to the same reference with mean coverage of 1,486.2X. It resulted in nuclear pseudogene reads mapped to the region at 1,496–1,665 bp showing ~40% increase in coverage in that 170 bp long region. (**c**) the non-competitive mapping to the Ae13CLOV028MT sequence with mean coverage of 1,500.3X

that many of the mitochondrial reads are improperly mapped when nuclear genome contigs are present during mapping. The coverage is also more variable in competitive mapping than non-competitive mapping (Figs. 1b vs. 3a). In non-competitive mapping, a portion of COX1 gene between 1,496 and 1,703 bp region had 40% increase in coverage (Figs. 1c and 3b). These are indeed extra nuclear pseudogene reads mapped to the mitogenome. However, not all pseudogene reads were properly paired, meaning that only one direction of paired-end reads was mapped to the mitogenome and their mate was unmapped. This issue can be resolved easily by filtering out "mate is unmapped" reads from bam files.

Visual inspection of mapped data using AaegL5 MT contig revealed a 138 bp-long region where no reads were mapped (Fig. 3a, b). This gap was present in all of the 173 *Ae. aegypti* specimens processed using Illumina NGS platform (Table 2). In addition, datasets from NCBI Sequencing reads (SRR6063610, SRR6063610, SRR871497) also show gap in coverage in this region (Schmidt et al. 2018b). Revised MT sequence was able to close the gap (Fig. 3c).

For nuclear genome analysis of *Ae. aegypti,* we recommend mapping the sequence to the Mt region first to filter out Mt reads as well as nuclear pseudogene reads and then use the remaining unmapped reads to map to the nuclear genome reference.

Table 2 Genotype of AX-93242921 SNP (genomic position 3:107766050; probe sequence – TCAGTTCGGTTTTTGTTGGCAAGTGCTGCATTGTT[C/T]CGATGAGGTGAGTAGCAGCA GTACGCTTTCGCGTT)

Platform	Location	Sample size	AX-93242921 SNP genotypes		
			CC	CT	TT
Illumina NGS	Central CA	68	100%		
Illumina NGS	Southern CA	74	100%		
Illumina NGS	Florida	4	100%		
Illumina NGS	Puerto Rico	6	83.3%	16.7%	
Illumina NGS	Dutch Caribbean	15	100%		
Illumina NGS	Mexico	3	100%		
Illumina NGS	South Africa	3	100%		
SNP chip	Australia	12			100%
SNP chip	French Polynesia	12			100%
SNP chip	Senegal	24		8.3%	91.7%
SNP chip	Texas	8			100%
SNP chip	Hawaii	6			100%
SNP chip	Brazil	40			100%
SNP chip	Uganda	12			100%
SNP chip	Florida	11		9.1%	90.9%
SNP chip	Puerto Rico	12			100%
SNP chip	Mexico	11			100%
SNP chip	Thailand	12			100%

11 Future Prospects for Sequencing Efforts

The paucity of WGS data for *Aedes aegypti* populations is likely due to the combination of interest and cost. As of May 2020, there are over 16,000 publications related to *An. gambiae,* while *Ae. aegypti* has just over 9,000 publications. The number of publications is not directly correlated with the level of interest or the number of people working on the subject matter, let alone the significance with respect to public health importance. However, the lower number of publications is a reflection of the research productivity and/or funding availability. There is a consortium called Ag1000G that started in 2014 and accumulated over 1,000 sequences from natural populations of *An. gambiae.* A similar effort for *Ae. aegypti* culminated in the completion of the *Ae. aegypti* genome assembly at the chromosome level (Matthews et al. 2018), and recently, a continental scale study of African *Ae. aegypti* population genomics produced 480 whole genome sequences at the individual level (Rose et al. 2020).

Recent advances in sequencing technology have significantly reduced the cost per base. A typical Illumina HiSeq 4000 lane was able to sequence ~7 *Ae. aegypti* samples to 10× depth in practice with the cost of $417/sample. With the new NovaSeq S4 platform, researchers can sequence up to 51 samples per lane which substantially lowers the sequencing cost per sample to $141 (Table 3). With the

Table 3 Sequencing cost comparison. PE stands for "paired-end." Note that price can vary depending on the read length. Total cost includes $60/sample for library prep as well as sequencing cost. The choice of read length was provided in the 2nd column. Prices are based on rate at the University of Florida Interdisciplinary Center for Biotechnology Research NextGen DNA Sequencing Core as of Jan 2021

Platform	PE read length (bp)	Cluster/ lane (M)	Output bases/ lane (PE)	Samples/ lane for 10×/ sample	Sequencing cost/lane	Total cost per sample	Total cost per lane
NovaSeq Sp500	250	365	182,500	14.3	$2,600	$242	$3,390.39
NovaSeq S1	150	730	219,000	17.1	$3,200	$247	$4,196.31
NovaSeq S2	150	1,850	555,000	43.4	$5,500	$187	$8,028.87
NovaSeq S4	150	2,200	660,000	51.6	$4,200	$141	$7,209.96

sequencing cost continuously decreasing, we expect more studies involving whole genome sequencing will be published.

In addition to lower-cost sequencing approaches facilitating increased availability of *Aedes aegypti* whole genome sequences, complementary fields of study are developing new and innovative approaches to analyze genomic data across space and time (Manel and Holderegger 2013; Balkenhol 2015). The field of landscape genomics offers a promising framework to investigate the functional connectivity and local adaptation of *Ae. Aegypti* across different geographic areas, providing the set of quantitative tools needed to bridge advanced molecular genomics with cutting-edge geospatial analyses (Manel and Holderegger 2013; Rellstab et al. 2015; Hemming-Schroeder et al. 2018). As noted above, thus far, a limited number of studies have investigated finer-scale *Ae. aegypti* dispersal patterns across space using a landscape genomics approach, but we anticipate an increase in these analyses and that there will be movement away from characterizing patterns toward quantifying contributions of environmental variables to population structure and insecticide resistance. In this context, the continued effort to synthesize genomic data across geographic areas will facilitate the exploration of *Ae. aegypti* biology and ecology in novel and potentially transformational ways.

12 Conclusions

In this chapter, we highlight key genomic features of the latest *Ae. aegypti* reference genome (AaegL5) necessary to analyze and interpret data. We also provide current updates on the use of population genomic data to understand the biology of *Ae. aegypti* important to arbovirus surveillance and vector control. As dengue transmission risk continues to increase globally, we expect increased research into novel control approaches, and we expect that utilizing information gained from the population genomics of *Ae. aegypti* will play a pivotal role in all aspects of these efforts. Recent advances in genetic control strategies for *Ae. aegypti* and the active efforts in field trials and deployment for Dengue control are one example of this potential, and we expect a steady increase in the number of studies utilizing these data assessing population connectivity, resistance potential, dispersal capacity, and population size to inform control strategies. With increasing data availability from multiple geographic locations, continued advances in landscape genomics approaches that quantify associations between population diversity, differentiation, or functional attributes and environmental variables will provide novel approaches toward informing intervention strategies. As these new studies continue to reveal new information regarding factors important to distinguishing subpopulations or the functional roles of additional loci, we anticipate new hypotheses will emerge contributing to a greater understanding of the biology and ecology of this important vector species.

References

Balkenhol N. Landscape genetics: concepts, methods, applications. Hoboken: Wiley Blackwell; 2015.

Bartholomay LC, Cho WL, Rocheleau TA, Boyle JP, Beck ET, Fuchs JF, et al. Description of the transcriptomes of immune response-activated hemocytes from the mosquito vectors *Aedes aegypti* and *Armigeres subalbatus*. Infect Immun. 2004;72(7):4114–26.

Batool K, Alam I, Wu S, Liu W, Zhao G, Chen M, et al. Transcriptomic analysis of *Aedes aegypti* in response to Mosquitocidal *Bacillus thuringiensis* LLP29 toxin. Sci Rep. 2018;8(1):12650.

Black WC 4th, Lanzaro GC. Distribution of genetic variation among chromosomal forms of *Anopheles gambiae* s.s: introgressive hybridization, adaptive inversions, or recent reproductive isolation? Insect Mol Biol. 2001;10(1):3–7.

Brelsfoard CL, Mains JW, Mulligan S, Cornel A, Holeman J, Kluh S, et al. *Aedes aegypti* males as vehicles for insecticide delivery. Insects. 2019;10(8)

Caicedo PA, Serrato IM, Sim S, Dimopoulos G, Coatsworth H, Lowenberger C, et al. Immune response-related genes associated to blocking midgut dengue virus infection in *Aedes aegypti* strains that differ in susceptibility. Insect Sci. 2019;26(4):635–48.

Campbell CL, Harrison T, Hess AM, Ebel GD. MicroRNA levels are modulated in *Aedes aegypti* after exposure to Dengue-2. Insect Mol Biol. 2014;23(1):132–9.

Campbell CL, Dickson LB, Lozano-Fuentes S, Juneja P, Jiggins FM, Black WC. Alternative patterns of sex chromosome differentiation in *Aedes aegypti* (L). BMC Genomics. 2017;18 (1):943.

Carballar-Lejarazu R, Ogaugwu C, Tushar T, Kelsey A, Pham TB, Murphy J, et al. Next-generation gene drive for population modification of the malaria vector mosquito, *Anopheles gambiae*. Proc Natl Acad Sci U S A. 2020;117(37):22805–14.

Cornel AJ, Brisco KK, Tadei WP, Secundino NF, Rafael MS, Galardo AK, et al. *Anopheles darlingi* polytene chromosomes: revised maps including newly described inversions and evidence for population structure in Manaus. Mem Inst Oswaldo Cruz. 2016a;111(5):335–46.

Cornel AJ, Holeman J, Nieman CC, Lee Y, Smith C, Amorino M, et al. Surveillance, insecticide resistance and control of an invasive *Aedes aegypti* (Diptera: Culicidae) population in California. F1000Res. 2016b;5:194.

Crawford JE, Alves JM, Palmer WJ, Day JP, Sylla M, Ramasamy R, et al. Population genomics reveals that an anthropophilic population of *Aedes aegypti* mosquitoes in West Africa recently gave rise to American and Asian populations of this major disease vector. BMC Biol. 2017;15 (1):16.

Crawford JE, Clarke DW, Criswell V, Desnoyer M, Cornel D, Deegan B, et al. Efficient production of male Wolbachia-infected *Aedes aegypti* mosquitoes enables large-scale suppression of wild populations. Nat Biotechnol. 2020;38(4):482–92.

Cruickshank TE, Hahn MW. Reanalysis suggests that genomic islands of speciation are due to reduced diversity, not reduced gene flow. Mol Ecol. 2014;23(13):3133–57.

David JP, Faucon F, Chandor-Proust A, Poupardin R, Riaz MA, Bonin A, et al. Comparative analysis of response to selection with three insecticides in the dengue mosquito *Aedes aegypti* using mRNA sequencing. BMC Genomics. 2014;15:174.

Dickson LB, Campbell CL, Juneja P, Jiggins FM, Sylla M, Black WC. Exon-enriched libraries reveal large genic differences between *Aedes aegypti* from Senegal, West Africa, and populations outside Africa. G3 (Bethesda). 2017;7(2):571–82.

Dudchenko O, Batra SS, Omer AD, Nyquist SK, Hoeger M, Durand NC, et al. De novo assembly of the *Aedes aegypti* genome using Hi-C yields chromosome-length scaffolds. Science. 2017;356 (6333):92–5.

Etebari K, Hegde S, Saldana MA, Widen SG, Wood TG, Asgari S, et al. Global transcriptome analysis of *Aedes aegypti* mosquitoes in response to Zika virus infection. mSphere. 2017;2(6)

Evans BR, Gloria-Soria A, Hou L, McBride C, Bonizzoni M, Zhao H, et al. A multipurpose high throughput SNP Chip for the dengue and yellow fever mosquito, *Aedes aegypti*. G3 (Bethesda). 2015;5(5):711–8.

Faucon F, Dusfour I, Gaude T, Navratil V, Boyer F, Chandre F, et al. Identifying genomic changes associated with insecticide resistance in the dengue mosquito *Aedes aegypti* by deep targeted sequencing. Genome Res. 2015;25(9):1347–59.

Filipović I, Hapuarachchi HC, Tien WP, Razak M, Lee C, Tan CH, et al. Using spatial genetics to quantify mosquito dispersal for control programs. BMC Biol. 2020;18(1):104.

Fontaine A, Filipovic I, Fansiri T, Hoffmann AA, Cheng C, Kirkpatrick M, et al. Extensive genetic differentiation between Homomorphic sex chromosomes in the mosquito vector, *Aedes aegypti*. Genome Biol Evol. 2017;9(9):2322–35.

Gantz VM, Jasinskiene N, Tatarenkova O, Fazekas A, Macias VM, Bier E, et al. Highly efficient Cas9-mediated gene drive for population modification of the malaria vector mosquito *Anopheles stephensi*. Proc Natl Acad Sci U S A. 2015;112(49):E6736–43.

Grisales N, Poupardin R, Gomez S, Fonseca-Gonzalez I, Ranson H, Lenhart A. Temephos resistance in *Aedes aegypti* in Colombia compromises dengue vector control. PLoS Negl Trop Dis. 2013;7(9):e2438.

Guerra CA, Reiner RC Jr, Perkins TA, Lindsay SW, Midega JT, Brady OJ, et al. A global assembly of adult female mosquito mark-release-recapture data to inform the control of mosquito-borne pathogens. Parasit Vectors. 2014;7:276.

Hall AB, Basu S, Jiang X, Qi Y, Timoshevskiy VA, Biedler JK, et al. Sex determination. A male-determining factor in the mosquito *Aedes aegypti*. Science. 2015;348(6240):1268–70.

Harrington LC, Fleisher A, Ruiz-Moreno D, Vermeylen F, Wa CV, Poulson RL, et al. Heterogeneous feeding patterns of the dengue vector, *Aedes aegypti*, on individual human hosts in rural Thailand. PLoS Negl Trop Dis. 2014;8(8):e3048.

Harris AF, McKemey AR, Nimmo D, Curtis Z, Black I, Morgan SA, et al. Successful suppression of a field mosquito population by sustained release of engineered male mosquitoes. Nat Biotechnol. 2012;30(9):828–30.

Hemming-Schroeder E, Lo E, Salazar C, Puente S, Yan G. Landscape genetics: a toolbox for studying vector-borne diseases. Front Ecol Evol. 21 p. 2018; https://doi.org/10.3389/fevo.2018.00021.

Hlaing T, Tun-Lin W, Somboon P, Socheat D, Setha T, Min S, et al. Mitochondrial pseudogenes in the nuclear genome of *Aedes aegypti* mosquitoes: implications for past and future population genetic studies. BMC Genet. 2009;10:11.

Ishak IH, Kamgang B, Ibrahim SS, Riveron JM, Irving H, Wondji CS. Pyrethroid resistance in Malaysian populations of dengue vector *Aedes aegypti* is mediated by CYP9 family of cytochrome P450 genes. PLoS Negl Trop Dis. 2017;11(1):e0005302.

Juarez JG, Garcia-Luna S, Chaves LF, Carbajal E, Valdez E, Avila C, et al. Dispersal of female and male *Aedes aegypti* from discarded container habitats using a stable isotope mark-capture study design in South Texas. Sci Rep. 2020;10(1):6803.

Jung KH, Lee J, Dardick C, Seo YS, Cao P, Canlas P, Phetsom J, Xu X, Ouyang S, An K, et al. Identification and functional analysis of light-responsive unique genes and gene family members in rice. PLoS Genet. 2008;4(8):e1000164.

Kang DS, Barron MS, Lovin DD, Cunningham JM, Eng MW, Chadee DD, et al. A transcriptomic survey of the impact of environmental stress on response to dengue virus in the mosquito, *Aedes aegypti*. PLoS Negl Trop Dis. 2018;12(6):e0006568.

Kasai S, Komagata O, Itokawa K, Shono T, Ng LC, Kobayashi M, et al. Mechanisms of pyrethroid resistance in the dengue mosquito vector, *Aedes aegypti*: target site insensitivity, penetration, and metabolism. PLoS Negl Trop Dis. 2014;8(6):e2948.

Koh C, Allen SL, Herbert RI, McGraw EA, Chenoweth SF. The transcriptional response of *Aedes aegypti* with variable extrinsic incubation periods for dengue virus. Genome Biol Evol. 2018;10 (12):3141–51.

Kreutzer R, Kitzmiller J, Ferreira E. Inversion polymorphism in the salivary gland chromosome of *Anopheles darlingi* root. Mosq News. 1972;32(4):555–65.

Lee Y, Schmidt H, Collier TC, Conner WR, Hanemaaijer MJ, Slatkin M, et al. Genome-wide divergence among invasive populations of *Aedes aegypti* in California. BioRXiv. 2017; https://doi.org/10.1101/166629.

Lee Y, Schmidt H, Collier TC, Conner WR, Hanemaaijer MJ, Slatkin M, et al. Genome-wide divergence among invasive populations of *Aedes aegypti* in California. BMC Genomics. 2019;20(1):204.

Lertkiatmongkol P, Pethuan S, Jirakanjanakit N, Rongnoparut P. Transcription analysis of differentially expressed genes in insecticide-resistant *Aedes aegypti* mosquitoes after deltamethrin exposure. J Vector Ecol. 2010;35(1):197–203.

Leta S, Beyene TJ, De Clercq EM, Amenu K, Kraemer MUG, Revie CW. Global risk mapping for major diseases transmitted by *Aedes aegypti* and *Aedes albopictus*. Int J Infect Dis. 2018;67:25–35.

Li M, Yang T, Kandul NP, Bui M, Gamez S, Raban R, et al. Development of a confinable gene drive system in the human disease vector *Aedes aegypti*. Elife. 2020;9

Lien NTK, Ngoc NTH, Lan NN, Hien NT, Tung NV, Ngan NTT, et al. Transcriptome sequencing and analysis of changes associated with insecticide resistance in the dengue mosquito (*Aedes aegypti*) in Vietnam. Am J Trop Med Hyg. 2019;100(5):1240–8.

Liew C, Curtis CF. Horizontal and vertical dispersal of dengue vector mosquitoes, *Aedes aegypti* and *Aedes albopictus*. Singapore Med Vet Entomol. 2005;18(4):351–60.

Manel S, Holderegger R. Ten years of landscape genetics. Trends Ecol Evol. 2013;28(10):614–21.

Marcantonio M, Reyes T, Barker CM. Quantifying *Aedes aegypti* dispersal in space and time: a modeling approach. Ecosphere. ee02977 p. 2019; https://doi.org/10.1002/ecs2.2977.

Marcombe S, Poupardin R, Darriet F, Reynaud S, Bonnet J, Strode C, et al. Exploring the molecular basis of insecticide resistance in the dengue vector *Aedes aegypti*: a case study in Martinique Island (French West Indies). BMC Genomics. 2009;10:494.

Marshall JM, Buchman A, Sanchez CH, Akbari OS. Overcoming evolved resistance to population-suppressing homing-based gene drives. Sci Rep. 2017;7(1):3776.

Marshall JM, Raban RR, Kandul NP, Edula JR, Leon TM, Akbari OS. Winning the tug-of-war between effector gene design and pathogen evolution in vector population replacement strategies. Front Genet. 2019;10:1072.

Matthews BJ, Dudchenko O, Kingan SB, Koren S, Antoshechkin I, Crawford JE, et al. Improved reference genome of *Aedes aegypti* informs arbovirus vector control. Nature. 2018;563 (7732):501–7.

Muir LE, Kay BH. *Aedes aegypti* survival and dispersal estimated by mark-release-recapture in northern Australia. Am J Trop Med Hyg. 1998;58(3):277–82.

Nene V, Wortman JR, Lawson D, Haas B, Kodira C, Tu ZJ, et al. Genome sequence of *Aedes aegypti*, a major arbovirus vector. Science. 2007;316(5832):1718–23.

Petrarca V, Beier JC. Intraspecific chromosomal polymorphism in the *Anopheles gambiae* complex as a factor affecting malaria transmission in the Kisumu area of Kenya. Am J Trop Med Hyg. 1992;46(2):229–37.

Pless E, Gloria-Soria A, Evans BR, Kramer V, Bolling BG, Tabachnick WJ, et al. Multiple introductions of the dengue vector, *Aedes aegypti*, into California. PLoS Negl Trop Dis. 2017;11(8):e0005718.

Poupardin R, Srisukontarat W, Yunta C, Ranson H. Identification of carboxylesterase genes implicated in temephos resistance in the dengue vector *Aedes aegypti*. PLoS Negl Trop Dis. 2014;8(3):e2743.

Raban RR, Marshall JM, Akbari OS. Progress towards engineering gene drives for population control. J Exp Biol. 2020;223(Pt Suppl 1)

Rašić G, Filipovic I, Weeks AR, Hoffmann AA. Genome-wide SNPs lead to strong signals of geographic structure and relatedness patterns in the major arbovirus vector, *Aedes aegypti*. BMC Genomics. 2014;15:275.

Redmond SN, Sharma A, Sharakhov I, Tu Z, Sharakhova M, Neafsey DE. Linked-read sequencing identifies abundant microinversions and introgression in the arboviral vector *Aedes aegypti*. BMC Biol. 2020;18(1):26.

Rellstab C, Gugerli F, Eckert AJ, Hancock AM, Holderegger R. A practical guide to environmental association analysis in landscape genomics. Mol Ecol. 2015;24(17):4348–70.

Rose NH, Sylla M, Badolo A, Lutomiah J, Ayala D, Aribodor OB, et al. Climate and urbanization drive mosquito preference for humans. Curr Biol. 2020;30(18):3570–3579.e6.

Saavedra-Rodriguez K, Suarez AF, Salas IF, Strode C, Ranson H, Hemingway J, et al. Transcription of detoxification genes after permethrin selection in the mosquito *Aedes aegypti*. Insect Mol Biol. 2012;21(1):61–77.

Saavedra-Rodriguez K, Campbell CL, Lenhart A, Penilla P, Lozano-Fuentes S, Black WC. Exome-wide association of deltamethrin resistance in *Aedes aegypti* from Mexico. Insect Mol Biol. 2019;28(5):591–604.

Sanchez CH, Bennett JB, Wu SL, Rasic G, Akbari OS, Marshall JM. Modeling confinement and reversibility of threshold-dependent gene drive systems in spatially-explicit *Aedes aegypti* populations. BMC Biol. 2020;18(1):50.

Schmidt TL, Filipovic I, Hoffmann AA, Rasic G. Fine-scale landscape genomics helps explain the slow spatial spread of Wolbachia through the *Aedes aegypti* population in Cairns, Australia. Heredity (Edinb). 2018a;120(5):386–95.

Schmidt H, Hanemaaijer MJ, Cornel AJ, Lanzaro GC, Braack L, Lee Y. Complete mitogenome sequence of *Aedes (Stegomyia) aegypti* derived from field isolates from California and South Africa. Mitochondrial DNA B. 2018b;3(2):994–5.

Schmidt TL, van Rooyen AR, Chung J, Endersby-Harshman NM, Griffin PC, Sly A, et al. Tracking genetic invasions: genome-wide single nucleotide polymorphisms reveal the source of pyrethroid-resistant *Aedes aegypti* (yellow fever mosquito) incursions at international ports. Evol Appl. 2019;12(6):1136–46.

Schmidt H, Collier TC, Hanemaaijer MJ, Houston PD, Lee Y, Lanzaro GC. Abundance of conserved CRISPR-Cas9 target sites within the highly polymorphic genomes of *Anopheles* and *Aedes* mosquitoes. Nat Commun. 2020;11(1):1425.

Stevenson BJ, Pignatelli P, Nikou D, Paine MJ. Pinpointing P450s associated with pyrethroid metabolism in the dengue vector, *Aedes aegypti*: developing new tools to combat insecticide resistance. PLoS Negl Trop Dis. 2012;6(3):e1595.

Strode C, Wondji CS, David JP, Hawkes NJ, Lumjuan N, Nelson DR, et al. Genomic analysis of detoxification genes in the mosquito *Aedes aegypti*. Insect Biochem Mol Biol. 2008;38 (1):113–23.

Sul JH, Martin LS, Eskin E. Population structure in genetic studies: confounding factors and mixed models. PLoS Genet. 2018;14(12):e1007309.

Tadesse FG, Ashine T, Teka H, Esayas E, Messenger LA, Chali W, et al. *Anopheles stephensi* Mosquitoes as vectors of Plasmodium vivax and falciparum, Horn of Africa, 2019. Emerg Infect Dis. 2021;27(2)

Terradas G, McGraw EA. Using genetic variation in *Aedes aegypti* to identify candidate anti-dengue virus genes. BMC Infect Dis. 2019;19(1):580.

Touré Y, Petrarca V, Traore S, Coulibaly A, Maiga H, Sankare O, et al. The distribution and inversion polymorphism of chromosomally recognized taxa of the *Anopheles gambiae* complex in Mali, West Africa. Parassitologia. 1998;40(4):477–511.

Vontas J, Kioulos E, Pavlidi N, Morou E, della Torre A, Ranson H. Insecticide resistance in the major dengue vectors *Aedes albopictus* and *Aedes aegypti*. Pestic Biochem Phys. 2012;104 (2):126–31.

Wang JM, Cheng Y, Shi ZK, Li XF, Xing LS, Jiang H, et al. *Aedes aegypti* HPX8C modulates immune responses against viral infection. PLoS Negl Trop Dis. 2019;13(4):e0007287.

WHO. WHO fact sheet on dengue. Geneva: World Health Organization; 2020. www.who.int. Accessed 12 Jan 2021. https://www.who.int/en/news-room/fact-sheets/detail/dengue-and-severe-dengue

Zhao L, Alto BW, Smartt CT, Shin D. Transcription profiling for Defensins of *Aedes aegypti* (Diptera: Culicidae) during development and in response to infection with Chikungunya and Zika viruses. J Med Entomol. 2018;55(1):78–89.

Zhao L, Alto BW, Jiang Y, Yu F, Zhang Y. Transcriptomic analysis of *Aedes aegypti* innate immune system in response to ingestion of Chikungunya virus. Int J Mol Sci. 2019;20(13)

The Population Genomics of *Anopheles gambiae* Species Complex: Progress and Prospects

Hanno Schmidt, Oscar D. Kirstein, Tse-Yu Chen, Lindsay P. Campbell, Travis C. Collier, and Yoosook Lee

Abstract *Anopheles gambiae* sensu *lato* is a species complex containing principal malaria vectors such as *An. gambiae* sensu stricto, *An. coluzzii,* and *An. arabiensis.* Numerous studies have shown dynamic species hybridization among member of this complex makes them an ideal model for studying evolution and speciation as well as for applied vector biology. Applying a population genomics approach to the *An. gambiae* and *An. coluzzii* species group has led to a number of important and epidemiologically relevant insights including: (1) organization of genomic divergence into "islands of speciation"; (2) competing models of population origin of *An. gambiae* and *An. Coluzzii*; (3) description of asymmetric introgression between diverged populations; (4) description of adaptive introgression as a mechanism for the rapid evolution of insecticide resistance; (5) gene/genomic region discovery associated with insecticide resistance and *Plasmodium* resistance; (6) assessment of gene drive resistance potential among natural populations of *An. gambiae* and *An. coluzzii.* With increasing availability of genomic tools and population genomic databases for this species group, it is poised to step further in utilizing genomics resources to answer traditionally challenging questions such as dispersal and population size estimation.

Keywords *Anopheles coluzzii* · *Anopheles gambiae* · Genomics · Hybridization · Introgression · Malaria vector · Species complex

H. Schmidt
Institute of Organismic and Molecular Evolution, University of Mainz, Mainz, Germany
e-mail: schmi014@uni-mainz.de

O. D. Kirstein
Emory University, Atlanta, GA, USA
e-mail: odkirstein@emory.edu

T.-Y. Chen · L. P. Campbell · Y. Lee (✉)
Florida Medical Entomology Laboratory, University of Florida, Vero Beach, FL, USA
e-mail: papilioninae@ufl.edu; lcampbell2@ufl.edu; yoosook.lee@ufl.edu

T. C. Collier
Vector Genetics Laboratory, University of California, Davis, Davis, CA, USA

Julian R. Dupuis and Om P. Rajora (eds.), *Population Genomics: Insects*, Population Genomics [Om P. Rajora (Editor-in-Chief)],
https://doi.org/10.1007/13836_2021_92,

1 Introduction

Anopheles gambiae belongs to a group of closely related species, the *An. gambiae* complex or *An. gambiae* sensu *lato*. This complex consists of around eight species, depending on the criteria applied. The status of these species was initially established based on F_1 hybrid sterility among crosses between different populations exhibiting difference in chromosome inversions (Davidson 1956, 1964a, b; Davidson and Hunt 1973; Coluzzi et al. 1979). Later studies utilizing molecular markers and ecological studies demonstrating the rarity of hybrids in natural setting supported the elevation of *An. coluzzii* as separate species from *An. gambiae* sensu stricto (Coetzee et al. 2013). Since *An. gambiae* sensu stricto is most important in a public health context, most of the research has been done on this species and its sister species *An. coluzzii*. The impact on public health is mainly due to these species being competent vectors of malaria, a disease responsible for millions of cases and hundreds of thousands of deaths every year, primarily in Africa.

In this chapter, we focus on population genomics of these two anopheline species, focusing on topics of hybridization, genetic isolation, introgression, and historical population divergence. Studying the population genomics of *An. gambiae s.l.* is especially challenging because of (1) inter-specific gene flow and (2) a huge, structured distribution range throughout sub-Saharan Africa. We describe the current spread of genetic resistance against common insecticides through the mechanism of adaptive introgression and selective sweeps. We also review the work thus far in understanding pathogen resistance and the most recent efforts to tackle malaria spread by genetically modified mosquito populations.

2 Characteristics of *Anopheles gambiae* Genome

Anopheles gambiae sensu stricto is a member of African malaria vector species complex, *Anopheles gambiae* sensu *lato* (Fig. 1). Due to the species' medical importance, its ~280 Mb reference genome became available in 2002 (Holt et al. 2002) ahead of any other mosquito species. Its estimated genome size using haploid DNA content (C-value) is 0.27 pg, and the AgamP4 reference genome length is 281.38 Mbps. Other *Anopheles* species show a similar range of C-values around 0.2 pg, and whole-genome sequencing based assembly lengths of 200–300 Mbps genome size (Fig. 2).

Like all other *Anopheles* species, *An. gambiae* has two autosomes, chromosome 2 and 3, and sex chromosomes X and Y. Female individuals have two X chromosomes and males have one X and one Y chromosome. While *An. gambiae* remains the only *Anopheles* species with a genome assembly at the chromosome-level (available through VectorBase[1]), about 20 *Anopheles* species' genomes have been

[1]Strictly speaking, AgamP4 should be considered *An. coluzzii* genome given the callability bias toward *An. coluzzii* in the islands of speciation (Turner et al. 2005).

Species complex	Species	Chromosome-level Reference Genome	Draft Reference (scaffold count)	Genome size (Mbp)	C-value
An. gambiae sensu lato	*An. arabiensis*		AaraD1 (1,214)	246.6	
	An. coluzzii	AgamP4	AcolM1 (10,521)	224.5	0.27
	An. gambiae sensu stricto	AgamP4	AgamS1 (13,042)	236.4	
	An. amharicus				
	An. bwambae				
	An. melas		AmelC2 (20,229)	224.2	
	An. merus		AmerM2 (2,027)	288.0	
	An. quadriannulatus		AquaS1 (2,823)	283.8	

Fig. 1 Species within *An. gambiae* complex and their reference genome availability. Scaffold count and Genome size information are obtained from VectorBase (Giraldo-Calderon et al. 2015), accessed in May 2020. C-value (Haploid DNA contents in picograms) is from Animal Genome Size Database (http://www.genomesize.com/), accessed in March 2021

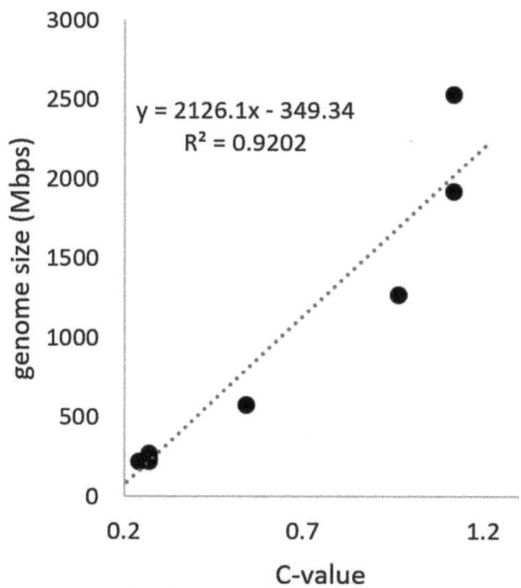

Fig. 2 Relationship between C-value and genome size. C-value data are from Animal Genome Size Database (http://www.genomesize.com/) and genome size are from VectorBase (http:/ vectorbase.org). Both data were accessed in March 2021

sequenced thus far (Marinotti et al. 2013; Jiang et al. 2014; Zhou et al. 2014; Ghurye et al. 2019) and efforts to increase genomic resources for other species are underway. Of note, the Y chromosome of *An. gambiae* is yet to be assembled into a chromosome-level scaffold due to its repeat-rich heterochromatin structure (Hall et al. 2016).

Formerly, *An. gambiae* was classified into two molecular forms, "M" and "S" forms (Favia et al. 2001). When the reference genome was initially sequenced, a hybrid strain was chosen to capture both forms. Later, *An. gambiae* was split into two separate species, *An. gambiae s.s.* (S form) and *An. coluzzii* (M form; Fig. 1) based on molecular and ecological evidence (Coetzee et al. 2013). Therefore, the

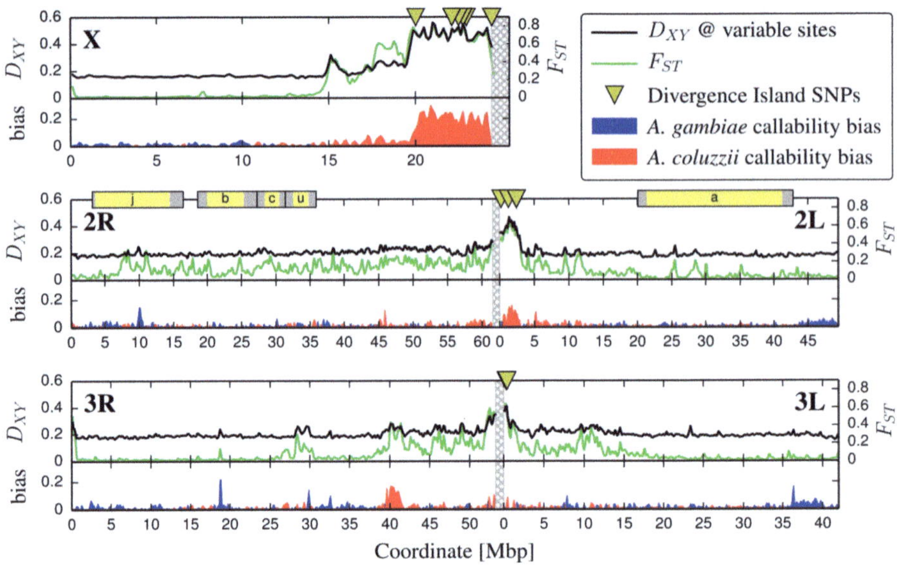

Fig. 3 Genome characteristics of AgamP4 reference genome. Differentiation between *An. coluzzii* and *An. gambiae s.s.* are elevated both in terms of D_{xy} (black line) and F_{ST} (green line) in centromeric regions. Divergence Island SNPs are fixed mutations used to distinguish two species in a multiplex genotyping assay called Divergence Island SNP (DIS) assay (Lee et al. 2014). A region containing more *coluzzii*-specific SNPs is represented in red. A region containing more gambiae-specific SNPs is marked in blue

reference genome *AgamP4* has characteristics of both species. The two species have very similar genomes with 49.1% of sliding windows having F_{ST} less than 0.1 between them (Fig. 3). Highly differentiated regions include regions adjacent to centromeres and most of the right arm of chromosome 2 (2R). Based on allopatric population genomes of *An. coluzzii* and *An. gambiae s.s.* in Mali, we measured the callability bias, which indicates the relative proportions of nucleotide sequence in AgamP4 containing species-specific SNPs. For example, bias is zero (0) if a stretch of reference sequence contains an equal number of gambiae-specific and coluzzii-specific SNPs. A region containing more *coluzzii*-specific SNPs is represented in red in Fig. 3. The elevated *An. coluzzii* callability bias near the known "speciation island" (Turner et al. 2005) in the centromeric regions suggests that AgamP4 is closer to *An. coluzzii* rather than *An. gambiae s.s.* Whether this bias may impact the mapping performance between the two species is unknown. However, other studies have reported a significant bias toward higher mapping rate of reference alleles compared to that of alternate alleles (Degner et al. 2009). It is a fairly standard practice to create a customized reference genome with population-specific alleles to reduce reference bias when conducting allele-specific expression experiments (Rozowsky et al. 2011; Zhuo et al. 2017). In a recent study it has been suggested that an ongoing process of purifying selection is occurring in the *An. coluzzii* population against DNA of *An. gambiae* ancestry, and stable introgression is limited to only specific genes involved in insecticide resistance (Hanemaaijer et al. 2018).

3 Geographical and Historical Population Structure

Anopheles gambiae is distributed across the sub-Saharan Africa from the Atlantic Ocean in the west to the Indian Ocean in the east, and from the Sahara Desert in the north to almost the very southern point of the continent (Fig. 4b). Its sibling species *An. coluzzii* is confined to the western part of the African continent but still occupies about a quarter of it (Fig. 4a). The geographic scale alone makes population studies challenging. Difference in flight activity between species and seasonal aptitude of flight activity related to long distance travel within a species (Dao et al. 2014; Faiman et al. 2020) may also influence population structure. A limited dispersal capacity (<5 km), as found in mark-release-recapture experiments (Costantini et al. 1996), could add to (meta-) populations' high structuring. Due to the extraordinary public health importance of the species, several extensive projects have been carried out to decipher historical and recent population genetic structures.

The current genetic structure of populations is shaped by the historical events and gene flow barriers. Several population genomic analyses on SNP distribution data have shown that large scale *An. gambiae* populations are strongly influenced by geographic barriers like the Congo Basin tropical rainforest and the East African rift system (*Anopheles gambiae* 1000 Genomes Consortium, Data Analysis Group, Partner Working Group, Sample Collections-Angola, Burkina Faso, Cameroon,

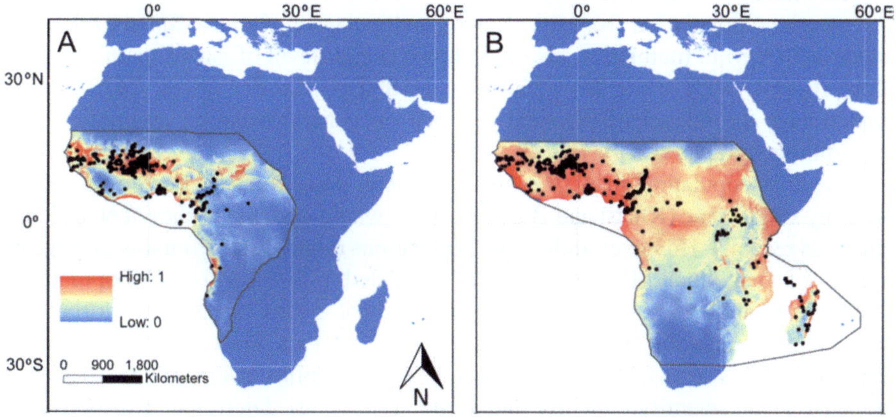

Fig. 4 Species distribution models for (**a**) *Anopheles coluzzii* and (**b**) *An. gambiae* s.s. Black dots indicate georeferenced observations, and red colors shaded to blue indicate high versus low predicted suitability values. All models were run using the Maxent program with the "cloglog" output and executed in the "kuenm" package in R (Phillips et al. 2006; Cobos et al. 2019) with WorldClim v2 bioclimatic data serving as environmental data layers (Fick and Hijmans 2017). The gray line indicates the model calibration and projection region. Models were run with two sets of environmental variables, for 500 iterations, with 10 bootstrap replicates, multiple regularization multiplier values to optimize parameterization, and all combinations of feature types (Merow et al. 2013). The best performing model was selected based on a modified area under the receiver operating characteristic value and lowest AICc score (Peterson et al. 2008; Warren and Seifert 2011)

Gabon, Guinea, Guinea Bissau, Kenya, et al. 2017; Schmidt et al. 2019). F_{ST} values of 0.001 and smaller for populations thousands of kilometers apart (*Anopheles gambiae* 1000 Genomes Consortium, Data Analysis Group, Partner Working Group, Sample Collections-Angola, Burkina Faso, Cameroon, Gabon, Guinea, Guinea Bissau, Kenya, et al. 2017) illustrate that neighboring populations in general are highly connected. However, the mentioned geographical structures serve as strong gene flow barriers increasing the genetic differentiation (e.g., $F_{ST} > 0.05$ for two populations of *An. gambiae* divided by Congo Basin, which is higher than between neighboring populations of *An. gambiae* and *An. coluzzii*; *Anopheles gambiae* 1000 Genomes Consortium, Data Analysis Group, Partner Working Group, Sample Collections-Angola, Burkina Faso, Cameroon, Gabon, Guinea, Guinea Bissau, Kenya, et al. 2017). This can be even more pronounced in island populations, as for those found in islands of the Comoros archipelago (Schmidt et al. 2019). The Mozambique Channel (Indian Ocean) served as a strong barrier to gene flow, resulting in F_{ST} values >0.2 between *An. gambiae* populations from these remote islands and from mainland sites (Schmidt et al. 2019). Microsatellite-based studies on less isolated islands in Lake Victoria revealed conflicting results with signs of a high degree of genetic isolation on one island (Lukindu et al. 2018) and high degree of gene flow ($F_{ST} < 0.006$) on another (Chen et al. 2004). Although we could not locate the F_{ST} data based on microsatellite markers from Lukindu et al. (2018), they provided the F_{ST} values based on the protein-coding mitogenome which indicates FST value range of 0.01–0.1 between mainland and islands. An *An. gambiae* population from the Bijagos archipelago in Guinea-Bissau, relatively close (<50 km) to mainlands, also has shown a high level of island-mainland gene flow ($F_{ST} < 0.02$) using 96 genome-wide SNPs (Marsden et al. 2013), again underlining the impact of physical distance on genetic isolation as predicted by island biogeography theory.

Effective population sizes (N_e) of *An. gambiae* have robustly been estimated to $\sim 10^7$ by single gene (Khatri and Burt 2019) and whole-genome (Schmidt et al. 2019) analyses recently. However, individual populations might differ from this average by orders of magnitude if geographically restricted ($\sim 10^6$ for island populations; (Schmidt et al. 2019)), and depending on seasonal cycles (up to $\sim 10^9$ during wet season in simulations; (Khatri and Burt 2019)). Given the complex population structure across a vast distribution area, the geographic origin of the species is both of great scientific interest and quite difficult to determine. Chromosomal inversion data makes a central African (Ayala and Coluzzi 2005) or eastern African origin (Kamali et al. 2012) plausible. Recent population genomic data indicate either a central African (*Anopheles gambiae* 1000 Genomes Consortium, Data Analysis Group, Partner Working Group, Sample Collections-Angola, Burkina Faso, Cameroon, Gabon, Guinea, Guinea Bissau, Kenya, et al. 2017) or a western African origin (Schmidt et al. 2019). However, these differing localizations based on population genomics do not contradict each other, as they both determine the origin to a region west of all major geographic barriers and thus a region in which population structure in *An. gambiae* is almost absent nowadays. Coalescent analyses point toward a stepwise migration from west to east that was shaped by multiple bottlenecks when small numbers of mosquitoes overcame one of the migration barriers and successfully established new populations behind it (Schmidt et al. 2019). New studies of *An. coluzzii* populations (Campos et al. 2020) also suggest a West African origin.

Overall, *An. gambiae* seems to have a highly complicated population structure that needs more attention for understanding the evolutionary dynamics and predicting effects of vector control strategies. Additional research into ecological and population genomics is needed to build a more accurate models of local/regional population dynamics of species within the *Anopheles gambiae* complex. Areas of future research include the seasonal change in species composition within the species complex (Touré et al. 1998; Lehmann et al. 2010), spatiotemporal dynamics in relation to other *Anopheles* species sharing similar habitats, genomic signatures of local adaptation, and location-specific genetic variants that may confer resistance to control measures.

4 Asymmetric Introgression Between *An. coluzzii* and *An. gambiae*

Hybridization events within *An. gambiae* complex is a dynamic process occurring gradually or spontaneously, with relatively high frequency in sympatric populations of *An. gambiae s.s.* and *An. coluzzii* (Marsden et al. 2011; Riehle et al. 2011; Lee et al. 2013). Environmental and anthropogenic factors such as use of insecticide treated bednets can alter hybridization dynamics by diminishing natural reproductive and ecological barriers due to habitat deterioration or fitness landscape change (Norris et al. 2015). Those factors work as external forces that pressure and challenge the ecological balance, triggering the breakdown of mating barriers. Thus, these factors not only increase mating rate between the different species of the complex, but also decrease assortative mating, resulting in the appearance of F1 hybrids and subsequently introgressed genotypes (Lee et al. 2013; Mancini et al. 2015). Episodes of natural interbreeding have mainly been observed across West Sahelian Africa, an area with clear characteristics of a hybrid zone but unclear ecotones. Hybrids of *An. gambiae s.s.* and *An. coluzzii* have been recorded in The Gambia (Caputo et al. 2008), Guinea-Bissau (Oliveira et al. 2008; Marsden et al. 2011; Nwakanma et al. 2013), Mali (Lee et al. 2013; Hanemaaijer et al. 2018), and other countries in West Africa (Riehle et al. 2011; Lee et al. 2013).

However, in the far western region of Africa such as Guinea-Bissau or the Gambia, however, hybridization appears to be particularly frequent and stable (Caputo et al. 2008; Oliveira et al. 2008; Marsden et al. 2011). In 2011, a study using a set of 96 SNPs from all three chromosomes clearly demonstrated the first case of asymmetric introgression in *An. gambiae* (Marsden et al. 2011). *Anopheles gambiae* ancestry values ranged from 0 (pure *An. coluzzii*) to 1 (pure *An. gambiae*) showing that hybrids between the two species are characterized by considerable

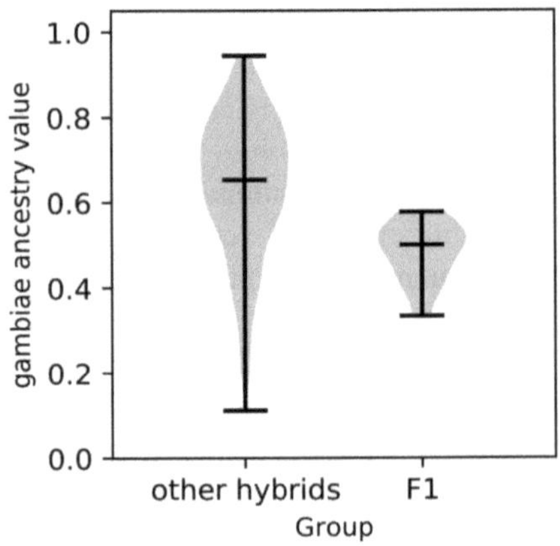

Fig. 5 Violin plot showing the distribution of ancestry contribution from *An. gambiae* s.s. among hybrid samples collected in Guinea-Bissau. The data is obtained from *PopI* database (Lee et al. 2004)

ancestry contribution from *An. gambiae* (Fig. 5). The cause of asymmetric introgression in Guinea-Bissau remains unknown.

Nonetheless, this uneven gene flow directionality does not appear to be always asymmetric. Studies showing the introgression of the islands of speciation from *An. gambiae* s.s., containing insecticide resistance mutation, into *An. coluzzii* in other West African countries like Mali (Norris et al. 2015), Burkina Faso (Hanemaaijer et al. 2019), Benin (Weill et al. 2000), and Ghana (Clarkson et al. 2014) suggest that both species can be recipient species accepting genetic material from other species but the direction of gene flow may differ depending on the spatiotemporal dynamics of the species.

5 Adaptive Introgression in *An. coluzzii*

Hybridization between sympatric or introduced related species can be seen as an adverse event for some areas as in agriculture, due to the fact that viable hybrids could carry specific traits that may be unfavorable for the ecological balance of a specific ecosystem. However, hybridization is a vital process for the evolution of species and speciation. Adaptive introgression is a highly dynamic process that contributes to the influx of foreign variants into a parental population. This process depends on the rate of hybridization events occurring in a specific location, the environment, the genetic similarity/compatibility between two hybridizing species, the differential fitness of potentially introgressing alleles within each species, and their genetic background (Norris et al. 2015). These factors contribute to the spread of beneficial traits under strong selection between closely related species where the costs of genetic incompatibility are more than compensated for by fitness gains (Seehausen 2004; Abbott et al. 2013; Hedrick 2013). Well-documented examples of adaptive introgression include the spread of wing color mimicking protective warning patterns within *Heliconius* butterflies (Pardo-Diaz et al. 2012) and the acquisition of resistance to anticoagulant rodenticide among mice and rats in Europe (Song et al. 2011). However, the pervasiveness of this process is still highly disputed for the fundamental implications it bears for evolutionary biology.

Episodes of ongoing hybridization in a natural sympatric population of *An. gambiae* s.s. and *An. coluzzii* in the village of Selinkenyi in Mali were observed at two time points (2002 and 2006) (Main B. J. et al. 2015; Norris et al. 2015; Hanemaaijer et al. 2018). However, earlier hybrids (2002) disappeared afterwards, suggesting hybrids suffer a fitness cost. Moreover, starting from 2010, a new type of *An. coluzzii* was observed, carrying the L1014F knockdown resistance (kdr) allele, known to increase resistance to Dichlorodiphenyltrichloroethane (DDT) and Pyrethroid insecticides (Martinez-Torres et al. 1998). In Mali, positive selection for the L1014F gene has arisen in *An. coluzzii* populations after the major hybridization event in 2006 by adaptive introgression from *An. gambiae* s.s. This event is correlated with the Insecticide-Treated Bed-nets campaigns in Mali (Main B. J. et al. 2015). These observations suggest that increased insecticide exposure increased the fitness benefit of L1014F to a point that it more than offset the fitness cost incurred by having the *An. gambiae* s.s. gene within an otherwise *An. coluzzii* genome, thereby driving the introgression of the entire 2L genomic island of divergence across the reproductive barrier.

The introgression of foreign genetic material from one species to another is not seen as beneficial to vector control programs. By spontaneous episodes of a permeable reproductive barrier due to external environmental pressures, an adaptive trait can be introgressed into sibling species, enhancing the fitness of hybrids by fixation of the new genetic material into the otherwise unfit species. The increment of insecticide resistance traits in some species of malaria vector mosquitoes occurred by hybridization with adaptive introgression of resistance trait to sibling susceptible species has been of extreme importance. Evidence of introgressive hybridization in mosquito evolution is widely demonstrated within the *An. gambiae* species complex (Weill et al. 2000; Clarkson et al. 2014; Norris et al. 2015).

6 Novel Gene Discovery Involved in Insecticide Resistance

The *An. gambiae* detox chip was the earliest tool available for population genomic studies (David et al. 2005). This microarray, containing unique fragments from 230 *Anopheles gambiae* genes, putatively involved in insecticide metabolism, has been applied to studying *An. gambiae* complex populations in Ghana (Muller et al. 2007), Cameroon (Muller et al. 2008), Nigeria (Djouaka et al. 2008; Awolola et al. 2009), and Benin (Djouaka et al. 2008). Some of the early stage experiments, however, suffer the flaw in experimental design of mixing multiple species in a pool (Djouaka et al. 2008; Muller et al. 2008) or comparing to a susceptible strain that has different genetic background (Muller et al. 2007). Given these flaws in experimental designs, the cytochrome P450 gene CYP6M2 appears consistently overexpressed in the resistant *An. gambiae* or *An. coluzzii* (Muller et al. 2007; Djouaka et al. 2008).

After comparing the whole-genome sequences of *An. coluzzii* populations prior to introgression (collected in Mali before 2006) with populations post-introgression (collected in Mali after 2006), Main B. J. et al. (2015) discovered a region on the X chromosome centered around the CYP9K1 gene that was significantly differentiated between the two time points. While 2L islands showed increased nucleotide diversity after introgression due to influx of new alleles from *An. gambiae*, the CYP9K1 region showed drastic decrease in nucleotide diversity indicating a selective sweep (Main B. J. et al. 2015). After targeted SNP genotyping surrounding the CYP9K1 region with increased population sample size, it was revealed that CYP9K1 had multiple haplotypes prior to the introgression event but only one of them increased its frequency to >90% post-introgression. A subsequent study uncovered a series of metabolic detoxification genes such as cytochrome P450 genes (i.e., CYP9K1 gene among others), carboxylesterases, and glutathione S-transferases (Main Bradley J. et al. 2018) that are undergoing selection. It appears that the development of insecticide resistance is a synergetic physiological process, involving both target site resistance gene and metabolic resistance in *An. coluzzii* from Mali.

7 *Plasmodium* Resistance Island

Principal malaria vectors like *An. gambiae s.s., An. coluzzii,* and *An. arabiensis* transmit *Plasmodium* parasites to humans, causing ~230 million malaria cases worldwide and ~400,000 death annually (WHO 2020). Five parasite species thus far have been reported to infect humans: *P. falciparum, P. vivax, P. ovale, P. malariae,* and *P. knowlesi* (Jongwutiwes et al. 2004; Cox-Singh et al. 2008). *Plasmodium falciparum* is the most virulent among these species accounting for ~85% of all malaria cases and nearly all of the mortality (Rich et al. 2009). The spatial distribution of the *P. falciparum* is largely concentrated in sub-Saharan Africa (Gething et al. 2011). The combination of the deadliest parasite and the presence of competent malaria vectors makes malaria control especially challenging in sub-Saharan Africa. Most research has focused on human-*Plasmodium* interactions. However, in the last two decades, efforts to develop a transmission blocking vaccine have recognized the critical importance of the sexual stage of the malaria parasite occurring within mosquitoes (Singh et al. 2021).

Early studies using paracentric chromosome inversion polymorphisms indicated that mosquitoes with inversions on the chromosome 2 showed higher infection rate of *P. falciparum* (Petrarca and Beier 1992; Riehle et al. 2011) when *An. gambiae* had standard arrangement (+) for 2La inversion. Follow-up studies using microsatellites showed evidence of a narrow genomic region controlling the resistance of mosquitoes against *Plasmodium* parasites. Quantitative loci are grouped into two major categories: (1) loci linked to *P. falciparum* infection intensity (Pfin) and (2) loci related to *P. falciparum* melanization (Pfmel). A quantitative trait locus (QTL) named pfin1 was found near AG2H603 (2L) in the East African population (Niare et al. 2002). Riehle et al. (2006) surveyed the West African *An. gambiae* population and noticed five loci linked to Pfin and two linked to Pfmel. Four loci were in a small region of chromosome 2L around AG2H325 and AG2H603 which were Pfin1, Pfin4, Pfin5, and Pfeml2 (Riehle et al. 2006). The same region on chromosome 2L was also found in East Africa *An. gambiae* population and identified a locus named EA_Pfin1 connected to strong *Plasmodium* resistance (Riehle et al. 2007). Due to the cluster of QTLs in a confined region of Chromosome 2L inversion region, the region was called *Plasmodium*-Resistance Island (PRI), which explained at least 89% of the variation related to *P. falciparum* resistance or susceptibility (Riehle et al. 2006).

The *Anopheles Plasmodium*-responsive leucine-rich repeat 1 gene (APL1; AGAP007033; 2L:41257850 – 41260185) was identified from PRI using sequencing (Rottschaefer et al. 2011). Among the three isoforms APL1A, APL1B, and APL1C, only APL1A showed the inhibition of *P. falciparum* in *An. gambiae* (Mitri et al. 2009). Moreover, from the three alleles of the APL1A gene, the APL1A2 allele appears to be highly protective against *P. falciparum* infection but not the other two alleles (Holm et al. 2012).

There are other genomic regions; however, they also indicated to be involved in parasite resistance of *An. gambiae.* In 1997, Zheng et al. found major quantitative trait loci (QTL) named Pen1 located at 2R near AG2H175 and two minor QTL named Pen2 (3L) and Pen3 (2R) which could be involved in the encapsulation of *Plasmodium cynomolgi* in *An. gambiae* (Zheng et al. 1997). The same study also indicated the QTL named pfin2 near AG2H290 (2R) and both pfin1 and pfin2 reduced the number of *P. falciparum* oocysts (Niare et al. 2002). Two SNPs were identified on one of the QTL, Pfin7, at Chromosome 3L. A C:A mutation between the intergenic region of Toll 11 (AGAP011186) and Toll 10 (AGAP011187) has found and confirmed the function from Toll11 to against *P. falciparum* through an RNAi experiment (Redmond et al. 2015).

Several studies have focused on the comparison between infected and uninfected midgut. ChIP-sequencing of *An. gambiae* from West Africa, after being infected with *P. falciparum*, showed that H3K9ac/H3K4me3/H3K27ac histone marks are associated with promoters of the active gene and that H3K9me3 are associated with silencing genes. Over 100 promoter or intragenic regions were involved including several immune-related pathways (Ruiz et al. 2019). In *An. gambiae* Keele laboratory strain, miRNA microarrays showed that miRNAs aga-miR-989 and aga-miR-305 were expressed at higher levels in *P. falciparum*-infected midguts. Antagomir, a small synthetic RNA that silences microRNA, caused inhibition of the aga-miR-305 and suppressed the *Plasmodium* infection. On the other hand, treated mosquitoes with artificial aga-miR-305 made them more susceptible. This suggested that miRNA aga-miR-305 might regulate the anti-*Plasmodium* response (Dennison et al. 2015).

Most parasite interaction studies have used *P. berghei*, the rodent malaria parasite (Clayton et al. 2014; Belachew 2018). However, it is not clear if the immune response involved in rodent malaria parasite is the same as human malaria parasite.

Some studies suggested that *P. berghei* may not be an ideal model for studying transcriptional immune responses due to the different pattern of gene expression as opposed to using *P. falciparum* (Tahar et al. 2002; Dong et al. 2006). Three *An. gambiae* genes, LRIM1 (leucine-rich repeat immune protein 1), and C-type lectin agonist gene CTL4 and CTLMA2 (CTL mannose binding 2), influence *P. berghei* development but did not show the same effects on *P. falciparum* infections in the same strain (Cohuet et al. 2006). Further studies on the immunity outcome in *P. falciparum*-infected mosquitoes, and the comparison between *P. berghei* and *P. falciparum* model system would illuminate if *P. berghei* system is a good alternative to study *P. falciparum* and *Anopheles* vector interactions.

Susceptibility to *Plasmodium* infection among African malaria vectors is typically less than 3% but can reach as high as ~25% (Beier et al. 1999; Riehle et al. 2006; Dia et al. 2009; Ndiath et al. 2011; Fryxell et al. 2012). It is assumed that any direct selection for *Plasmodium* resistance in these populations is weak due to the generally low infection rate among mosquito populations (Lee et al. 2020), a variety of different immune signaling pathways involved in resistance (Corby-Harris et al. 2010; Surachetpong et al. 2011; Clayton et al. 2014), and the fact that the innate immune system of these mosquitoes is co-evolving with many other microbial pressures present in the environment (Weiss and Aksoy 2011). *Anopheles* mosquitoes experience different immune challenges over the course of their development (e.g., larval density, larval nutrition (Muturi et al. 2012; Telang et al. 2012) and during blood feeding as adults (Okech et al. 2007; Hillyer and Estevez-Lao 2010)). A recent study reported extensive levels of divergence in immune genes compared to non-immune genes between *An. gambiae s.s.* and *An. coluzzii*. The two species are often sympatric during the adult stage but have some degree of separation between larval habitats (Gimonneau et al. 2012; Sanford et al. 2013). Based on these results, the authors concluded that immune challenges in the larval habitat, possibly reflected in the microbiome, would be the most important driver of adaptation/selection shaping the evolution of immune genes (Rottschaefer et al. 2015; Lee et al. 2020).

8 Application of Genetic Modification and Gene Drive

Malaria remains difficult to manage partly due to drug resistance of parasites, insecticide resistance of mosquito vectors, difficulties in distributing drugs and resources to rural regions, and lack of effective vaccines. One strategy currently being actively studied as a disease control tool is the use of genetically engineered mosquitoes (GEM, (Gantz et al. 2015; Kyrou et al. 2018; Carballar-Lejarazu et al. 2020)). GEM carry "effector" genes that suppress mosquito reproduction (Kyrou et al. 2018) or block pathogen development in mosquitoes (Isaacs et al. 2012; Gantz et al. 2015). Effector genes are introduced into natural mosquito populations upon GEM release and spread through wild-type mosquito populations using a super-Mendelian gene drive system (Gantz and Bier 2015). Latest efforts led to highly efficient, ready-to-use strains of *An. gambiae* that have good prospects of complementing and enhancing existing vector control strategies (Carballar-Lejarazu et al. 2020; Simoni et al. 2020).

A CRISPR-Cas9-based gene drive system is designed upon a specific 23 base pair stretch of genomic target DNA for the insertion of an engineered genetic element via the cut-and-paste-mechanism of CRISPR-Cas9. This "construct" is then spread into natural populations by the gene drive functionality. Standing genetic variation within a genomic CRISPR-Cas9 target site may lead to population resistance to the particular drive system because specimens with the alternative allele may not be susceptible due to a lack of target site recognition. The concern about the drive resistance has since emerged as a key issue to address in developing a successful GEM (Marshall et al. 2017, 2019; Raban et al. 2020), especially for population suppression approaches that have an inherently high positive selection on resistance alleles (Fig. 6).

Contrary to intuition, the highly polymorphic genome from *An. gambiae* was shown to be a promising target for the CRISPR-Cas9 approach (Schmidt et al. 2020). Whole-genome sequences of the two primary African malaria vectors, *An. gambiae* and *An. coluzzii* were examined to locate possible CRISPR/Cas9 target sites in all protein-coding genes. Sites were filtered for high specificity and GC content. Then, variants within the potential target sites were identified from whole-genome sequences over thousands of field collected mosquitos. In this study, if one sets the threshold for tolerable drive resistance allele frequency at 1%, then still ~90% of protein-coding genes contain at least one usable CRISPR-Cas9 target (Fig. 6) due to the extremely high number of potential target sites overall. The drive resistance allele's general occurrence and frequency are expected to have different degrees of impact depending on the type of the GEM strategies (e.g., population suppression vs population replacement; Fig. 6).

9 Conclusions and Future Prospects

Anopheles gambiae is among the rare cases with excellent resources in both molecular genetics and genomics, as well as extensive studies on natural populations. It is a heavily studied species in various fields, with investigations spanning fundamental to applied research questions, including emerging technologies like gene drive. There is no doubt that the current efforts sequencing thousands of specimens from all over the species' distribution range will enable researchers to address even more challenging questions, such as association mapping and haplotype sequencing for recent demographic history and population size estimation. For now, the species complex appears as a large mosaic of populations with both intra- and inter-specific gene flow at varying degrees. The incomplete isolation between populations allows for genetic connectedness over huge distances and even introgression of beneficial genes into sympatric sister species. This connectivity could be detrimental or advantageous for future public health programs, but should be taken into account

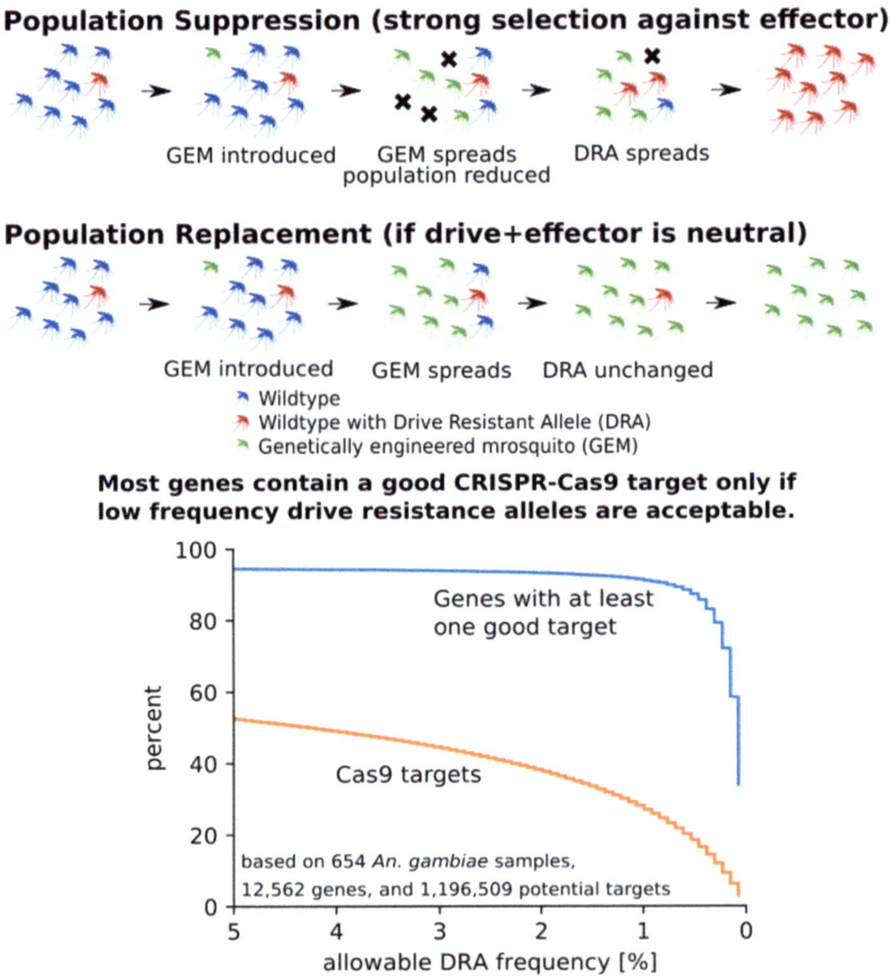

Fig. 6 Decrease in Cas9 targets (orange) vs genes with at least one good target site (blue) over various allowable drive resistance frequencies. Presence of drive-resistant alleles would have greater impact in compromising the population suppression strategy than population modification strategy. DRA = drive resistance allele. Adapted from Schmidt et al. (2020)

when considering the dynamic landscape of disease risk across the continent of Africa.

The broad scope of genomic studies for *Anopheles gambiae* resulted in additional studies of population genomics that were not discussed in detail here. These studies include the use of genomic data to understand chromosome inversions and their role in ecotypification (Love et al. 2016), and the exploration of more recent, local population size fluctuations by analyses of identity-by-descent regions (Bergey et al. 2020) and soft sweeps (Khatri and Burt 2019). There are also efforts to expand the geographic coverage of available *Anopheles gambiae* genome sequences (Malaria Genomic Epidemiology Network 2019). Moreover, some efforts are underway to improve existing references of many of *Anopheles* species using improved chromosome-level assembly techniques (Kingan et al. 2019). We look forward to more chromosome-level assemblies becoming available, which will facilitate research on structural variations and their role in hybridization and adaptation of this important family of mosquitoes of high public health significance.

References

Abbott R, Albach D, Ansell S, Arntzen JW, Baird SJ, Bierne N, et al. Hybridization and speciation. J Evol Biol. 2013;26(2):229–46.

Anopheles gambiae 1000 Genomes Consortium, Data Analysis Group, Partner Working Group, Sample Collections-Angola, Burkina Faso, Cameroon, Gabon, Guinea, Guinea Bissau, Kenya et al. Genetic diversity of the African malaria vector *Anopheles gambiae*. Nature. 2017;552 (7683):96–100.

Awolola TS, Oduola OA, Strode C, Koekemoer LL, Brooke B, Ranson H. Evidence of multiple pyrethroid resistance mechanisms in the malaria vector *Anopheles gambiae* sensu stricto from Nigeria. Trans R Soc Trop Med Hyg. 2009;103(11):1139–45.

Ayala FJ, Coluzzi M. Chromosome speciation: humans, *Drosophila*, and mosquitoes. Proc Natl Acad Sci U S A. 2005;102(Suppl 1):6535–42.

Beier JC, Killeen GF, Githure JI. Short report: entomologic inoculation rates and *Plasmodium falciparum* malaria prevalence in Africa. Am J Trop Med Hyg. 1999;61(1):109–13.

Belachew EB. Immune response and evasion mechanisms of *Plasmodium falciparum* parasites. J Immunol Res. 2018;2018:6529681.

Bergey CM, Lukindu M, Wiltshire RM, Fontaine MC, Kayondo JK, Besansky NJ. Assessing connectivity despite high diversity in island populations of a malaria mosquito. Evol Appl. 2020;13(2):417–31.

Campos M, Hanemaaijer J, Gripkey H, Collier T, Lee Y, Cornel A, et al. The origin of island populations of the African malaria mosquito, *Anopheles coluzzii*. Authorea. https://www.authorea.com/users/352717/articles/476853-the-origin-of-island-populations-of-the-african-malaria-mosquito-anopheles-coluzzii. 2020; https://doi.org/10.22541/au.159819143.34774248.

Caputo B, Nwakanma D, Jawara M, Adiamoh M, Dia I, Konate L, et al. *Anopheles gambiae* complex along The Gambia river, with particular reference to the molecular forms of *An. gambiae* s.s. Malar J. 2008;7:182.

Carballar-Lejarazu R, Ogaugwu C, Tushar T, Kelsey A, Pham TB, Murphy J, et al. Next-generation gene drive for population modification of the malaria vector mosquito, *Anopheles gambiae*. Proc Natl Acad Sci U S A. 2020;117(37):22805–14.

Chen H, Minakawa N, Beier J, Yan G. Population genetic structure of *Anopheles gambiae* mosquitoes on Lake Victoria islands, West Kenya. Malar J. 2004;3:48.

Clarkson CS, Weetman D, Essandoh J, Yawson AE, Maslen G, Manske M, et al. Adaptive introgression between *Anopheles* sibling species eliminates a major genomic island but not reproductive isolation. Nat Commun. 2014;5:4248.

Clayton AM, Dong Y, Dimopoulos G. The *Anopheles* innate immune system in the defense against malaria infection. J Innate Immun. 2014;6(2):169–81.

Cobos ME, Peterson AT, Barve N, Osorio-Olvera L. Kuenm: an R package for detailed development of ecological niche models using Maxent. PeerJ. 2019;7:e6281.

Coetzee M, Hunt RH, Wilkerson RC, della Torre A, Coulibaly MB, Besansky NJ. *Anopheles coluzzii* and *Anopheles amharicus,* new members of the *Anopheles gambiae* complex. Zootaxa. 2013;3619(2):246–74.

Cohuet A, Osta MA, Morlais I, Awono-Ambene PH, Michel K, Simard F, et al. *Anopheles* and *Plasmodium*: from laboratory models to natural systems in the field. EMBO Rep. 2006;7 (12):1285–9.

Coluzzi M, Sabatini A, Petrarca V, Di Deco MA. Chromosomal differentiation and adaptation to human environments in the *Anopheles gambiae* complex. Trans R Soc Trop Med Hyg. 1979;73 (5):483–97.

Corby-Harris V, Drexler A, Watkins de Jong L, Antonova Y, Pakpour N, Ziegler R, et al. Activation of Akt signaling reduces the prevalence and intensity of malaria parasite infection and lifespan in *Anopheles stephensi* mosquitoes. PLoS Pathog. 2010;6(7):e1001003.

Costantini C, Li SG, Della Torre A, Sagnon N, Coluzzi M, Taylor CE. Density, survival and dispersal of *Anopheles gambiae* complex mosquitoes in a west African Sudan savanna village. Med Vet Entomol. 1996;10(3):203–19.

Cox-Singh J, Davis TM, Lee KS, Shamsul SS, Matusop A, Ratnam S, et al. *Plasmodium knowlesi* malaria in humans is widely distributed and potentially life threatening. Clin Infect Dis. 2008;46 (2):165–71.

Dao A, Yaro AS, Diallo M, Timbine S, Huestis DL, Kassogue Y, et al. Signatures of aestivation and migration in Sahelian malaria mosquito populations. Nature. 2014;516(7531):387–90.

David JP, Strode C, Vontas J, Nikou D, Vaughan A, Pignatelli PM, et al. The *Anopheles gambiae* detoxification chip: a highly specific microarray to study metabolic-based insecticide resistance in malaria vectors. Proc Natl Acad Sci U S A. 2005;102(11):4080–4.

Davidson G. Insecticide resistance in *Anopheles gambiae* Giles: a case of simple mendelian inheritance. Nature. 1956;178(4538):863–4.

Davidson G. *Anopheles Gambiae*, a complex of species. Bull World Health Organ. 1964a;31:625–34.

Davidson G. The five mating-types in the *Anopheles Gambiae* complex. Riv Malariol. 1964b;43:167–83.

Davidson G, Hunt RH. The crossing and chromosome characteristics of a new, sixth species in the *Anopheles gambiae* complex. Parassitologia. 1973;15(1):121–8.

Degner JF, Marioni JC, Pai AA, Pickrell JK, Nkadori E, Gilad Y, et al. Effect of read-mapping biases on detecting allele-specific expression from RNA-sequencing data. Bioinformatics. 2009;25(24):3207–12.

Dennison NJ, BenMarzouk-Hidalgo OJ, Dimopoulos G. MicroRNA-regulation of *Anopheles gambiae* immunity to *Plasmodium falciparum* infection and midgut microbiota. Dev Comp Immunol. 2015;49(1):170–8.

Dia I, Ba H, Mohamed SA, Diallo D, Lo B, Diallo M. Distribution, host preference and infection rates of malaria vectors in Mauritania. Parasit Vectors. 2009;2(1):61.

Djouaka RF, Bakare AA, Coulibaly ON, Akogbeto MC, Ranson H, Hemingway J, et al. Expression of the cytochrome P450s, CYP6P3 and CYP6M2 are significantly elevated in multiple pyrethroid resistant populations of *Anopheles gambiae* s.s. from Southern Benin and Nigeria. BMC Genomics. 2008;9:538.

Dong Y, Taylor HE, Dimopoulos G. AgDscam, a hypervariable immunoglobulin domain-containing receptor of the *Anopheles gambiae* innate immune system. PLoS Biol. 2006;4(7): e229.

Faiman R, Yaro AS, Diallo M, Dao A, Djibril S, Sanogo ZL, et al. Quantifying flight aptitude variation in wild *Anopheles gambiae* in order to identify long-distance migrants. Malar J. 2020;19(1):263.

Favia G, Lanfrancotti A, Spanos L, Siden-Kiamos I, Louis C. Molecular characterization of ribosomal DNA polymorphisms discriminating among chromosomal forms of *Anopheles gambiae* s.s. Insect Mol Biol. 2001;10(1):19–23.

Fick SE, Hijmans RJ. Worldclim 2: new 1-km spatial resolution climate surfaces for global land areas. Int J Climatol. 2017;37(12):4302–15.

Fryxell RT, Nieman CC, Fofana A, Lee Y, Traore SF, Cornel AJ, et al. Differential *Plasmodium falciparum* infection of *Anopheles gambiae* s.s. molecular and chromosomal forms in Mali. Malar J. 2012;11:133.

Gantz VM, Bier E. Genome editing. The mutagenic chain reaction: a method for converting heterozygous to homozygous mutations. Science. 2015;348(6233):442–4.

Gantz VM, Jasinskiene N, Tatarenkova O, Fazekas A, Macias VM, Bier E, et al. Highly efficient Cas9-mediated gene drive for population modification of the malaria vector mosquito *Anopheles stephensi*. Proc Natl Acad Sci U S A. 2015;112(49):E6736–43.

Gething PW, Patil AP, Smith DL, Guerra CA, Elyazar IR, Johnston GL, et al. A new world malaria map: *Plasmodium falciparum* endemicity in 2010. Malar J. 2011;10:378.

Ghurye J, Koren S, Small ST, Redmond S, Howell P, Phillippy AM, et al. A chromosome-scale assembly of the major African malaria vector *Anopheles funestus*. Gigascience. 2019;8(6)

Gimonneau G, Pombi M, Choisy M, Morand S, Dabire RK, Simard F. Larval habitat segregation between the molecular forms of the mosquito *Anopheles gambiae* in a rice field area of Burkina Faso, West Africa. Med Vet Entomol. 2012;26(1):9–17.

Giraldo-Calderon GI, Emrich SJ, MacCallum RM, Maslen G, Dialynas E, Topalis P, et al. VectorBase: an updated bioinformatics resource for invertebrate vectors and other organisms related with human diseases. Nucleic Acids Res. 2015;43(Database issue):D707–13.

Hall AB, Papathanos PA, Sharma A, Cheng C, Akbari OS, Assour L, et al. Radical remodeling of the Y chromosome in a recent radiation of malaria mosquitoes. Proc Natl Acad Sci U S A. 2016;113(15):E2114–23.

Hanemaaijer MJ, Collier TC, Chang A, Shott CC, Houston PD, Schmidt H, et al. The fate of genes that cross species boundaries after a major hybridization event in a natural mosquito population. Mol Ecol. 2018;27(24):4978–90.

Hanemaaijer MJ, Higgins H, Eralp I, Yamasaki Y, Becker N, Kirstein OD, et al. Introgression between *Anopheles gambiae* and *Anopheles coluzzii* in Burkina Faso and its associations with kdr resistance and *Plasmodium* infection. Malar J. 2019;18(1):127.

Hedrick PW. Adaptive introgression in animals: examples and comparison to new mutation and standing variation as sources of adaptive variation. Mol Ecol. 2013;22(18):4606–18.

Hillyer JF, Estevez-Lao TY. Nitric oxide is an essential component of the hemocyte-mediated mosquito immune response against bacteria. Dev Comp Immunol. 2010;34(2):141–9.

Holm I, Lavazec C, Garnier T, Mitri C, Riehle MM, Bischoff E, et al. Diverged alleles of the *Anopheles gambiae* leucine-Rich repeat gene APL1A display distinct protective profiles against *Plasmodium falciparum*. PLoS One. 2012;7(12):e52684.

Holt RA, Subramanian GM, Halpern A, Sutton GG, Charlab R, Nusskern DR, et al. The genome sequence of the malaria mosquito *Anopheles gambiae*. Science. 2002;298(5591):129–49.

Isaacs AT, Jasinskiene N, Tretiakov M, Thiery I, Zettor A, Bourgouin C, et al. Transgenic *Anopheles stephensi* coexpressing single-chain antibodies resist *Plasmodium falciparum* development. Proc Natl Acad Sci U S A. 2012;109(28):E1922–30.

Jiang X, Peery A, Hall AB, Sharma A, Chen XG, Waterhouse RM, et al. Genome analysis of a major urban malaria vector mosquito, *Anopheles stephensi*. Genome Biol. 2014;15(9):459.

Jongwutiwes S, Putaporntip C, Iwasaki T, Sata T, Kanbara H. Naturally acquired *Plasmodium knowlesi* malaria in human, Thailand. Emerg Infect Dis. 2004;10(12):2211–3.

Kamali M, Xia A, Tu Z, Sharakhov IV. A new chromosomal phylogeny supports the repeated origin of vectorial capacity in malaria mosquitoes of the *Anopheles gambiae* complex. PLoS Pathog. 2012;8(10):e1002960.

Khatri BS, Burt A. Robust estimation of recent effective population size from number of independent origins in soft sweeps. Mol Biol Evol. 2019;36(9):2040–52.

Kingan SB, Urban J, Lambert CC, Baybayan P, Childers AK, Coates B, et al. A high-quality genome assembly from a single, field-collected spotted lanternfly (*Lycorma delicatula*) using the PacBio sequel II system. Gigascience. 2019;8(10)

Kyrou K, Hammond AM, Galizi R, Kranjc N, Burt A, Beaghton AK, et al. A CRISPR-Cas9 gene drive targeting doublesex causes complete population suppression in caged *Anopheles gambiae* mosquitoes. Nat Biotechnol. 2018;36(11):1062–6.

Lee Y, Collier TC, Manoukis NC, Lozano-Fuentes S, Vallejo EE, Yan W, Cornel AJ, Lanzaro GC, Taylor CE. 2004. An individual-level population genomics database for arthropod disease vectors (PopI). Davis, CA.; updated 2020; accessed 2020 August. http://popi.ucdavis.edu.

Lee Y, Marsden CD, Norris LC, Collier TC, Main BJ, Fofana A, et al. Spatiotemporal dynamics of gene flow and hybrid fitness between the M and S forms of the malaria mosquito, *Anopheles gambiae*. Proc Natl Acad Sci U S A. 2013;110(49):19854–9.

Lee Y, Marsden CD, Nieman C, Lanzaro GC. A new multiplex SNP genotyping assay for detecting hybridization and introgression between the M and S molecular forms of *Anopheles gambiae*. Mol Ecol Resour. 2014;14(2):297–305.

Lee Y, Souvannaseng L, Collier TC, Main BJ, Norris LC, Fofana A, et al. Evidence for divergent selection on immune genes between the African Malaria vectors, *Anopheles coluzzii* and *A gambiae*. Insects. 2020;11(12)

Lehmann T, Dao A, Yaro AS, Adamou A, Kassogue Y, Diallo M, et al. Aestivation of the African malaria mosquito, *Anopheles gambiae* in the Sahel. Am J Trop Med Hyg. 2010;83(3):601–6.

Love RR, Steele AM, Coulibaly MB, Traore SF, Emrich SJ, Fontaine MC, et al. Chromosomal inversions and ecotypic differentiation in *Anopheles gambiae*: the perspective from whole-genome sequencing. Mol Ecol. 2016;25(23):5889–906.

Lukindu M, Bergey CM, Wiltshire RM, Small ST, Bourke BP, Kayondo JK, et al. Spatio-temporal genetic structure of *Anopheles gambiae* in the northwestern Lake Victoria Basin, Uganda: implications for genetic control trials in malaria endemic regions. Parasit Vectors. 2018;11 (1):246.

Main BJ, Lee Y, Collier TC, Norris LC, Brisco K, Fofana A, et al. Complex genome evolution in *Anopheles coluzzii* associated with increased insecticide usage in Mali. Mol Ecol. 2015;24 (20):5145–57.

Main BJ, Everitt A, Cornel AJ, Hormozdiari F, Lanzaro GC. Genetic variation associated with increased insecticide resistance in the malaria mosquito, *Anopheles coluzzii*. Parasit Vectors. 2018;11(1):225.

Malaria Genomic Epidemiology Network. Insights into malaria susceptibility using genome-wide data on 17,000 individuals from Africa, Asia and Oceania. Nat Commun. 2019;10(1):5732.

Mancini E, Spinaci MI, Gordicho V, Caputo B, Pombi M, Vicente JL, et al. Adaptive potential of hybridization among Malaria vectors: introgression at the immune locus TEP1 between *Anopheles coluzzii* and *A. gambiae* in 'Far-West' Africa. PLoS One. 2015;10(6):e0127804.

Marinotti O, Cerqueira GC, de Almeida LG, Ferro MI, Loreto EL, Zaha A, et al. The genome of *Anopheles darlingi*, the main neotropical malaria vector. Nucleic Acids Res. 2013;41 (15):7387–400.

Marsden CD, Lee Y, Nieman CC, Sanford MR, Dinis J, Martins C, et al. Asymmetric introgression between the M and S forms of the malaria vector, *Anopheles gambiae*, maintains divergence despite extensive hybridization. Mol Ecol. 2011;20(23):4983–94.

Marsden CD, Cornel A, Lee Y, Sanford MR, Norris LC, Goodell PB, Nieman CC, Han S, Rodrigues A, Denis J, et al. An analysis of two island groups as potential sites for trials of transgenic mosquitoes for malaria control. Evol Appl. 2013;6(4):706–20.

Marshall JM, Buchman A, Sanchez CH, Akbari OS. Overcoming evolved resistance to population-suppressing homing-based gene drives. Sci Rep. 2017;7(1):3776.

Marshall JM, Raban RR, Kandul NP, Edula JR, Leon TM, Akbari OS. Winning the tug-of-war between effector gene design and pathogen evolution in vector population replacement strategies. Front Genet. 2019;10:1072.

Martinez-Torres D, Chandre F, Williamson M, Darriet F, Berge JB, Devonshire AL, et al. Molecular characterization of pyrethroid knockdown resistance (kdr) in the major malaria vector *Anopheles gambiae ss*. Insect Mol Biol. 1998;7(2):179–84.

Merow C, Smith MJ, Silander JA. A practical guide to MaxEnt for modeling species' distributions: what it does, and why inputs and settings matter. Ecography. 2013;36(10):1058–69.

Mitri C, Jacques JC, Thiery I, Riehle MM, Xu J, Bischoff E, et al. Fine pathogen discrimination within the APL1 gene family protects *Anopheles gambiae* against human and rodent malaria species. PLoS Pathog. 2009;5(9):e1000576.

Muller P, Donnelly MJ, Ranson H. Transcription profiling of a recently colonised pyrethroid resistant *Anopheles gambiae* strain from Ghana. BMC Genomics. 2007;8:36.

Muller P, Chouaibou M, Pignatelli P, Etang J, Walker ED, Donnelly MJ, et al. Pyrethroid tolerance is associated with elevated expression of antioxidants and agricultural practice in *Anopheles arabiensis* sampled from an area of cotton fields in northern Cameroon. Mol Ecol. 2008;17 (4):1145–55.

Muturi EJ, Blackshear M Jr, Montgomery A. Temperature and density-dependent effects of larval environment on *Aedes aegypti* competence for an alphavirus. J Vector Ecol. 2012;37 (1):154–61.

Ndiath MO, Cohuet A, Gaye A, Konate L, Mazenot C, Faye O, et al. Comparative susceptibility to *Plasmodium falciparum* of the molecular forms M and S of *Anopheles gambiae* and *Anopheles arabiensis*. Malar J. 2011;10:269.

Niare O, Markianos K, Volz J, Oduol F, Toure A, Bagayoko M, et al. Genetic loci affecting resistance to human malaria parasites in a west African mosquito vector population. Science. 2002;298(5591):213–6.

Norris LC, Main BJ, Lee Y, Collier TC, Fofana A, Cornel AJ, et al. Adaptive introgression in an African malaria mosquito coincident with the increased usage of insecticide-treated bed nets. Proc Nat Acad Sci U S A. 2015;112(3):815–20.

Nwakanma DC, Neafsey DE, Jawara M, Adiamoh M, Lund E, Rodrigues A, et al. Breakdown in the process of incipient speciation in *Anopheles gambiae*. Genetics. 2013;193(4):1221–31.

Okech BA, Gouagna LC, Yan G, Githure JI, Beier JC. Larval habitats of *Anopheles gambiae* s.s. (Diptera: Culicidae) influences vector competence to *Plasmodium falciparum* parasites. Malar J. 2007;6:50.

Oliveira E, Salgueiro P, Palsson K, Vicente JL, Arez AP, Jaenson TG, et al. High levels of hybridization between molecular forms of *Anopheles gambiae* from Guinea Bissau. J Med Entomol. 2008;45(6):1057–63.

Pardo-Diaz C, Salazar C, Baxter SW, Merot C, Figueiredo-Ready W, Joron M, et al. Adaptive introgression across species boundaries in *Heliconius* butterflies. PLoS Genet. 2012;8(6): e1002752.

Peterson AT, Pepes M, Soberón J. Rethinking receiver operating characteristic analysis applications in ecological niche modelling. Ecol Model. 2008;213:63–72.

Petrarca V, Beier JC. Intraspecific chromosomal polymorphism in the *Anopheles gambiae* complex as a factor affecting malaria transmission in the Kisumu area of Kenya. Am J Trop Med Hyg. 1992;46(2):229–37.

Phillips SJ, Anderson RP, Schapire RE. Maximum entropy modeling of species geographic distributions. Ecol Model. 2006;190(3–4):231–59.

Raban RR, Marshall JM, Akbari OS. Progress towards engineering gene drives for population control. J Exp Biol. 2020;223(Pt Suppl 1)

Redmond SN, Eiglmeier K, Mitri C, Markianos K, Guelbeogo WM, Gneme A, et al. Association mapping by pooled sequencing identifies TOLL 11 as a protective factor against *Plasmodium falciparum* in *Anopheles gambiae*. BMC Genomics. 2015;16(1):779.

Rich SM, Leendertz FH, Xu G, LeBreton M, Djoko CF, Aminake MN, et al. The origin of malignant malaria. Proc Natl Acad Sci U S A. 2009;106(35):14902–7.

Riehle MM, Markianos K, Niare O, Xu J, Li J, Toure AM, et al. Natural malaria infection in *Anopheles gambiae* is regulated by a single genomic control region. Science. 2006;312 (5773):577–9.

Riehle MM, Markianos K, Lambrechts L, Xia A, Sharakhov I, Koella JC, et al. A major genetic locus controlling natural *Plasmodium falciparum* infection is shared by east and west African *Anopheles gambiae*. Malar J. 2007;6:87.

Riehle MM, Guelbeogo WM, Gneme A, Eiglmeier K, Holm I, Bischoff E, et al. A cryptic subgroup of *Anopheles gambiae* is highly susceptible to human malaria parasites. Science. 2011;331 (6017):596–8.

Rottschaefer SM, Riehle MM, Coulibaly B, Sacko M, Niare O, Morlais I, et al. Exceptional diversity, maintenance of polymorphism, and recent directional selection on the APL1 malaria resistance genes of *Anopheles gambiae*. PLoS Biol. 2011;9(3):e1000600.

Rottschaefer SM, Crawford JE, Riehle MM, Guelbeogo WM, Gneme A, Sagnon N, et al. Population genetics of Anopheles coluzzii immune pathways and genes. G3. 2015;5(3):329–39.

Rozowsky J, Abyzov A, Wang J, Alves P, Raha D, Harmanci A, et al. AlleleSeq: analysis of allele-specific expression and binding in a network framework. Mol Syst Biol. 2011;7:522.

Ruiz JL, Yerbanga RS, Lefevre T, Ouedraogo JB, Corces VG, Gomez-Diaz E. Chromatin changes in *Anopheles gambiae* induced by *Plasmodium falciparum* infection. Epigenetics Chromatin. 2019;12(1):5.

Sanford MR, Ramsay S, Cornel AJ, Marsden CD, Norris LC, Patchoke S, et al. A preliminary investigation of the relationship between water quality and *Anopheles gambiae* larval habitats in Western Cameroon. Malar J. 2013;12:225.

Schmidt H, Lee Y, Collier TC, Hanemaaijer MJ, Kirstein OD, Ouledi A, et al. Transcontinental dispersal of *Anopheles gambiae* occurred from west African origin via serial founder events. Commun Biol. 2019;2:473.

Schmidt H, Collier TC, Hanemaaijer MJ, Houston PD, Lee Y, Lanzaro GC. Abundance of conserved CRISPR-Cas9 target sites within the highly polymorphic genomes of *Anopheles* and *Aedes* mosquitoes. Nat Commun. 2020;11(1):1425.

Seehausen O. Hybridization and adaptive radiation. Trends Ecol Evol. 2004;19(4):198–207.

Simoni A, Hammond AM, Beaghton AK, Galizi R, Taxiarchi C, Kyrou K, et al. A male-biased sex-distorter gene drive for the human malaria vector *Anopheles gambiae*. Nat Biotechnol. 2020;38(9):1054–60.

Singh M, Suryanshu K, Singh G, Dubey A, Chaitanya RK. *Plasmodium*'s journey through the *Anopheles* mosquito: a comprehensive review. Biochimie. 2021;181:176–90.

Song Y, Endepols S, Klemann N, Richter D, Matuschka FR, Shih CH, et al. Adaptive introgression of anticoagulant rodent poison resistance by hybridization between old world mice. Curr Biol. 2011;21(15):1296–301.

Surachetpong W, Pakpour N, Cheung KW, Luckhart S. Reactive oxygen species-dependent cell signaling regulates the mosquito immune response to *Plasmodium falciparum*. Antioxid Redox Signal. 2011;14(6):943–55.

Tahar R, Boudin C, Thiery I, Bourgouin C. Immune response of *Anopheles gambiae* to the early sporogonic stages of the human malaria parasite *Plasmodium falciparum*. EMBO J. 2002;21 (24):6673–80.

Telang A, Qayum AA, Parker A, Sacchetta BR, Byrnes GR. Larval nutritional stress affects vector immune traits in adult yellow fever mosquito *Aedes aegypti* (*Stegomyia aegypti*). Med Vet Entomol. 2012;26(3):271–81.

Touré Y, Petrarca V, Traore S, Coulibaly A, Maiga H, Sankare O, et al. The distribution and inversion polymorphism of chromosomally recognized taxa of the *Anopheles gambiae* complex in Mali, West Africa. Parassitologia. 1998;40(4):477–511.

Turner TL, Hahn MW, Nuzhdin SV. Genomic islands of speciation in *Anopheles gambiae*. PLoS Biol. 2005;3(9):e285.

Warren DL, Seifert SN. Ecological niche modeling in Maxent: the importance of model complexity and the performance of model selection criteria. Ecol Appl. 2011;21(2):335–42.

Weill M, Chandre F, Brengues C, Manguin S, Akogbeto M, Pasteur N, et al. The kdr mutation occurs in the Mopti form of *Anopheles gambiae* s.s. through introgression. Insect Mol Biol. 2000;9(5):451–5.

Weiss B, Aksoy S. Microbiome influences on insect host vector competence. Trends Parasitol. 2011;27(11):514–22.

WHO. World malaria report 2020: 20 years of global progress and challenges. https://www.who.int/teams/global-malaria-programme/reports/world-malaria-report-2020. Geneva: World Health Organization; 2020.

Zheng L, Cornel AJ, Wang R, Erfle H, Voss H, Ansorge W, et al. Quantitative trait loci for refractoriness of *Anopheles gambiae* to *Plasmodium cynomolgi* B. Science. 1997;276 (5311):425–8.

Zhou D, Zhang D, Ding G, Shi L, Hou Q, Ye Y, et al. Genome sequence of *Anopheles sinensis* provides insight into genetics basis of mosquito competence for malaria parasites. BMC Genomics. 2014;15:42.

Zhuo Z, Lamont SJ, Abasht B. RNA-Seq analyses identify frequent allele specific expression and no evidence of genomic imprinting in specific embryonic tissues of chicken. Sci Rep. 2017;7 (1):11944.

Index

Julian R. Dupuis and Om P. Rajora (eds.), *Population Genomics: Insects*, Population Genomics [Om P. Rajora (Editor-in-Chief)], https://doi.org/10.1007/978-3-032-10103-7

MIX
Papier aus verantwortungsvollen Quellen
Paper from responsible sources
FSC® C105338

If you have any concerns about our products,
you can contact us on
ProductSafety@springernature.com

In case Publisher is established outside the EU,
the EU authorized representative is:
Springer Nature Customer Service Center GmbH
Europaplatz 3, 69115 Heidelberg, Germany

Printed by Libri Plureos GmbH
in Hamburg, Germany